Emerging Applications of Differential Equations and Game Theory

Sırma Zeynep Alparslan Gök
Süleyman Demirel University, Turkey

Duygu Aruğaslan Çinçin
Süleyman Demirel University, Turkey

A volume in the Advances in Computer and Electrical Engineering (ACEE) Book Series

IGI Global
DISSEMINATOR OF KNOWLEDGE

Published in the United States of America by
IGI Global
Engineering Science Reference (an imprint of IGI Global)
701 E. Chocolate Avenue
Hershey PA, USA 17033
Tel: 717-533-8845
Fax: 717-533-8661
E-mail: cust@igi-global.com
Web site: http://www.igi-global.com

Library of Congress Cataloging-in-Publication Data

Names: Alparslan-Gök, Sırma Zeynep, 1980- editor. | Aruğaslan-Çinçin, Duygu, 1979- editor.
Title: Emerging applications of differential equations and game theory / Sırma Zeynep Alparslan-Gök and Duygu Aruğaslan-Çinçin, editors.
Description: Hershey, PA : Engineering Science Reference, [2020] | Includes bibliographical references.
Identifiers: LCCN 2019016789| ISBN 9781799801344 (hardcover) | ISBN 9781799801351 (softcover) | ISBN 9781799801368 (ebook)
Subjects: LCSH: Differential equations. | Game theory.
Classification: LCC QA372 .A3945 2020 | DDC 515/.35--dc23 LC record available at https://lccn.loc.gov/2019016789

This book is published in the IGI Global book series Advances in Computer and Electrical Engineering (ACEE) (ISSN: 2327-039X; eISSN: 2327-0403)

British Cataloguing in Publication Data
A Cataloguing in Publication record for this book is available from the British Library.

Advances in Computer and Electrical Engineering (ACEE) Book Series

Srikanta Patnaik
SOA University, India

ISSN:2327-039X
EISSN:2327-0403

MISSION

The fields of computer engineering and electrical engineering encompass a broad range of interdisciplinary topics allowing for expansive research developments across multiple fields. Research in these areas continues to develop and become increasingly important as computer and electrical systems have become an integral part of everyday life.

The **Advances in Computer and Electrical Engineering (ACEE) Book Series** aims to publish research on diverse topics pertaining to computer engineering and electrical engineering. **ACEE** encourages scholarly discourse on the latest applications, tools, and methodologies being implemented in the field for the design and development of computer and electrical systems.

COVERAGE

- Power Electronics
- VLSI Fabrication
- Qualitative Methods
- Chip Design
- Algorithms
- Microprocessor Design
- Applied Electromagnetics
- VLSI Design
- Computer Architecture
- Sensor Technologies

IGI Global is currently accepting manuscripts for publication within this series. To submit a proposal for a volume in this series, please contact our Acquisition Editors at Acquisitions@igi-global.com or visit: http://www.igi-global.com/publish/.

The Advances in Computer and Electrical Engineering (ACEE) Book Series (ISSN 2327-039X) is published by IGI Global, 701 E. Chocolate Avenue, Hershey, PA 17033-1240, USA, www.igi-global.com. This series is composed of titles available for purchase individually; each title is edited to be contextually exclusive from any other title within the series. For pricing and ordering information please visit http://www.igi-global.com/book-series/advances-computer-electrical-engineering/73675. Postmaster: Send all address changes to above address.

Titles in this Series

For a list of additional titles in this series, please visit:
https://www.igi-global.com/book-series/advances-computer-electrical-engineering/73675

Novel Practices and Trends in Grid and Cloud Computing
Pethuru Raj (Reliance Jio Infocomm Ltd. (RJIL), India) and S. Koteeswaran (Vel Tech, India)
Engineering Science Reference • © 2019 • 374pp • H/C (ISBN: 9781522590231) • US $255.00 (our price)

Blockchain Technology for Global Social Change
Jane Thomason (University College London, UK) Sonja Bernhardt (ThoughtWare, Australia) Tia Kansara (Replenish Earth Ltd, UK) and Nichola Cooper (Blockchain Quantum Impact, Australia)
Engineering Science Reference • © 2019 • 243pp • H/C (ISBN: 9781522595786) • US $195.00 (our price)

Contemporary Developments in High-Frequency Photonic Devices
Siddhartha Bhattacharyya (RCC Institute of Information Technology, India) Pampa Debnath (RCC Institute of Information Technology, India) Arpan Deyasi (RCC Institute of Information Technology, India) and Nilanjan Dey (Techno India College of Technology, India)
Engineering Science Reference • © 2019 • 369pp • H/C (ISBN: 9781522585312) • US $225.00 (our price)

Applying Integration Techniques and Methods in Distributed Systems and Technologies
Gabor Kecskemeti (Liverpool John Moores University, UK)
Engineering Science Reference • © 2019 • 351pp • H/C (ISBN: 9781522582953) • US $245.00 (our price)

Handbook of Research on Cloud Computing and Big Data Applications in IoT
B. B. Gupta (National Institute of Technology Kurukshetra, India) and Dharma P. Agrawal (University of Cincinnati, USA)
Engineering Science Reference • © 2019 • 609pp • H/C (ISBN: 9781522584070) • US $295.00 (our price)

IGI Global
DISSEMINATOR OF KNOWLEDGE

701 East Chocolate Avenue, Hershey, PA 17033, USA
Tel: 717-533-8845 x100 • Fax: 717-533-8661
E-Mail: cust@igi-global.com • www.igi-global.com

Table of Contents

Detailed Table of Contents

For a nonempty set G, the authors define an operation * reckoned with closeness property (i.e.,* is an operation which is not a binary operation). Then they define the partial group as a generalisation of a group. A partial group G which is a non empty set satisfies following conditions hold for all a,b and c∈d:(PG1) If ab,(ab) c, bc and a (bc) is defined, then,(ab) c=a (bc)(PG2) . For every,a∈G there exists an e∈G such that ae and ea are defined and, ae=ea=a (PG3) . For every,a∈G there exists an a∈G such that aa and aa are defined and aa=aa=e. The * operation effects and changes the properties of group axioms. So that lots of group theoretic theorems and conclusions do not work in partial groups. Thus, this description gives us some fundemental and important properties and analogous to group theory. Also the authors have some differences from group theory.

In this chapter, the authors study the exponential rational function method to find new exact solutions for the time-fractional fifth-order Sawada-Kotera equation, the space-time fractional Whitham-Broer-Kaup equations, and the space-time fractional generalized Hirota-Satsuma coupled KdV equations. These fractional differential equations are converted into ordinary differential equations by using the fractional complex transform. The fractional derivatives are defined in the sense of Jumarie's modified Riemann-Liouville. The proposed method is direct and effective for solving different kind of nonlinear fractional equations in mathematical physics.

Society often relies on information disclosed by enterprises and verified by auditors to decide on an efficient allocation of capital. Auditing sector serves as a means of verification to protect investors from making decisions based on inaccurate information. However, auditors can use their superior information for extracting additional rents. This study explores an economy where entrepreneurs choose their financial reporting quality considering incentives imposed by the society, and rent-seeking auditors may manipulate their reports to extract gains in the expense of public interest. The analysis captures the dynamics of strategy changes among different actors by introducing a population game framework. The steady-state equilibrium analysis shows that there is a pure state and mixed states whose stability is affected by policy parameters such as subsidies, taxes, competitive auditor fee, and rate of adjustment of different behavioral dynamics. It appears that corruption in auditing sector and poor quality in financial reporting may arise as a temporally persistent outcome.

A reproducing kernel Hilbert space (shorty, RKHS) $H=H(\Omega)$ on some set Ω is a Hilbert space of complex valued functions on Ω such that for every $\lambda\in\Omega$ the linear functional (evaluation functional) $f\rightarrow f(\lambda)$ is bounded on H. If H is RKHS on a set Ω, then, by the classical Riesz representation theorem for every $\lambda\in\Omega$ there is a unique element $kH,\lambda\in H$ such that $f(\lambda)=\langle f,kH,\lambda\rangle$; for all $f\in H$. The family $\{kH,\lambda:\lambda\in\Omega\}$ is called the reproducing kernel of the space H. The Berezin set and the Berezin number of the operator A was respectively given by Karaev in [26] as following $Ber(A)=\{A(\lambda):\lambda\in\Omega\}$ and $ber(A):=sup\lambda\in\Omega|A(\lambda)|$. In this chapter, the authors give the Berezin number inequalities for an invertible operator and some other related results are studied. Also, they obtain some inequalities of the slater type for convex functions of selfadjoint operators in reproducing kernel Hilbert spaces and examine related results.

In this chapter, the features of a continuous time GARCH (COGARCH) process is discussed since the process can be applied as an explicit solution for the stochastic differential equation which is defined for the volatility of unequally spaced time series. COGARCH process driven by a Lévy process is an analogue of discrete time GARCH process and is further generalized to solutions of Lévy driven stochastic differential equations. The Compound Poisson and Variance Gamma processes are defined and used to derive the increments for the COGARCH process. Although there are various parameter estimation methods introduced for COGARCH, this study is focused on two methods which are Pseudo Maximum Likelihood Method and General Methods of Moments. Furthermore, an example is given to illustrate the findings.

In this chapter, the authors extend transportation situations under uncertainty by using grey numbers. Further, they try in this research building models for grey game problems on transportation situations proposing the ideas of grey solutions and their corresponding structures. They introduce cooperative grey games and grey solutions. They focus on the grey Shapley value and the grey core of the modeled game arising from transportation situations. Moreover, they prove the nonemptiness of the grey core for the transportation grey games, and some results on the relationship between the grey core.

Discrete q-Hermite I polynomials are a member of the q-polynomials of the Hahn class. They are the polynomial solutions of a second order difference equation of hypergeometric type. These polynomials are one of the q-analogous of the Hermite polynomials. It is well known that the q-Hermite I polynomials approach the Hermite polynomials as q tends to 1. In this chapter, the orthogonality property of the discrete q-Hermite I polynomials is considered. Moreover, the orthogonality relation for the k-th order q-derivatives of the discrete q-Hermite I polynomials is obtained. Finally, it is shown that, under a suitable transformation, these relations give the corresponding relations for the Hermite polynomials in the limiting case as q goes to 1.

A boundary value problem for the second order differential equation -y′′+∑_{m=0} N−1λ^{m}q_{m}(x)y=λ2Ny with two boundary conditions a_{i1}y(0)+a_{i2} y′(0)+a_{i3}y(π)+a_{i4}y′(π)=0, i=1,2 is considered. Here n>1, λ is a complex parameter, q0(x),q1(x),...,q_{n-1}(x) are summable complex-valued functions, a_{ik} (i=1,2; k=1,2,3,4) are arbitrary complex numbers. It is proved that the system of eigenfunctions and associated eigenfunctions is complete in the space and using elementary asymptotical metods asymptotic formulas for the eigenvalues are obtained.

The authors consider a nonlinear epidemic equation by modeling it with generalized piecewise constant argument (GPCA). The authors investigate invariance region for the considered model. Sufficient conditions guaranteeing the existence and uniqueness of the solutions of the model are given by creating integral equations. An important auxiliary result giving a relation between the values of the unknown function solutions at the deviation argument and at any time t is indicated. By using Lyapunov-Razumikhin method developed by Akhmet and Aruğaslan for the differential equations with generalized piecewise constant argument (EPCAG), the stability of the trivial equilibrium is investigated in addition to the stability examination of the positive equilibrium transformed into the trivial equilibrium. Then sufficient conditions for the uniform stability and the uniform asymptotic stability of trivial equilibrium and the positive equilibrium are given.

This work studies the singular Hahn-Dirac system. Here μ is a complex spectral parameter, p(.) and r(.) are real-valued continuous functions at ω0, defined on [ω0,∞) and q∈(0,1), ω0:=ω/1-q, ω>0, x∈[ω0,∞). The existence of a spectral function for this system is proved. Further, a Parseval equality and an expansion formula in eigenfunctions are proved in terms of the spectral function.

The authors propose a nonlinear epidemic model by developing it with generalized piecewise constant argument (GPCA) introduced by Akhmet. The authors investigate invariance region for the considered model. For the taken model into consideration, they obtain a useful inequality concerning relation between the values of the solutions at the deviation argument and at any time for the epidemic model. The authors reach sufficient conditions for the existence and uniqueness of the solutions. Then, based on Lyapunov-Razumikhin method developed by Akhmet and Aruğaslan for the differential equations with generalized piecewise constant argument (EPCAG), sufficient conditions for the stability of the trivial equilibrium and the positive equilibrium are investigated. Thus, the theoretical results concerning the uniform stability of the equilibriums are given.

Preface

Differential equations and game theory approaches have attracted a growing number of scientists, decision makers and practicing researchers in recent years. They are used for solving complex problems in real-world applications of economics, biology, neuroscience, medicine and engineering.

Differential equations play a key role to understand the real-world problems. Therefore, the theory of differential equations has been studied by many authors in recent years. The obtained results help us to solve several problems in applied fields of science such as mathematical biology, neural networks, economics, mechanics, physics, medicine and differential game theory. As a result, it is very important to stimulate more studies on this area.

Game theory serves as a new core perspective of the compendium and text book. It deals with multi-person decision making, in which each decision maker tries to maximize own utility. In such situations, decisions regarding whether (or not) to cooperate within the grand coalition rely on estimations of individual benefits/costs gains importance. Game theory is a mathematical theory dealing with the modeling and the analysis of conflict and cooperation and has broad applicability in Operational Research, economy, modern finance, climate negotiations and policy, environmental management and pollution control, etc.

With this book the editors aim as follows:

- to offer a *compendium* for all researchers who will become able to familiarize with the emerging research subjects that are of common practical and methodological interest for representatives of engineering, economics and biology,
- to provide a *dictionary and encyclopedia* that will enable scientists and practitioners to quickly access the key notions of their domains and, via suitable methods and references, to advance ahead and to branch further on the way to handle their own and their institutions' problems and challenges,
- to create an *atmosphere* in which young researchers can grow when reading the book and develop further towards becoming recognized experiments,

- and initiate a *momentum* of excitement and encouragement on the way of preparing this book and of reading it later on, that will strongly support interdisciplinary science, the *solution* of striking real-world problems and a creative and fruitful collaboration between experts from all over the world,
- to further introduce IGI as a premium publisher in new regions and as a Center of Excellence.

This book provides an excellent reference to graduate, postgraduate students, decision makers and researchers in private sectors, universities, and industries in the field of various sciences, engineering and management such as mathematics/applied mathematics, game theory, biology, neuroscience, computer science, economics and finance wherever one wants to model their uncertain practical and real-life problems. This book aims to become significant and to become very fruitful for humankind.

The book is organized into 11 chapters, prepared by experts and scholars from all over the world. A brief description of each of the contents of the chapters is given as follows:

Chapter 1: This chapter defines an operation reckoned with closeness property for a nonempty set. Then the authors define the partial group as a generalization of a group. The operation effects and changes the properties of group axioms. So that lots of group theoretic theorems and conclusions do not work in partial groups. Thus, this description gives some fundamental and important properties from group theory. Also some differences from group theory are given.

Chapter 2: In this chapter, the authors study the exponential rational function method to find new exact solutions for the time-fractional fifth-order Sawada-Kotera equation, the space-time fractional Whitham-Broer-Kaup equations and the space-time fractional generalized Hirota-Satsuma coupled KdV equations. These fractional differential equations are converted into ordinary differential equations by using the fractional complex transform. The fractional derivatives are defined in the sense of Jumarie's modified Riemann-Liouville method. The proposed method is direct and effective for solving different kinds of nonlinear fractional equations in mathematical physics.

Chapter 3: This chapter explores an economy where entrepreneurs choose their financial reporting quality considering incentives imposed by the society, and rent-seeking auditors may manipulate their reports to extract gains in the expense of public interest. The analysis captures the dynamics of strategy changes among different actors, by introducing a population game framework. The steady-state equilibrium analysis shows that there is a pure state and mixed states whose stability is affected by policy parameters such as subsidies, taxes, competitive auditor fee and rate of adjustment of different behavioral dynamics. It appears that corruption in auditing sector and poor quality in financial reporting may arise as a temporally persistent outcome.

Chapter 4: This chapter gives the Berezin number inequalities for an invertible operator and some other related results are. Also, some inequalities of the slater type for convex functions of self adjoint operators in reproducing kernel Hilbert spaces are obtained and related results are examined.

Chapter 5: In this chapter, the features of a continuous time GARCH (COGARCH) process is discussed since the process can be applied as an explicit solution for the stochastic differential equation which is defined for the volatility of unequally spaced time series. COGARCH process driven by a Lévy process, which is an analogue of discrete time GARCH process and is further generalized to solutions of Lévy driven stochastic differential equations. The Compound Poisson and Variance Gamma processes are defined and used to derive the increments for the COGARCH process. Although there are various parameter estimation methods introduced for COGARCH this study is focused on two methods: Pseudo Maximum Likelihood Method and General Methods of Moments. Furthermore, an example is given to illustrate the findings.

Chapter 6: The authors extend transportation situations under uncertainty by using grey numbers. Further, building models for grey game problems on transportation situations; proposing the ideas of grey solutions and their corresponding structures are given. In the sequel, cooperative grey games and grey solutions are introduced. The grey Shapley value and the grey core of the modeled game arising from transportation situations are handled. Moreover, the nonemptiness of the grey core for the transportation grey games, and some results on the relationship between the grey core are proven.

Chapter 7: This chapter is concerned with the orthogonality property of the discrete q-Hermite I polynomials. Moreover, the orthogonality relation for the k-th order q-derivatives of the discrete q-Hermite I polynomials is obtained. Finally, it is shown that, under a suitable transformation, these relations give the corresponding relations for the Hermite polynomials in the limiting case as q goes to 1.

Chapter 8: The objective of this chapter is to investigate a boundary value problem for the second order differential equation with two boundary conditions. It is proved that the system of eigenfunctions and associated eigenfunctions is complete in the space and using elementary asymptotical methods asymptotic formulas for the eigenvalues are obtained.

Chapter 9: In this chapter, the authors consider a nonlinear epidemic equation by modeling it with generalized piecewise constant argument (GPCA). The authors investigate invariance region for the considered model. Sufficient conditions guaranteeing the existence and uniqueness of the solutions of the model are given by creating integral equations. An important auxiliary result giving a relation between the values of the unknown function solutions at the deviation argument and at any time t is indicated. By using Lyapunov-Razumikhin method for the differential equations with generalized piecewise constant argument (EPCAG), the stability of the trivial equilibrium is investigated in addition to the stability examination of the positive equilibrium transformed into the trivial equilibrium. Then, sufficient conditions for the uniform stability and the uniform asymptotic stability of trivial equilibrium and the positive equilibrium are given.

Chapter 10: This chapter aims to study the singular Hahn-Dirac system. In the sequel, the existence of a spectral function for this system is proved. Further, a Parseval equality and an expansion formula in eigenfunctions are proved in terms of the spectral function.

Chapter 11: This chapter proposes a nonlinear epidemic model by developing it with generalized piecewise constant argument (GPCA). The authors investigate invariance region for the considered model. For the model taken into consideration, they obtain a useful inequality concerning relation between the values of the solutions at the deviating argument and at any time for the epidemic model. The authors reach sufficient conditions for the existence and uniqueness of the solutions. Then, based on Lyapunov-Razumikhin method for the differential equations with generalized piecewise constant argument (EPCAG), sufficient conditions for the stability of the trivial equilibrium and the positive equilibrium are investigated. Thus, the theoretical results concerning the uniform stability of the equilibriums are given.

This book will be an excellent reference for the global research scholars across the planet in the research areas on application of differential equations and game theory. Firstly, we would like to sincerely thank all the authors for their marvelous contribution to the book in submitting their valuable book chapters. Secondly, thanks to all the referees for their valuable time and great effort in reviewing all the book chapters. Lastly but not least, we also thank to the editors for their strong support and motivation in making this book publication very successful.

Sirma Zeynep Alparslan-Gök
Süleyman Demirel University, Turkey

Duygu Aruğaslan-Çinçin
Süleyman Demirel University, Turkey

Chapter 1
A Generalization of Groups:
Partial Groups

Zekiye Çiloğlu Şahin
Süleyman Demirel University, Turkey

Yılmaz Çeven
https://orcid.org/0000-0002-2968-1546
Süleyman Demirel University, Turkey

ABSTRACT

*For a nonempty set G, the authors define an operation * reckoned with closeness property (i.e., * is an operation which is not a binary operation). Then they define the partial group as a generalisation of a group. A partial group G which is a non empty set satisfies following conditions hold for all a,b and c∈d:(PG1) If ab,(ab)c, bc and a (bc) is defined, then,(ab) c=a (bc)(PG2) . For every,a∈G there exists an e∈G such that ae and ea are defined and, ae=ea=a (PG3) . For every,a∈G there exists an a∈G such that aa and aa are defined and aa=aa=e. The * operation effects and changes the properties of group axioms. So that lots of group theoretic theorems and conclusions do not work in partial groups. Thus, this description gives us some fundemental and important properties and analogous to group theory. Also the authors have some differences from group theory.*

INTRODUCTION

Groupoid is an algebraic structure which is first introduced by Brandt in "Über eine Verallgemeinerung des Gruppengriffes". The definition of a groupoid is given as the following:

DOI: 10.4018/978-1-7998-0134-4.ch001

Let G be a nonempty set. A binary operation on G is a mapping from $G \times G$ to G. If $*$ is a binary operation on G, then it is known that $a*b \in G$ for all $a,b \in G$. ($*(a,b)$ is denoted by $a*b$).

A partially defined binary operation on G is a function from D to G i.e. $*: D \rightarrow G$, where D is a subset of $G \times G$.

If an operation $*$ is a partially defined binary operation on G, then $a*b$ is undefined or $a*b$ is not contained in G for some $a,b \in G$.

A semigroupoid (which is also called as partial semigroup by Bergelson V., Blass A. and Hindman N., in "Partition theorems for spaces of variable words") is a nonempty set G equipped with a partially defined binary operation $*$ on G and this operation is associative in the following sense:

If either $(a*b)*c$ or $a*(b*c)$ is defined, then so is the other and $(a*b)*c = a*(b*c)$.

Let $(G,*)$ is a semigroupoid. Then G is called a groupoid if it satisfies the following conditions:

1. For each a in G, there are left and right identity elements e_r and e_l such that $e_l * a = a = a * e_r$,
2. Each a in G has an inverse a^{-1} for which $aa^{-1} = e_l$ and $a^{-1}a = e_r$.

An algebraic structure called as an effect algebra, has recently been introduced for investigations in the foundations of quantum mechanics by Foulis D., and Bennett M. K., in the paper "Effect algebras and unsharp quantum logics".

Moreover, effect algebras play a fundamental role in recent investigations of fuzzy probability theory. The connection between effect algebras and probablity theory can be found in "Fundamentals of fuzzy probability theory" by Bugajski S.

The definition of an effect algebra is given as the following:

An effect algebra is an algebraic system $(E,0,1,*)$ where 0,1 are distinct elements of E and $*$ is a partially defined binary operation on E that satisfies the following conditions for all a,b and $c \in E$:

($E1$): If $a*b$ is defined then $b*a$ is defined and $a*b=b*a$,

($E2$): If $a*b$ and $(a*b)*c$ are defined, then $b*c$ and $a*(b*c)$ are defined and $a*(b*c) = (a*b)*c$,

($E3$): For every $a \in E$, there exists a unique $a' \in E$ such that $a*a'$ is defined and $a*a' = 1$,

($E4$): If $a*1$ is defined, then $a=0$.

This study occured by inspiration of the definitions of effect algebra and groupoid. In this paper, we introduced the notion named partial group as a generalization of definition of group. We also defined the notions of partial subgroup, cyclic partial subgroup and partial group homomorphism and obtained some fundamental properties.

PARTIAL GROUPS

In the following we will define the partial group. For brevity, if $a*b \in G$ for any $a,b \in G$, $a*b$ will be written as ab.

Definition 2.1: Let G be a nonempty set. G is called a partial group if the following conditions hold for all a,b and $c \in G$:

($PG1$): If ab, $(ab)c$, bc and $a(bc)$ is defined, then $(ab)c=a(bc)$,

($PG2$): For every $a \in G$, there exists an $e \in G$ such that ae and ea are defined and $ae=ea=a$,

($PG3$): For every $a \in G$, there exists an $a' \in G$ such that aa' and $a'a$ are defined and $aa' = a'a = e$.

The element $e \in G$ satisfies ($PG2$) in Definition 2.1 is called identity element of G and the element $a' \in G$ satisfies ($PG3$) in Definition 2.1 is called the inverse of a and denoted by a^{-1}, in generally.

If the non empty set G satisfies only the axiom ($PG1$) then it is called semi partial group. Note that semi partial group differs from partial semigroup.

A partial group G is said to be abelian or commutative if $ab=ba$ for all $a,b \in G$ such that ab and ba are defined.

Proposition 2.2 Let G be a semi partial group. Then G is a partial group if and only if the following conditions hold:

1. There exists an element $e \in G$ such that ea and ae are exist and $ea=ae=a$.
2. For each $a \in G$, there exists an element $a^{-1} \in G$ such that $a^{-1}a$ and aa^{-1} are defined and $a^{-1}a=aa^{-1}=e$.

Proof. The proof of proposition can be done straight forward from the definition of partial group and semi partial group.

Remark 2.3. Note that the definition of associativity is different from the definitons in effect algebra and groupoid.

Proposition 2.4. Every group is a partial group. Conversely it is not true. A partial group which is closed under its operation is a group.

Proof. It is clear by Definition 2.1.

Remark 2.5. In the above descriptions some concepts on group theory and partial groups have similarity. But if we want to give some counter descriptions we can look at *Generalized Associative Law* and *Generalized Commutative Law*. These properties give us some important results for group theory. But for partial groups these results do not work in general. Let remind these two law and think about why they do not work on partial groups:

Generalized Associative Law: If G is a semigroup and $a_1,a_2,\ldots,a_n \in G$ then any two meaningful products of $a_1,a_2,\ldots,a_n$ in this order are equal.

But in partial groups this law is not valid, because of the * operation is not closed and associativity property differs from group theory.

Similarly, we can state the following theorem:

Generalized Commutative Law: If G is a commutative semigroup and $a_1,a_2,\ldots,a_n \in G$, then for any permutation $i_1,i_2,\ldots,i_n$ of $1,2,\ldots,n$

$$a_1 a_2 \ldots a_n = a_{i_1} a_{i_2} \ldots a_{i_n} .$$

But unfortunately, this is not possible in commutative partial groups too. These explanations states that; partial group structure is weaker than group structure. In the following some examples are given for partial groups:

Example 2.6. Let $G = \{0, \mp 1, \ldots, \mp n\}$ where $n \in \mathbb{Z}^+$ and $+$ be known addition operation on $\mathbb{Z}$. Then it is easily seen that G is a partial group but it is not a group.

Example 2.7. Let $G = \mathbb{Z}^* \cup \left\{ \frac{1}{n} \middle| n \in \mathbb{Z}^* \right\}$ where $\mathbb{Z}^* = \mathbb{Z} - \{0\}$. Then it is easily seen that G is a partial group but it is not a group by the known multiplication on $\mathbb{R}$.

Example 2.8. Let $G=[-r,r]$ where $\mathbb{R}^+$ and $+$ be known addition operation on $\mathbb{R}^+$. Then it is easily seen that G is a partial group but is not a group.

Example 2.9. The set $G = \left\{ 1, -1, i, -i, 2i, \frac{-i}{2} \right\}$, where i is a complex number with the multiplication operation on $\mathbb{C}$ is a partial group.

Proposition 2.10. Let G be a partial group. Then the following conditions are hold:

1. The identity element e is unique,
2. For each $a \in G$, a^{-1} unique,
3. For each $a \in G$, $(a^{-1})^{-1}=a$,
4. İf ab $(ab)b^{-1}$, $b^{-1}a^{-1}$, $(ab)b^{-1}a^{-1}$, $a^{-1}(ab)$ and $b^{-1}a^{-1}(ab)$ are defined, then $(ab)^{-1}=b^{-1}a^{-1}$,

5. İf ab, ac, $a^{-1}(ab)$ and $a^{-1}(ac)$ are defined and $ab=ac$, then $b=c$,
6. İf $ab=ac$, ba, ca, $(ba)a^{-1}$ and $(ca)a^{-1}$ are defined and $ba=ca$ then $b=c$.

Proof.

1. Suppose G has two identity elements such as e_1,e_2. Then by the definition of identity element e_1e_2, e_2e_1 are defined and $e_1e_2= e_2e_1= e_1=e_2$Thus the identity element e is unique.
2. Similar to the proof of (i).
3. This comes from the axiom ($PG3$).
4. Since ab, $(ab)b^{-1}$ are defined and from ($PG3$), bb^{-1} is defined then by ($PG1$) we have $(ab)b^{-1} = a(bb^{-1}) = ae = a$. In this case, since $b^{-1}a^{-1}$ and $(ab)b^{-1}a^{-1}$ are defined, we get $e = aa^{-1} = \left(\left(ab\right)b^{-1}\right)a^{-1} = \left(ab\right)\left(b^{-1}a^{-1}\right)$. Similarly, we get $a^{-1}(ab) = (a^{-1}a)b = eb = b$. Then we obtain

$$e = b^{-1}b = \left(b^{-1}\left(ab\right)\right) = e = b^{-1}b = b^{-1}\left(a^{-1}\left(ab\right)\right) = \left(b^{-1}a^{-1}\right)\left(ab\right)$$

Hence $(ab)^{-1}=b^{-1}a^{-1}$.

5. Let $ab=ac$. Then we have $a^{-1}(ab)=a^{-1}(ac)$. Hence we get $(a^{-1}a)b=(a^{-1}a)c$ that is $b=c$.
6. Similar to the prof of (v).

Theorem 2.11. Let G is a partial group and $a,b\in G$. Then the following conditions are valid:

1. If $a^{-1}b$ and $a(a^{-1}b)$ are defined then the equation $ax=b$ has a unique solution x and $x=a^{-1}b$.
2. If ba^{-1} and $(ba^{-1})a$ are defined then the equation $ya=b$ has a unique solution y and $y=ba^{-1}$.

Proof. The proof of the theorem comes from the Proposition 2.10.(v).

Example 2.12. For the partial group in Example 2.7. the equations $4x=2$ and $\left(1/3\right)x = 5$ have solutions but the equation $4x=3$ has not got any solution in G.

Lemma 2.13. Let G and H be two partial groups. Then the Cartesian product of G and H which is defined as

$$G\times H = \{\left(g,h\right) \mid g\in G,\ \mathrm{h}\in H\}$$

is also a partial group.

Proof. Let the operation * on $G\times H$ defined as

$(g_1,h_1)*(g_2,h_2)=(g_1g_2,h_1h_2)$.

Then we need to show that the axioms of partial group are satisfied:

($PG1$): If xy, $(xy)z$ and $x(yz)$ is defined on $G\times H$ then we need to show $(xy)z=x(yz)$. Let $x=(g_1,h_1)$, $y=(g_2,h_2)$ and $z=(g_3,h_3)$. Since xy is defined then $xy=(g_1g_2,h_1h_2)$ is defined. This means that g_1g_2 and h_1h_2 is defined. Similarly, since $(xy)z$ and $x(yz)$ is defined on $G\times H$ we obtain $(g_1g_2)g_3$, $(h_1h_2)h_3$ and $g_1(g_2g_3)$, $h_1(h_2h_3)$ are defined. Since G and H are partial groups we handle that $(xy)z=x(yz)$.

($PG2$): For each $(g,h)\in G\times H$, let $(e_G,e_H)\in G\times H$ is candidate for the identity element of $G\times H$. Since G and H are partial groups e_G and e_H are identity elements of G and H. So that the equality $(g,h)(e_G,e_H)=(g,h)=(e_Gg,e_Hg)$ holds.

($PG3$): For every $(g,h)\in G\times H$ let $(g^{-1},h^{-1})\in G\times H$ candidate for the inverse element of $G\times H$. Since G and H are partial groups g^{-1} and h^{-1} are inverse elements of g and h in G,H. So that the equality $(g,h)(g^{-1},h^{-1})=(e_G,e_H)=(g^{-1},h^{-1})(g,h)$

Definition 2.14. Let G be a partial group and $a_1,a_2,\ldots,a_n\in G$. If a_1a_2 and $(a_1a_2)a_3$ are defined then we can define

$$\prod_{i=1}^{3}a_i=(a_1a_2)a_3 .$$

By recursion for $n\geq 4$, if $\prod_{i=1}^{n-1}a_i$ and $\left(\prod_{i=1}^{n-1}a_i\right)a_n$ are defined then we can define

$$\prod_{i=1}^{n}a_i=\left(\prod_{i=1}^{n-1}a_i\right)a_n .$$

Specially, for any $a\in G$, we can define $a^0=e$, $a^1=a$. For $n\geq 2$ if $\prod_{i=1}^{n-1}a$ and $\left(\prod_{i=1}^{n-1}a\right)a$ are defined then we can define $a^n=\left(\prod_{i=1}^{n-1}a\right)a$ for $a\in G$, $a^{-1}\in G$. If $(a^{-1})^n$ is defined then we define $(a^{-1})^n=a^{-n}$. Also,

Remark 2.15. In a partial group, we defined a^n as $a^n = \left(\prod_{i=1}^{n-1} a\right)a$ for $n \geq 2$. Note that in generally, in a partial group, $a^n = \left(\prod_{i=1}^{n-1} a\right)a$ can not be equal to $a^n = a\left(\prod_{i=1}^{n-1} a\right)$ as we can see in the following example for a partial group.

Example 2.16. Let $G = \{e, a, b, c\} \subset S = \{e, a, b, c, d\}$ and "•" be a partial defined operation on G as the following table:

$\cdot$	e	a	b	c
e	e	a	b	c
a	a	b	c	e
b	b	c	e	a
c	c	e	d	b

Note that $c.b$ is undefined. Then G is not a group but it is a partial group. We can see that $c^2 \cdot c = b \cdot c = a$ and $c \cdot c^2 = c \cdot b = d$. So that $c^2 \cdot c \neq c \cdot c^2$.

Corollary 2.17. Let G be a partial group and n,i, j and k be nonnegative integers. Then the following statements are satisfied with;

(i) if a^n is defined, then a^k is defined for all $k \leq n$,

(ii) if for $2 \leq i+j \leq n$, $a^i a^j$ is defined, $a^i a^j = a^{i+j}$..

Definition 2.18. Let G be a partial group, $m \in Z^+$ and $a \in G$. If a^m is defined and m is the least integer such that $a^m = e$, the number m is called the order of a. In this case it is called that a has finite order. If there does not exist an $m \in Z^+$ such that $a^m = e$ (if only $a^0 = e$), then it is called that a has infinite order. The order of a is denoted by $|a|$.

Example 2.19. For the partial group in Example 2.9., $|i| = 4$, $|2i| = \infty$.

Definition 2.20. Let G be a partial group.

$$Z(G) = \{x \in G \mid \text{If } ax \text{ and } xa \text{ are defined for all } a \in A \text{ and } ax = xa\}$$

is called the center of G.

Remark 2.21. A partial group is centerless if $Z(G)$ is trivial, i.e., it consists of only the identity element. If G is commutative then $G = Z(G)$.

Definition 2.22. Let G be a partial group and H be a nonempty subset of G. If H is a partial group with the operation on G then H is called a partial subgroup of G.

Example 2.23. Let G be the partial group in Example 2.6. and $H=\{0,\mp1,\ldots,\mp k\}$ where $0\leq k\leq n$ and $k\in Z$. Then H is a partial subgroup of G.

Theorem 2.24. Let G be a partial group and H be a nonempty subset of G. Then H is a partial subgroup of G if and only if the following conditions are hold with;

(i) $e\in H$,

(ii) $a^{-1}\in H$ for all $a\in H$.

Proof. The proof of the theorem is straightforward by definition.

Proposition 2.25. The center $Z(G)$ of a partial group G is a partial subgroup of G.

Proof. We need to show that;

1. Is $e\in G$ contained in $Z(G)$?

Since e is an identity element of the partial group G by $(PG2)$ for all $a\in G$, $ae=ea$. Then e is contained in $Z(G)$.

2. If $x\in Z(G)$, we need to show that $x^{-1}\in Z(G)$.

Since $x\in Z(G)$, for all a in G, ax and xa are defined and $ax=xa$. At the same time $xx^{-1}=x^{-1}x=e\in Z(G)$, sc. xx^{-1}, $x^{-1}x\in Z(G)$. Then for all $a\in G$

$$\begin{aligned}
ax^{-1} &= (ea)x^{-1} \\
&= \left(\left(xx^{-1}\right)a\right)x^{-1} \\
&\underset{PG1}{=} \left(x\left(x^{-1}a\right)\right)x^{-1} \\
&\underset{x\in Z(G)}{=} \left(\left(x^{-1}a\right)x\right)x^{-1} \\
&\underset{PG1}{=} (x^{-1}a)(xx^{-1}) \\
&= \left(x^{-1}a\right)e \\
&= x^{-1}a
\end{aligned}$$

and so that $x^{-1}\in Z(G)$. This completes the proof.

Example 2.26. Let G be a partial group, A and B be partial subgroups of G. Then the sets $A\cap B$ and $A\cup B$ are partial subgroups of G. In generally, if $\{H_i{:}i\in I\}$ is a family of partial subgroups of Gwhich are finite or infinite, then the sets $\bigcap_{i\in I}H_i$ and $\bigcup_{i\in I}H_i$ are partial subgroups of G.

Let G be a partial group and $a \in G$. The smallest partial subgroup which contains a is $\{e,a,a^{-1}\}$. Assume that X be a subset of G. Then the smallest partial subgroup of G contains X is the set $X \cup X^{-1} \cup \{e\}$ where the set X^{-1} is the set of inverses of elements of the set X.

Example 2.27. For the partial group $G = \{1,-1,i,-i,2i,-\frac{i}{2}\}$ in Example 2.9, the smallest partial subgroup contains i is given by $\{1,i,-i\}$. The smallest partial subgroup contains $\{-1,i\}$ is defined by $\{1,-1,i,-i\}$.

Let G be a partial group and let a be an element of G such that the elements a^k for all $k \in Z$ are defined. Assume taht the set is denoted by $\{a^k : k \in \mathrm{Z}\} = \langle a \rangle$. It is clear that the set $\langle a \rangle$ is a partial subgroup of G. The partial subgroup $\langle a \rangle$ of G is called the cyclic partial subgroup generated by a. If there exists an element a in G such that $\langle a \rangle = G$, then G is called a cyclic partial group.

Example 2.28. For the partial group in Example 2.7., the set $\langle 2 \rangle = \{2^k : k \in \mathrm{Z}\}$ is a cyclic partial subgroup of G.

Example 2.29. In the partial group in Example 2.17, since $\langle a \rangle = G$, G is a cyclic partial group.

Remark 2.30. All partial subgroups of a cyclic partial group can not be cyclic. For example, in Example 2.17, the partial subgroup $H=\{e,a,c\}$ is not cyclic partial subgroup.

Up to this part we gave some partial subgroup examples and explained some related structures. At this point we thought about Lagrange's theorem which is not valid for partial groups. Let us remind Lagrange's theorem in group theory:

Lagrange's Theorem: If H is a subgroup of a group G then $|G|=[G:H].|H|$. In particular, if G is finite the order of subgroup of G divides the order of G.

But in partial groups, let consider the example 2.6: We can say that the subset

$$S = \{0,\pm1,\pm2,\pm3,\pm4,\pm5\} \subseteq G = \{0,\pm1,\pm2,\ldots,\pm n\}$$

is also a partial subgroup of G. The subset S consists of 11 elements and the set G consists of $2n+1$ elements. But the division algorithm in Lagrange's theorem does not work in general except specific numbers. So that Lagrange's theorem is not valid for partial groups under these assumptions.

Definition 2.31. Let G and H be partial groups. A function $\phi:G\to H$ is called a partial group homomorphism if for all $a,b\in G$ such that ab is defined in G, $\phi(a)\ \phi(b)$ is defined in H and $\phi(ab)= \phi(a)\ \phi(b)$.

If ϕ is injective as a map of sets, ϕ is said to be a monomorphism. If ϕ is surjective, ϕ is called an epimorphism. If ϕ is bijective, ϕ is called by an isomorphism. In this case G and H are said to be isomorphic and written as $G\cong$H.

Example 2.32. Let G be partial group in Example 2.7. The map f: $G\to G$ f given by $x\mapsto\frac{1}{x}$ is an isomorphism. But the map given by $x\mapsto x^2$ is a homomorphism, it is not an isomorphism.

Definition 2.33. Let G and G be partial groups and ϕ: $G\to H$ be a partial group homomorphism. Then,

1. $Ker(\phi)=\{g\in G|\phi(g)=e_H\}$ is called the kernel of ϕ.
2. If $\varnothing\neq A\subseteq G$ then the set $\phi(A)=\{\phi(g)\mid g\in A\}$ is called the image of A under ϕ and $\phi(G)$ is the image of ϕ.
3. If $\varnothing\neq B\subseteq H$ then the set $\phi^{-1}(B)=\{g\in G|\ \phi(g)\in B\}$ is called the preimage of B.

Proposition 2.34. Let G and H be partial groups and φ: $G\to H$ be a homomorphism of partial groups. Then the following conditions are satisfied:

1. If A is a partial subgroup of G, then $\varphi(A)$ is a partial subgroup of H.
2. If B is a partial subgroup of H, then $\varphi^{-1}(B)$ is a partial subgroup of G.
3. $Ker(\varphi)$ is a partial subgroup of G.

Proof. Since A is a partial subgroup of G, then $e_G\in A$. Using φ is a homomorphism $\varphi(e_G)\in\varphi(A)$, and we obtain $e_H=\varphi(e_G)\in\varphi(A)$. After that let us show if $x\in\varphi(A)$ whether or not $x^{-1}\in\varphi(A)$. If $x\in\varphi(A)$ then there exists an element a in A such that $x=\varphi(A)$ and then since A is partial subgroup of G, when $a\in A$, also $a^{-1}\in A$. Since φ is a homomorphism we get $\varphi(a^{-1})\in\varphi(A)$. We know from the definition of partial group that aa^{-1} and $a^{-1}a$ are defined. Then the following equality

$$e_H=\varphi(e_G)=\varphi(aa^{-1})=\varphi(a^{-1}a)=\varphi(a)\varphi(a^{-1})=\varphi(a^{-1})\varphi(a)$$

is valid. So that $\varphi(a^{-1})=\varphi(a)^{-1}$ and $x^{-1}\in\varphi(A)$.

Proofs of axioms (ii) and (iii) are held in similar a way of the proof of (i).

CONCLUSION

In this paper, we introduced a notion named partial groups using a nonempty set and an operation which is not a binary operation on this set as a generalization of group theory. We also defined the notions of partial subgroup, cyclic partial group, cyclic partial subgroup and partial group homomorphism. We also obtained some fundamental properties about these notions. This work is an entry to a new concept and applicable to applied mathematics, such as fuzzy theory, probabilistic theory. We will continue to consider the properties of partial group and then introduce partial ring in the future work. Also, by reviewrs request we can prepare a new section as a second chapter of this partial groups such as partial free groups, partial quotient groups, isomorphism theories in partial groups, etc... by considering different properties and structers of groups.

ADDITIONAL READING

Beltrametti, E. G., & Bugajski, S. (1995). A classical extensions of quantum mechanics. *Journal of Physics. A, Mathematical and General*, *28*(:12), 3329–3343. doi:10.1088/0305-4470/28/12/007

Bergelson, V., Blass, A., & Hindman, N. (1994). Partition theorems for spaces of variable words. *Proceedings of the London Mathematical Society*, *68*(3), 449–476. doi:10.1112/plms3-68.3.449

Brandt, W. (1926). Über eine Verallgemeinerung des Gruppengriffes. *Mathematische Annalen*, *96*(1), 360–366. doi:10.1007/BF01209171

Brown, R. (1987). From groups to groupoids: A brief survey. *Bulletin of the London Mathematical Society*, *19*(2), 113–134. doi:10.1112/blms/19.2.113

Bugajski, S. (1996). Fundamentals of fuzzy probability theory. *International Journal of Theoretical Physics*, *35*(11), 2229–2244. doi:10.1007/BF02302443

Chevalley, C. (1956). *Fundemental Concepts of Algebra*. New York: Academic Press, Inc.

Dokuchaev, M., Exel, R., & Piccione, P. (2000). Partial Representations and Partial Group Algebras. *Journal of Algebra, 226*, 505–532. doi: 10.1006/jabr.1999.8204

Foulis, D., & Bennett, M. K. (1994). Effect algebras and unsharp quantum logics. *Foundations of Physics*, *24*(10), 1331–1352. doi:10.1007/BF02283036

Fuchs, L. (1970). *Infinite Abelian Groupd*. New York: Academic Press, Inc.

Goldhaber, J. (1970). *And Ehrlich, G., Algebra*. New York: The Macmillan Company.

Gorenstein, D. (1968). *Finite Groups*. New York: Harper and Row, Publishers.

Hall, M. (1970). *The Theory of Groups*. New York: Academic Press, Inc.

Hazewinkel, M. (1994). Groupoid. In *Encyclopedia of Mathematics*. Kluwer Academic Publishers.

Herstein, I. (1965). *Topics in Algebra*. Waltham, MA: Addison-Wesley, Publishing Company, Inc.

Hungerford, T. W. (1974). Algebra. Springer-Verlag.

Jekel, S. (2013). *Partial Groups*. Northeastern University. Doi: 10.13140/2.1.1409.3127

Kurosh, A. G. (1960). *The Theory of Groups* (Vols. 1–2). New York: Chelsea Publishing Company.

Lang, S. (1965). *Algebra*. Reading, MA: Addison-Wesley, Publishing Company, Inc.

MacLane, S., & Birkhoff, G. (1967). *Algebra*. New York: The Macmillan Company.

Rotman, J. (1973). *The Theory of Groups* (2nd ed.). Boston: Allyn and Bacon, Inc.

Scott, W. R. (1964). *Group Theory*. Englewood Cliffs, NJ: Prentice-Hall, Inc.

Van der Waerden, B. L. (n.d.). *Algebra* (7th ed.; 2 vols.). New York: Frederick Ungar Publishing Co.

Zassenhaus, H. (1958). *The Theory of Groups*. New York: Chelsea Publishing Company.

Chapter 2
Application of the Exponential Rational Function Method to Some Fractional Soliton Equations

Mustafa Ekici
https://orcid.org/0000-0003-2494-8229
Usak University, Turkey

Metin Ünal
Usak University, Turkey

ABSTRACT

In this chapter, the authors study the exponential rational function method to find new exact solutions for the time-fractional fifth-order Sawada-Kotera equation, the space-time fractional Whitham-Broer-Kaup equations, and the space-time fractional generalized Hirota-Satsuma coupled KdV equations. These fractional differential equations are converted into ordinary differential equations by using the fractional complex transform. The fractional derivatives are defined in the sense of Jumarie's modified Riemann-Liouville. The proposed method is direct and effective for solving different kind of nonlinear fractional equations in mathematical physics.

DOI: 10.4018/978-1-7998-0134-4.ch002

INTRODUCTİON

Fractional calculus deals with the derivatives and integrals of any real or complex order (Greenberg, 1978; Podlubny, 1999; Hilfer, 2000). This type of calculus was initiated and developed by many great scientists like Leibniz, L'Hospital, Euler, Bernoulli, Riemann, Liouville, and many others (Kilbas, Srivastava, & Trujillo, 2006). Many researchers have been working to develop the fractional calculus and make use of its applicability to various areas of mathematics, physics and engineering. It has been found that fractional calculus can be extensively used for many physical phenomena as a strong and effective tool to describe mathematical modeling (Caputo, & Fabrizio, 2015; Erbe, Goodrich, Jia, & Peterson, 2016).

In recent years, fractional differential equations (FDEs) have been popular among the researchers and gained much attention. FDEs are generalization of classical differential equations of integer order. They are widely used to describe various complex phenomena such as in fluid flow, control theory, signal processing, systems identification, viscoelasticity, acoustic waves etc. The exact solutions of nonlinear fractional differential equations by using various different methods have been investigated by many researchers. Many useful method for finding exact solutions of FDEs have been proposed. Such as the tanh-function expansion method (Fan, 2000), the Jacobi elliptic function expansion (Liu, Fu, Liu, & Zhao, 2001), the homogeneous balance method (Wang, 1995; Wang, 1996; Wang, Zhou, &Li, 1996), the trial function method (Kudryashov, 1990), the exponential function method (He & Wu, 2006), the (G'/G)-expansion method (Zhang, Tong, & Wang, 2008), the sub-ODE method (Zhang, Wang, & Li, 2006; Wang, Li, & Zhang, 2007), Adomian decomposition method (El-Sayed & Gaber, 2006; El-Sayed, Behiry, & Raslan, 2010), the homotopy analysis method (Arafa, Rida, & Mohamed, 2011), the differential transformation method (Odibat & Momani, 2008), the fractional sub-equation method (S. Zhang & H. Q. Zhang, 2011; Guo, Mei, Li, & Sun, 2012; Lu, 2012) and so on.

There are several types of fractional derivatives, among the popular of them are given in the following sections.

Riemann-Liouville Fractional Derivatives

The Riemann-Liouville fractional integration operator is defined as

$$I^{\alpha} f(t)=\frac{1}{\Gamma(\alpha)}\int_{a}^{t}(t-\xi)^{\alpha-1} f(\xi)\,d\xi$$

where $a>0$, $t>\alpha$, $a,t,\alpha \in R$ (Gorenflo & Mainardi, 2000). The Riemann-Liouville fractional differential operator is defined as

$$D^{\alpha} f(t) = \begin{cases} \dfrac{1}{\Gamma(n-\alpha)} \dfrac{d^n}{dt^n} \displaystyle\int_a^t \dfrac{f(\xi)}{(t-\xi)^{\alpha+1-n}} d\xi, & n-1<\alpha<n \in N \\ \dfrac{d^n}{dt^n} f(t), & \alpha = n \in N \end{cases}$$

where $a>0$, $t>\alpha$, $a,t,\alpha \in R$ (Gorenflo & Mainardi, 2000). The well-known fractional integral is the Riemann-Liouville type which is based on the generalisation of the usual Riemann integral $\int_a^x f(t)\,dt$ (Jarad, Abdeljawad, & Baleanu, 2012).

Caputo's Fractional Derivatives

An alternative definition of the fractional derivative is the Caputo's fractional derivative, originally introduced by Caputo (Caputo, 1967) in the following form,

$${}^{C}D^{\alpha} f(t) = \frac{1}{\Gamma(n-\alpha)} \int_a^t \frac{f^{(n)}(\xi)}{(t-\xi)^{\alpha+1-n}} d\xi, \quad n-1<\alpha<n$$

where $n \in N$ and $f(t)$ has $n+1$ continuous bounded derivatives in every finite interval $[a,t]$.

The main advantage of Caputo's approach is that the initial conditions for fractional differential equations with Caputo derivatives take on the same form as for integer-order differential equations. Although the Riemann-Liouville fractional integrals and derivatives contributed immensely to the development of the theory of fractional calculus, it turns out that this approach has certain disadvantages when trying to model real-world phenomena with fractional differential equations. Another difference between the Riemann-Liouville definition and the Caputo definition is that the Caputo derivative of a constant is zero, whereas in the case of a finite value of the lower terminal 'a' the Riemann-Liouville fractional derivative of a constant C is not zero.

Hadamard Fractional Derivative

Hadamard proposed a fractional power of the form $\left(x\frac{d}{dx}\right)^{\alpha}$. This fractional derivative is invariant with respect to dilation on the whole axis. The Hadamard approach to fractional integral was based on the generalisation of the n-th integral

$$(J^n f)(t)=\int_a^x \frac{dt_1}{t_1}\int_a^{t_1}\frac{dt_2}{t_2}\ldots\int_a^{t_{n-1}} f(t_n)\frac{dt_n}{t_n}.$$

Just like Riemann-Liouville, Hadamard derivative has its own disadvantages as well, one of which is the fact that the derivative of a constant is not equal to zero in general. The authors in (Jarad, Abdeljawad, & Baleanu, 2012) resolved these problems by modifying the derivative into a more suitable one having physically interpretable initial conditions similar to the ones in the Caputo settings.

Jumarie's Modified Riemann-Liouville Derivative

The Jumarie's modified Riemann-Liouville derivative of order α is defined as in the following:

$$D_t^a f(t)=\begin{cases}\dfrac{1}{r(1-a)}\displaystyle\int_0^t (t-\xi)^{-a-1}\left[f(\xi)-f(0)\right]d\xi, & a<0\\ \dfrac{1}{r(1-a)}\dfrac{d}{dt}\displaystyle\int_0^t (t-\xi)^{-a}\left[f(\xi)-f(0)\right]d\xi, & 0<a<1\\ (f^{(n)}(t))^{(a-n)}, & n\le a<n+1, n\ge 1\end{cases}$$

where $f(t)$ is a continuous $f:\mathbb{R}\to\mathbb{R},\ t\to f(t)$ function and $\Gamma(\alpha)$ is the gamma function defined as:

$$\Gamma(\alpha)=\lim_{n\to\infty}\frac{n^{\alpha}n!}{\alpha(\alpha+1)(\alpha+2)\ldots(\alpha+n)}.$$

The Jumarie's modified Riemann-Liouville derivative has many interesting properties such as;

- $D_t^{\alpha} t^r = \dfrac{\Gamma(1+r)}{\Gamma(1+r-\alpha)} t^{r-\alpha}$,
- $D_t^{\alpha}\left(f(t)g(t)\right) = g(t)D_t^{\alpha} f(t) + f(t)D_t^{\alpha} g(t)$
- $D_t^{\alpha} f\left(g(t)\right) = f_g'\left[g(t)\right]D_t^{\alpha} g(t) = D_g^{\alpha} f\left[g(t)\right]\left(g'(t)\right)^{\alpha}$

The α-order derivative of a constant is zero, and it can be applied to the functions whether are differentiable or not.

In this chapter we consider the exponential rational function method (Demiray, 2004) for solving fractional partial differential equations in the sense of modified Riemann-Liouville derivative by Jumarie (Jumarie, 2006). The rest of this chapter is organized as follows. In Section 2, we describe the exponential rational functional method for solving fractional partial differential equations. In Section 3, we make use of this method to find new exact solutions for some space-time fractional partial differential equations and some discussions are given in the conclusion.

DESCRIPTION OF THE EXPONANTIAL RATIONAL FUNCTION METHOD

In this section we give the description of the exponential rational functional method for solving FDEs. Suppose that a nonlinear FDEs are in the following form:

$$P\left(u, D_t^{\alpha} u, D_x^{\beta} u, D_t^{2\alpha} u, D_t^{\alpha} D_x^{\beta} u, D_x^{2\beta} u, \ldots\right) = 0, \quad 0 < \alpha, \ \beta < 1 \tag{1}$$

where $u=u(x,t)$ is an unknown function and P is a polynomial in u and its various partial fractional derivatives. This method can be summerized in the following steps (Yusufoğlu & Bekir, 2007).

Step 1: The fractional complex transform converts fractional differential equations into ordinary differential equations, so that all anlytical methods devoted to the advanced calculus can be easily applied fractional calculus (Li & He, 2010). Assign a compound variable ξ with the real variables x and t by the following transformation:

$$u(x,t) = U(\xi), \quad \xi = \frac{kx^{\beta}}{\Gamma(1+\beta)} + \frac{st^{\alpha}}{\Gamma(1+\alpha)} + \xi_0 \tag{2}$$

where k and s are arbitrary constants. The wave variable assigned in Equation 2 transforms Equation 1 into the following ordinary differential equation (ODE);

$$Q(U,U',U'',U''',\ldots)=0 \tag{3}$$

where Q is a polynomial of U and its various derivatives. The superscripts stand for ordinary derivatives with respect to ξ. Integrate Equation 3 one or more as possible and for simplicity, set the constant(s) of integration to zero.

Step 2: We express the exact solution of Equation 3 in the following form:

$$U(\xi)=\sum_{n=0}^{N}\frac{a_n}{\left(1+e^{\xi}\right)^n} \tag{4}$$

where $a_i(i=0,1,2,\ldots,N)$ are arbitrary constants to be determined later, such that $a_N \neq 0$. This formulation plays a significant and fundamental part for finding the exact solutions of mathematical problems.

Step 3: To determine the positive integer N, substitute Equation 4 into Equation 3 and take the homogeneous balance between the highest order derivatives and the highest order nonlinear terms. If the degree of $U(\xi)$ is $\deg[U(\xi)]=N$, then, the degree of the other expressions will be as follows:

$$deg\left[\frac{d^mU(\xi)}{d\xi^m}\right]=N+m, \quad deg\left[U^m\left(\frac{d^lU(\xi)}{d\xi^l}\right)^p\right]=Nm+p(N+l)$$

Step 4: Collect all terms with the same power of $e^{i\xi}$ $(i=0,1,2,\ldots)$ and equate the coefficients of $e^{i\xi}$ to zero in Equation 3. This procedure yields a system of algebraic equations which can be easily solved with the help of mathematical software programme such as Maple, Mathematica etc.. This completes the determination of the solutions.

In the following section, we present three examples to illustrate the applicability of the exponantial rational function method and fractional complex transform to solve nonlinear fractional differential equations.

APPLICATION OF THE EXPONANTIAL RATIONAL FUNCTION METHOD

In this section we apply the exponential rational function method for solving for some FDEs.

The Time-Fractional Fifth-Order Sawada–Kotera Equation

The time-fractional fifth-order Sawada–Kotera equation is given

$$D_t^{\alpha} u + u_{xxxxx} + 45u_x u^2 + 15\left(u_x u_{xx} + u u_{xxx}\right) = 0, \quad 0 < \alpha \leq 1 \tag{5}$$

where D_t^{α} and D_x^{α} are Jumarie's modified Riemann–Liouville derivative of order α defined in Section 1. Equation 5 is the variation of the fifth-order Sawada–Kotera equation (Liu & Dai, 2008). There are a lot of studies for the classical Sawada–Kotera equation and some profound results have been established. The Sawada–Kotera equation is an important nonlinear evolution equation which arise in many different physical phenomena to describe the motion of long waves in shallow water under gravity and has wide applications in nonlinear optics. It is well known that wave phenomena of plasma media and fluid dynamics are modelled by kink shaped tanh solution or by bell shaped sech solution. On the other hand, it is often preferable in many physical situations to have an equation which allows us to model waves that propagate in opposite directions, and it belongs to the completely integrable hierarchy of higher-order KdV equations, and has many sets of conservation laws (Goktas, & Hereman, 1997). Many properties of Equation 5 have been researched intensively by other authors. For example, It has multisoliton solutions, conserved quantities, Bäcklund transformation, Darboux transformation and so on (Wazwaz, 2010).

Now we apply the exponential rational function method to the Equation 5. Suppose that

$$u(x,t) = U(\xi), \quad \xi = kx^{\alpha} + \frac{st^{\alpha}}{\Gamma(1+\alpha)} + \xi_0 \tag{6}$$

where k, s, ξ_0 are all constants with $k, s \neq 0$. Substituting Equation 6 into Equation 5 reduces to the nonlinear ODE

$$sU' + k^5 U^{(5)} + 45kU^2U' + 15k^3\left(U'U'' + UU'''\right) = 0 \tag{7}$$

where $U' = \frac{dU}{d\xi}$. Balancing the highest order derivative term and the nonlinear term in Equation 7, we have

$$U(\xi) = a_0 + \frac{a_1}{\left(1+e^{\xi}\right)} + \frac{a_2}{(1+e^{\xi})^2}. \tag{8}$$

Next substitute Equation 8 into Equation 7, we get a polynomial equation of $e^{i\xi}\left(i = 1, 2, ..., 6\right)$, and equating all the coefficients of this polynomial to zero, we get a set of linear equations. Using the mathematical software programme Maple to solve these equations, yields the following solutions for $k, s, a_i (i=0,1,2)$:

Case 1

$$a_0 = -\frac{k^2}{3}, a_1 = 2k^2, a_2 = -2k^2, s = -k^5$$

Using these values with the Equation 8, we obtain the solitary wave solution of Equation 5:

$$U(\xi) = -\frac{k^2}{3} + \frac{k^2}{2}\operatorname{sech}^2\frac{\xi}{2}$$

where $\xi = kx^{\alpha} - \frac{k^5 t^{\alpha}}{\Gamma(1+\alpha)} + \xi_0$.

Case 2

$$a_0 = -\frac{k^2}{3}, a_1 = 4k^2, a_2 = -4k^2, s = -k^5$$

Using these values with the Eq. (8), we obtain the solitary wave solution of Eq. (5)

$$U(\xi) = -\frac{k^2}{3} + k^2\operatorname{sech}^2\frac{\xi}{2}$$

where $\xi = kx^{\alpha} - \frac{k^5 t^{\alpha}}{\Gamma(1+\alpha)} + \xi_0$.

Case 3

$$a_0 = \frac{-5k^3 + \sqrt{5k^6 - 20sk}}{30k}, a_1 = 2k^2, a_2 = -2k^2, s = -k^5$$

Using these values with the Equation 8, we obtain the solitary wave solution of Equation 5

$$U(\xi) = \frac{-5k^3 + \sqrt{5k^6 - 20sk}}{30k} + \frac{k^2}{2}\operatorname{sech}^2\frac{\xi}{2}$$

where $\xi = kx^{\alpha} - \frac{k^5 t^{\alpha}}{\Gamma(1+\alpha)} + \xi_0$.

Case 4

$$a_0 = -\frac{5k^3 + \sqrt{5k^6 - 20sk}}{30k}, a_1 = 2k^2, a_2 = -2k^2, s = -k^5$$

Using these values with the Equation 8, we obtain the solitary wave solution of Equation 5

$$U(\xi) = -\frac{5k^3 + \sqrt{5k^6 - 20sk}}{30k} + \frac{k^2}{2}\operatorname{sech}^2\frac{\xi}{2}$$

where $\xi = kx^{\alpha} - \frac{k^5 t^{\alpha}}{\Gamma(1+\alpha)} + \xi_0$.

The Space-Time Fractional Whitham-Broer-Kaup Equations

The space-time fractional Whitham-Broer-Kaup equations in the following form

$$\begin{cases} D_t^{\alpha} u + u D_x^{\alpha} u + D_x^{\alpha} v + \beta D_x^{2\alpha} u = 0 \\ D_t^{\alpha} v + D_x^{\alpha}(uv) - \beta D_x^{2\alpha} v + \gamma D_x^{3\alpha} u = 0 \end{cases}, \quad 0 < \alpha \leq 1 \tag{9}$$

where D_t^α and D_x^α are Jumarie's modified Riemann–Liouville derivative of order α defined in Section 1. Equation 9 is the generalization of the Whitham-Broer-Kaup (WBK) equations (Xu et al., 2007). The WBK equations, describe the dispersive long wave in shallow water in physical context. $u=u(x,t)$ is the field of horizontal velocity, and $v=v(x,t)$ is the height deviating from equilibrium position of liquid, β and γ are real constants that represent different diffusion powers. If $\alpha=1$, $\beta\neq 0$, $\gamma=0$, Equation 9 is the classical long-wave equations that describe shallow water wave with diffusion. If $\alpha=1$, $\beta=0$, $\gamma=1$, Equation 9 reduces to the modified Boussinesq equations (Ablowitz & Clarkson, 1990).

We apply the exponential rational function method to Equation 9. Suppose that $u(x,t)=U(\xi)$, $v(x,t)=V(\xi)$, and

$$\xi = \frac{kx^\alpha}{\Gamma(1+\alpha)} + \frac{st^\alpha}{\Gamma(1+\alpha)} + \xi_0$$

where k,s,ξ_0 are all constants with $k,s\neq 0$. These transformations reduce the Equation 9 to the following nonlinear ODEs

$$sU' + kUU' + kV' + \beta k^2 U'' = 0$$
$$sV' + k(UV)' - \beta k^2 V'' + \gamma k^3 U''' = 0$$

integrating these equations and taking integrating constants to zero, gives

$$\begin{aligned} &sU + \frac{1}{2}kU^2 + kV + \beta k^2 U' = 0 \\ &sV + kUV - \beta k^2 V' + \gamma k^3 U'' = 0 \end{aligned} \tag{10}$$

where $U' = \frac{dU}{d\xi}$. Balancing the highest order derivative term and the nonlinear term in Equation 10, we get the following formal solutions

$$\begin{aligned} U(\xi) &= a_0 + \frac{a_1}{(1+e^\xi)} \\ V(\xi) &= b_0 + \frac{b_1}{(1+e^\xi)} + \frac{b_2}{(1+e^\xi)^2}. \end{aligned} \tag{11}$$

We substitute Equation 11 into Equation 10, we get a polynomial equation of $e^{i\xi}\,(i=0,1,2,3)$, and equating all the coefficients of this polynomial to zero, we get a set of linear equations. Using the mathematical software programme Maple to solve these equations, yields the following solutions for $k,s,a_i,b_i(i=0,1,2)$:

Case 1

$$a_0=\mp\frac{2s}{k},\quad a_1=\pm\frac{2s}{k},$$

$$b_0=0,\quad b_1=2s\left(\beta+\frac{s}{k^2}\right),\quad b_2=-2s\left(\beta+\frac{s}{k^2}\right)$$

Using these values with the Equation 11, we obtain the solitary wave solutions of Equation 9

$$U(\xi)=\mp\frac{s}{k}\left(1+\tanh\frac{\xi}{2}\right)$$

$$V(\xi)=\pm\frac{1}{2}s\left(\beta+\frac{s}{k^2}\right)\left(\operatorname{sech}^2\frac{\xi}{2}\right)$$

where $\xi=\dfrac{kx^\alpha}{\Gamma(1+\alpha)}+\dfrac{st^\alpha}{\Gamma(1+\alpha)}+\xi_0$.

Case 2

$$a_0=0,\quad a_1=\mp\frac{2s}{k},$$

$$b_0=0,\quad b_1=2s\left(\frac{s}{k^2}-\beta\right),\quad b_2=2s\left(\beta-\frac{s}{k^2}\right)$$

Using these values with the Equation 11, we obtain the solitary wave solutions of Equation 9

$$U(\xi)=\mp\frac{s}{k}e^{-\frac{\xi}{2}}\operatorname{sech}\frac{\xi}{2}$$

$$V(\xi)=\pm\frac{1}{2}s\left(\frac{s}{k^2}-\beta\right)\left(\operatorname{sech}^2\frac{\xi}{2}\right)$$

where $\xi=\frac{kx^\alpha}{\Gamma(1+\alpha)}+\frac{st^\alpha}{\Gamma(1+\alpha)}+\xi_0$.

The Space-Time Fractional Generalized Hirota–Satsuma Coupled KdV Equations

The space-time fractional generalized Hirota–Satsuma coupled KdV equations [16] in the following form;

$$\begin{cases} D_t^\alpha u-\frac{1}{2}D_x^{3\alpha}u+3uD_x^\alpha u-3D_x^\alpha(vw)=0 \\ D_t^\alpha v+D_x^{3\alpha}v-3uD_x^\alpha v=0 \\ D_t^\alpha w+D_x^{3\alpha}w-3uD_x^\alpha w=0 \end{cases},\quad 0<\alpha\le 1 \tag{12}$$

where $u=u(x,t)$, $v=v(x,t)$ and $w=w(x,t)$. Equation 12 describe the interaction of two long waves with different dispersion relations (Abazari & Abazari, 2012). D_t^α and D_x^α are Jumarie's modified Riemann–Liouville derivative of order α defined in Section 1. When we take α=1, it reduces to the generalized Hirota-Satsuma coupled KdV equations (Wu, Geng, Hu, & Zhu, 1999). With the compatibility condition of a 4×4 matrix spectral problem, taking $w=v^*$ (where $v^*$is the conjugate of v) the generalized Hirota-Satsuma coupled KdV equations can be reduced to the complex coupled KdV equations, and taking $w=v$, it reduces to the Hirota-Satsuma equations.

Next we apply the exponential rational function method to the Equation 12. Suppose that $u(x,t)=U(\xi)$, $v(x,t)=V(\xi)$, $w(x,t)=W(\xi)$, and

$$\xi=\frac{kx^\alpha}{\Gamma(1+\alpha)}+\frac{st^\alpha}{\Gamma(1+\alpha)}+\xi_0$$

where k,s,ξ_0 are all constants with $k,s\neq 0$. These transformations reduce the Equation 12 to the following nonlinear ODEs

$$sU' - \frac{1}{2}k^3U''' + 3kUU' - 3k(VW)' = 0$$
$$sV' + k^3V''' - 3kUV' = 0 \qquad (13)$$
$$sW' + k^3W''' - 3kUW' = 0$$

where $U' = \frac{dU}{d\xi}$. Balancing the highest order derivative term and the nonlinear term in Equation 13, we get the following formal solutions

$$U(\xi) = a_0 + \frac{a_1}{(1+e^{\xi})} + \frac{a_2}{(1+e^{\xi})^2}$$
$$V(\xi) = b_0 + \frac{b_1}{(1+e^{\xi})} + \frac{b_2}{(1+e^{\xi})^2} \qquad (14)$$
$$W(\xi) = c_0 + \frac{c_1}{(1+e^{\xi})} + \frac{c_2}{(1+e^{\xi})^2}.$$

We substitute Equation 14 into Equation 13 we get a polynomial equation of $e^{i\xi}\,(i=1,2,3,4)$, and equating all the coefficients of this polynomial to zero, we get a set of linear equations. Using the mathematical software programme Maple to solve these equations, yields the following solutions for $k, s, a_i, b_i, c_i (i=0,1,2)$:

Case 1

$$a_0 = \frac{k^3+s}{3k}, \quad a_1 = -4k^2, \quad a_2 = 4k^2$$

$$b_0 = -\frac{k^3+3kc_0+4s}{3k}, \quad b_1 = 2k^2, \quad b_2 = -2k^2$$

$$c_1 = 2k^2, \quad c_2 = -2k^2$$

Using these values with the Equation 14, we obtain the solitary wave solutions of Equation 12

$$U(\xi)=\frac{k^3+s}{3k}-k^2\text{sech}^2\frac{\xi}{2}$$

$$V(\xi)=-\frac{k^3+3kc_0+4s}{3k}+\frac{k^2}{2}\text{sech}^2\frac{\xi}{2}$$

$$W(\xi)=c_0+\frac{k^2}{2}\text{sech}^2\frac{\xi}{2}$$

where $\xi=\frac{kx^\alpha}{\Gamma(1+\alpha)}+\frac{st^\alpha}{\Gamma(1+\alpha)}+\xi_0.$

Case 2

$$a_0=\frac{k^3+s}{3k},\quad a_1=-4k^2,\quad a_2=4k^2$$

$$b_0=\frac{k^3-3kc_0+4s}{3k},\quad b_1=-2k^2,\quad b_2=2k^2$$

$$c_1=-2k^2,\quad c_2=2k^2$$

Using these values with the Equation 14, we obtain the solitary wave solutions of Equation 12

$$U(\xi)=\frac{k^3+s}{3k}-k^2\text{sech}^2\frac{\xi}{2}$$

$$V(\xi)=\frac{k^3-3kc_0+4s}{3k}-\frac{k^2}{2}\text{sech}^2\frac{\xi}{2}$$

$$\xi=\frac{kx^\alpha}{\Gamma(1+\alpha)}+\frac{st^\alpha}{\Gamma(1+\alpha)}+\xi_0.$$

where $\xi=\frac{kx^\alpha}{\Gamma(1+\alpha)}+\frac{st^\alpha}{\Gamma(1+\alpha)}+\xi_0.$

Case 3

$$a_0 = \frac{k^3+s}{3k}, \quad a_1 = -2k^2, \quad a_2 = 2k^2$$

$$b_0 = -\frac{k^4 + k\sqrt{3k^4+12sk}c_0 + 4sk}{\sqrt{3k^4+12sk}}, \quad b_1 = \sqrt{\frac{k^4+4sk}{3}}, \quad b_2 = 0$$

$$c_1 = \sqrt{\frac{k^4+4sk}{3}}, \quad c_2 = 0$$

Using these values with the Equation 14, we obtain the solitary wave solutions of Equation 12

$$U(\xi) = \frac{k^3+s}{3k} - \frac{k^2}{2}\operatorname{sech}^2\frac{\xi}{2}$$

$$V(\xi) = -\frac{1}{2\sqrt{3}}\sqrt{k^4+4sk}\left(1 \mp \sqrt{1-\operatorname{sech}^2\frac{\xi}{2}}\right) - kc_0$$

$$W(\xi) = c_0 + \frac{1}{2\sqrt{3}}\sqrt{k^4+4sk}\,e^{-\frac{\xi}{2}}\operatorname{sech}\frac{\xi}{2}$$

where $\xi = \frac{kx^\alpha}{\Gamma(1+\alpha)} + \frac{st^\alpha}{\Gamma(1+\alpha)} + \xi_0$.

Case 4

$$a_0 = \frac{k^3+s}{3k}, \quad a_1 = -2k^2, \quad a_2 = 2k^2$$

$$b_0 = \frac{k^4 - k\sqrt{3k^4+12sk}c_0 + 4sk}{\sqrt{3k^4+12sk}}, \quad b_1 = -\sqrt{\frac{k^4+4sk}{3}}, \quad b_2 = 0$$

$$c_1 = -\sqrt{\frac{k^4 + 4sk}{3}}, \quad c_2 = 0$$

Using these values with the Equation 14, we obtain the solitary wave solutions of Equation 12

$$U(\xi) = \frac{k^3 + s}{3k} - \frac{k^2}{2}\operatorname{sech}^2\frac{\xi}{2}$$

$$V(\xi) = \frac{1}{2\sqrt{3}}\sqrt{k^4 + 4sk}\left(3 \pm \sqrt{1 - \operatorname{sech}^2\frac{\xi}{2}}\right) - kc_0$$

$$W(\xi) = c_0 - \frac{1}{2\sqrt{3}}\sqrt{k^4 + 4ske^{-\frac{\xi}{2}}}\operatorname{sech}\frac{\xi}{2}$$

where $\xi = \frac{kx^{\alpha}}{\Gamma(1+\alpha)} + \frac{st^{\alpha}}{\Gamma(1+\alpha)} + \xi_0$.

Case 1 and Case 2 are similar solutions, so Case 3 and Case 4 are also similar solutions.

CONCLUSİON

In this chapter we have applied the exponential rational function method to the fractional partial differential equations namely the time-fractional fifth-order Sawada-Kotera equation, the space-time fractional Whitham-Broer-Kaup equations, the space-time fractional generalized Hirota-satsuma coupled KdV equations and succesfully found abundant new exact solutions and new hyperbolic solutions, which may be useful to further understand the nonlinear physical phenomena. In order to reduce these nonlinear fractional partial differential equations to the corresponding ODEs, we use the nonlinear fractional complex transformation. This transformation guaranties that a given fractional partial differential equation reduces to another ordinary differential equation of integer order, and the solutions can be expressed by a polynomial in e^{ξ}. The proposed method is shown to be easy, useful, direct, and powerful method for dealing with the systems of FDEs.

REFERENCES

Abazari, R., & Abazari, M. (2012). Numerical simulation of generalized Hirota–Satsuma coupled KdV equation by RDTM and comparison with DTM. *Communications in Nonlinear Science and Numerical Simulation*, *17*(2), 619–629. doi:10.1016/j.cnsns.2011.05.022

Ablowitz, M. J., & Clarkson, P. A. (1990). *Nonlinear evolution equations and inverse scattering transform*. Cambridge University Press.

Arafa, A. A. M., Rida, S. Z., & Mohamed, H. (2011). Homotopy analysis method for solving biological population model. *Communications in Theoretical Physics*, *56*(5), 797–800. doi:10.1088/0253-6102/56/5/01

Caputo, M. (1967). Linear models of dissipation whose Q is almost frequency independent—II. *Geophysical Journal International*, *13*(5), 529–539. doi:10.1111/j.1365-246X.1967.tb02303.x

Caputo, M., & Fabrizio, M. (2015). A new definition of fractional derivative without singular kernel. *Progress in Fractional Differentiation and Applications*, *1*(2), 73–85.

Demiray, H. (2004). A travelling wave solution to the KdV–Burgers equation. *Applied Mathematics and Computation*, *154*(3), 665–670. doi:10.1016/S0096-3003(03)00741-0

El-Sayed, A. M. A., Behiry, S. H., & Raslan, W. E. (2010). Adomian's decomposition method for solving an intermediate fractional advection–dispersion equation. *Computers & Mathematics with Applications (Oxford, England)*, *59*(5), 1759–1765. doi:10.1016/j.camwa.2009.08.065

El-Sayed, A. M. A., & Gaber, M. (2006). The Adomian decomposition method for solving partial differential equations of fractal order in finite domains. *Physics Letters. [Part A]*, *359*(3), 175–182. doi:10.1016/j.physleta.2006.06.024

Erbe, L., Goodrich, C. S., Jia, B., & Peterson, A. (2016). Survey of the qualitative properties of fractional difference operators: Monotonicity, convexity, and asymptotic behavior of solutions. *Advances in Difference Equations*, *2016*(1), 43. doi:10.118613662-016-0760-3

Fan, E. (2000). Extended tanh-function method and its applications to nonlinear equations. *Physics Letters. [Part A]*, *277*(4), 212–218. doi:10.1016/S0375-9601(00)00725-8

Goktas, U., & Hereman, W. (1997). Symbolic computation of conserved densities for systems of nonlinear evolution equations. *Journal of Symbolic Computation, 24*(5), 591–622. doi:10.1006/jsco.1997.0154

Gorenflo, R., & Mainardi, F. (2000). *Essentials of fractional calculus*. Academic Press.

Greenberg, M. D. (1978). *Foundation of Applied Mathematics*. Prentice–Hall Inc.

Guo, S., Mei, L., Li, Y., & Sun, Y. (2012). The improved fractional sub-equation method and its applications to the space–time fractional differential equations in fluid mechanics. *Physics Letters. [Part A], 376*(4), 407–411. doi:10.1016/j.physleta.2011.10.056

He, J. H., & Wu, X. H. (2006). Exp-function method for nonlinear wave equations. *Chaos, Solitons, and Fractals, 30*(3), 700–708. doi:10.1016/j.chaos.2006.03.020

Hilfer, R. (2000). *Applications of Fractional Calculus in Physics* (Vol. 128). Singapore: World Scientific Publishing. doi:10.1142/3779

Jarad, F., Abdeljawad, T., & Baleanu, D. (2012). Caputo-type modification of the Hadamard fractional derivatives. *Advances in Difference Equations, 2012*(1), 142. doi:10.1186/1687-1847-2012-142

Jumarie, G. (2006). Modified Riemann-Liouville derivative and fractional Taylor series of nondifferentiable functions further results. *Computers & Mathematics with Applications (Oxford, England), 51*(9-10), 1367–1376. doi:10.1016/j.camwa.2006.02.001

Kilbas, A. A., Srivastava, H. M., & Trujillo, J. J. (2006). *Theory of appliciations of fractional differential equations*. Academic Press.

Kudryashov, N. A. (1990). Exact solutions of the generalized Kuramoto-Sivashinsky equation. *Physics Letters. [Part A], 147*(5-6), 287–291. doi:10.1016/0375-9601(90)90449-X

Li, Z. B., & He, J. H. (2010). Fractional complex transform for fractional differential equations. *Mathematical and Computational Applications, 15*(5), 970–973. doi:10.3390/mca15050970

Liu, C., & Dai, Z. (2008). Exact soliton solutions for the fifth-order Sawada–Kotera equation. *Applied Mathematics and Computation, 206*(1), 272–275. doi:10.1016/j.amc.2008.08.028

Liu, S., Fu, Z., Liu, S., & Zhao, Q. (2001). Jacobi elliptic function expansion method and periodic wave solutions of nonlinear wave equations. *Physics Letters. [Part A]*, *289*(1), 69–74. doi:10.1016/S0375-9601(01)00580-1

Lu, B. (2012). Bäcklund transformation of fractional Riccati equation and its applications to nonlinear fractional partial differential equations. *Physics Letters. [Part A]*, *376*(28-29), 2045–2048. doi:10.1016/j.physleta.2012.05.013

Odibat, Z., & Momani, S. (2008). A generalized differential transform method for linear partial differential equations of fractional order. *Applied Mathematics Letters*, *21*(2), 194–199. doi:10.1016/j.aml.2007.02.022

Podlubny, I. (1999).. . *Fractional Differential Equations, Academic Press*, *198*, 10–200.

Wang, M. (1995). Solitary wave solutions for variant Boussinesq equations. *Physics Letters. [Part A]*, *199*(3-4), 169–172. doi:10.1016/0375-9601(95)00092-H

Wang, M. (1996). Exact solutions for a compound KdV-Burgers equation. *Physics Letters. [Part A]*, *213*(5-6), 279–287. doi:10.1016/0375-9601(96)00103-X

Wang, M., Li, X., & Zhang, J. (2007). Various exact solutions of nonlinear Schrödinger equation with two nonlinear terms. *Chaos, Solitons, and Fractals*, *31*(3), 594–601. doi:10.1016/j.chaos.2005.10.009

Wang, M., Zhou, Y., & Li, Z. (1996). Application of a homogeneous balance method to exact solutions of nonlinear equations in mathematical physics. *Physics Letters. [Part A]*, *216*(1-5), 67–75. doi:10.1016/0375-9601(96)00283-6

Wazwaz, A. M. (2010). *Partial differential equations and solitary waves theory*. Springer Science & Business Media.

Wu, Y., Geng, X., Hu, X., & Zhu, S. (1999). A generalized Hirota–Satsuma coupled Korteweg–de Vries equation and Miura transformations. *Physics Letters. [Part A]*, *255*(4-6), 259–264. doi:10.1016/S0375-9601(99)00163-2

Xu, T., Li, J., Zhang, H. Q., Zhang, Y. X., Yao, Z. Z., & Tian, B. (2007). New extension of the tanh-function method and application to the Whitham–Broer–Kaup shallow water model with symbolic computation. *Physics Letters. [Part A]*, *369*(5-6), 458–463. doi:10.1016/j.physleta.2007.05.047

Yusufoğlu, E., & Bekir, A. (2007). A travelling wave solution to the Ostrovsky equation. *Applied Mathematics and Computation*, *186*(1), 256–260. doi:10.1016/j.amc.2006.07.099

Zhang, J. L., Wang, M. L., & Li, X. Z. (2006). The subsidiary ordinary differential equations and the exact solutions of the higher order dispersive nonlinear Schrödinger equation. *Physics Letters. [Part A]*, *357*(3), 188–195. doi:10.1016/j.physleta.2006.03.081

Zhang, S., Tong, J. L., & Wang, W. (2008). A generalized-expansion method for the mKdV equation with variable coefficients. *Physics Letters. [Part A]*, *372*(13), 2254–2257. doi:10.1016/j.physleta.2007.11.026

Zhang, S., & Zhang, H. Q. (2011). Fractional sub-equation method and its applications to nonlinear fractional PDEs. *Physics Letters. [Part A]*, *375*(7), 1069–1073. doi:10.1016/j.physleta.2011.01.029

Chapter 3
Auditors in the Economy and the Impact of Rent-Seeking Behaviour and Penalties

Bilge Öztürk-Göktuna
https://orcid.org/0000-0003-2992-4648
Galatasaray University, Turkey

Mert Erinç
https://orcid.org/0000-0002-4498-5812
Bocconi University, Italy

ABSTRACT

Society often relies on information disclosed by enterprises and verified by auditors to decide on an efficient allocation of capital. Auditing sector serves as a means of verification to protect investors from making decisions based on inaccurate information. However, auditors can use their superior information for extracting additional rents. This study explores an economy where entrepreneurs choose their financial reporting quality considering incentives imposed by the society, and rent-seeking auditors may manipulate their reports to extract gains in the expense of public interest. The analysis captures the dynamics of strategy changes among different actors by introducing a population game framework. The steady-state equilibrium analysis shows that there is a pure state and mixed states whose stability is affected by policy parameters such as subsidies, taxes, competitive auditor fee, and rate of adjustment of different behavioral dynamics. It appears that corruption in auditing sector and poor quality in financial reporting may arise as a temporally persistent outcome.

DOI: 10.4018/978-1-7998-0134-4.ch003

INTRODUCTION

Investors and governments require financial information to evaluate a company's performance because they make their economic decisions based on their interpretations of this financial information. However, the information does not serve much if it is not reliable. Today's competitive market conditions put a great pressure on companies' managements to meet their investors' expectations. Due to this pressure, under-performing companies become much more inclined to manipulate their financial statements (Carruth, 2011). There are significant losses due to investments based on inaccurate financial information, as in the cases of Enron and WorldCom (Brickey, 2003). These experiences have highlighted the importance of obtaining reliable attestation over financial statements.

Self-interested actions of individuals in an economy may eventually result in market failures. In the context of this research, an example is an economy where some public companies report manipulated financial information, extracting additional gains through overstated performance, which undermines the efficient allocation of capital. Although the vast majority of companies avoid engaging in any sort of financial misconduct and play by the rules, even a couple of consecutive corporate scandals is enough to collapse the market, as in the cases of early and late 2000s of the U.S. Public authorities (i.e. governments) often intend to solve market failures by intervention in the market mostly through the use of agents (or bureaucrats) with relevant expertise and skills for collecting required information. However, this process inevitably creates room for other potential problems for the economy: inefficient spending, information asymmetry and principal-agent problem. Thus, there is a trade-off for public to consider when opting for and out of market interventions (Dollery & Wallis, 1997).

The independent auditors are considered to be the "watchdogs" of public interest. Their existence in the economy is for ensuring that all information published by public companies, are accurately stated, so that the investors supply their funds and capital to companies, conditional on their actual performance (Miller, 2006). Given the difficulty of understanding the complex nature of business models and the reporting standards that should be used to interpret the accuracy of the financial information, it would not be feasible for the investors to audit all the financial information regarding the performance of companies they plan to invest in, without the professional intermediacy of auditors who possess the required expertise and skills. From the perspective of the companies being audited, the intermediacy of auditors helps them in contractual terms, protecting their confidential private information from potential breaches in the absence of bilateral legally binding agreements. Therefore, the auditors have an important role in achieving a higher output for the society through a more efficient allocation of funds.

As a matter of fact, there is also some evidence that the auditors' existence causes a downside impact on the output through waste of resources, let alone helping to achieve a better one[1]. For example, many banks and investors have lost billions of dollars due to the accounting scandal of Enron in 2001. It was a huge energy company audited by Arthur Andersen, one of the largest audit firms in the world at the time. However, the auditors failed to reveal Enron's fraudulent financial information, and investors' trust for their assurance caused them to wastefully transfer their capital in a company in the edge of bankruptcy (Brickey, 2003).

Such audit failures can result from the conflicts of interest as well as the auditors' incompetence. There are various ways that auditors may pursue their self-interest. An auditor, for instance, may gain from being more cost-averse by performing an audit with lower effort or by decreasing the size of its engagement teams, both of which would certainly decrease the quality of the audit. Alternatively, auditors may gain from seeking additional rents from the clients they audit by charging extra fees for reporting an opinion more favorable to their clients. In turn, such behavioral problems may damage the public interest by creating an inefficient capital allocation environment. In this paper, the auditors' incentives to report dishonestly and how they are affected by deterrence policies are examined.

The public demand for auditing basically arises from the need for verified information. Using their professional skills and going through an exhaustive process, the auditors gather sufficient evidence to conclude whether the company's financial statements are reported in accordance with some pre-established reporting standards, and whether they reflect its true financial position. This evidence provides the auditor superior information regarding the financial position of the entities they are auditing. It is expected from an auditor to use this superior information to report on any material misstatements by the company's management that may influence the investors' decisions. However, auditors, like all other actors in the economy, are self-interested. Thus, they might have incentives to diverge from acting in accordance with public expectations. In this study, the benefits and costs of the existence of auditors in the economy -with and without dishonest behavior- are analyzed. In addition to this analysis, an evolutionary interpretation is made with an attempt to observe the dynamics of strategy choices of agents in the economy.

The rest of the chapter is organized as follows. After a brief background on the objective and review of the literature, we set up the model; define the terms used throughout the chapter and give the static equilibrium results. Then, we extend the model using a population game setup to consider the dynamic aspects of dishonesty in the auditing sector and provide with the description of the dynamically stable outcomes of the model. We give examples of populations with some dishonesty in the auditing sector as a dynamically stable outcome. The chapter ends with the concluding remarks.

BACKGROUND

In one of her speeches, Lynn Turner, a former chief accountant of the SEC, addresses the importance of financial information reliability by saying "The success of capital markets of the USA derived from people's willingness to invest more capital there since they receive higher quality financial information than is available in any other place in the world" (Turner, 2001). So, if the information does not reflect the company's financial position and performance fairly, there emerges a risk for investors that they invest in underperforming or insolvent companies due to misleading inaccurate financial information. The quality of financial reporting affects the reliability of financial statements by confirming companies' affairs and reducing the information asymmetry (Ali, 2008; Fairchild, 2008). Gul et al. (2013) examines the effects of informational problems, the quality of financial reporting on firms' cost of debt and suggest that the cost of debt is lower when financial audit is of higher quality and firms with more informational problems benefit more from the quality of reporting.

A major contribution of the auditing sector is considered to be its role in protecting investors from making decisions based on such inaccurate information, and thereby enhancing efficient allocation of capital in the economy (Doty, 2013; Zimmerman, 2015). This is also the reason why the investors and regulators have high expectations of auditors' work, reflected in their disappointment and criticisms in cases of audit failures (Bonner et al. 1998; Marriage, 2019). However, the sector brings its costs together with its benefits. In a model developed by Acemoglu and Verdier (2000), the costs of allocating some of the agents in the economy for verifying the information are simply categorized as follows: (1) the cost of withdrawing individuals (verifiers) from production sector to use them for monitoring, (2) the costs resulting from dishonest behavior of rent-seeking verifiers[2]. An auditor may gain from seeking additional rents from the clients he audits, that is, he can charge extra fees for reporting an opinion favorable to his client. Such behavioral problems would damage the public interest by creating an inefficient capital allocation environment. The regulators may try to prevent such behavior by imposing sanctions and punishments on auditors. Public Company Accounting Oversight Board (PCAOB), the regulator of the U.S. auditors, take action against non-complying engagement teams. These actions involve disciplinary proceedings including a censure, monetary penalties, revocation of a firm's registration, and a bar on an individual's association with registered accounting firms. The findings of Ye and Simunic (2016) suggest that when the legal regime is weak, regulatory oversight can improve social surplus by incentivizing auditors to comply with standards, and it can substitute for a weak legal system in disciplining auditors if regulatory penalty is sufficiently high. However, the sanctions are conditional on detection of auditors' misconduct, and the inspections to detect misconduct are sample based and costly. This clearly shows that there is a trade-off for the public authorities to consider while deciding whether to use auditors for monitoring financial information accuracy in the market.

The prevention of dishonest rent seeking behavior is costly and this would entail a certain share of dishonesty in the auditing sector. The main objective of this paper is to investigate the optimal and dynamically stable size of dishonest auditors in the auditing sector. The model illustrated in this chapter was inspired by Acemoglu and Verdier (2000) who suggest that because the corruption is costly to prevent, the regulators may find it more optimal to allow a certain fraction of corrupt officials as a second-best solution. Furthermore, using a similar setting Infante and Smirnova (2009) show that the existence of the dishonest verifiers are especially useful in an economy under weak institutional environment, where the proportion of incentives withheld by the verifiers are high. This chapter contributes to their findings by adapting this framework to a setting in which the regulator uses the non-producer market participants (i.e., auditors) instead of government bureaucrats to encourage the use of good technology over bad technology. This new setting not only allows examining the evolutionary dynamics of technology compliance in the production sector and honesty in the non-producing assurance sector, but also makes the findings more relevant for policy-making purposes as it remedies some of the prior setting's abstractions from reality.

This remedy is essentially provided through two main adjustments: First, while the regulator's objective of encouraging good technology usage is maintained in our setting, the responsibility of proving compliance is passed on to the market itself and thereby bearing the necessary costs. This property reflects the companies' obligation to have their financial statements audited by independent professionals in today's economy. The resulting opinion of the auditors becomes the input for critical decisions to be made by investors and regulators regarding the companies in question. Second, the present model introduces the auditors' probability of being inspected by a bureaucrat and the respective punishment in case of detection. This property, in turn, reflects the fact that the independent auditors have to go through periodic inspections by their regulators over a small sample of their audit engagements. As a result of these inspections, the regulators infer whether there has been any violation of professional standards due to obvious conflict of interest concerns.

THE MODEL

The setting involves two main types of agents in the economy: entrepreneurs and auditors. The mass of the population of entrepreneurs and auditors are $n=1$ and $m=1$ respectively. Entrepreneurs have a uniform production of y and choose between two types of financial reporting: high quality financial reporting which accurately reflects the company's performance and low quality financial reporting which is likely to include inaccurate information regarding the company's performance. The fraction of those choosing high quality financial reporting will be denoted by x. There is an additional cost of providing high quality financial information, $0<e<y$, for the entrepreneurs making it. However, the choice of high quality financial reporting provides all individuals in the economy, irrespective of their job and type of financial reporting choice, $\beta>e$ units of positive externality for each entrepreneur using high quality reporting, that is βx. Consequently, the social surplus is:

$$SS = y + (\beta - e)nx \quad (1)$$

In the case of a decentralized economy, where there are no auditors and where all individuals are entrepreneurs ($n=1$) and their reporting choices are unknown to public, the social surplus is $SS=y+(\beta-e)x$ and the entrepreneurs making a high quality financial reporting and low quality financial reporting will face payoffs of π_h and π_l respectively:

$$\pi_h = y\beta x - e \quad (2)$$

$$\pi_l = y + \beta x$$

The first best for this economy is given by $n=1$ and $x=1$ where all agents are entrepreneurs and provide high quality financial reports (Acemoglu & Verdier, 2000). This outcome, however, is not an equilibrium when agents decide simultaneously and non-cooperatively. It is evident that there is only one equilibrium here, which is $n=1$ and $x=0$. The simple reason is that the individuals are self-interested and there is no incentive for a single agent to contribute and face a cost of e. This equilibrium yields a social surplus of $SS=y$. It is important to note that the setting considers the production with high quality financial reporting creating a positive externality instead of the one which may use a low-quality financial reporting creating a negative externality. But this does not affect the model's findings since the reasoning goes both ways. This equilibrium implies that no enterprise would be willing to report fairly in the absence of auditors as their integrity brings them to a disadvantaged position in the competitive market.

Hereafter, *honest auditors* will be introduced into this economy and their existence will be analyzed with respect to their impact on the utilization of positive externalities for reaching a better equilibrium. In this case, society imposes the entrepreneurs to hire an external auditor to monitor the quality choices of entrepreneurs and implement policies accordingly. The entrepreneurs using low quality financial reporting are punished with a penalty[3] of $\tau<y$ and those using high quality financial reporting are rewarded[4] with a subsidy of $s<y$ both being applicable given the results of the audit. The parameters τ and s are incentives imposed by society, for entrepreneurs to diverge from undesired behavior towards the desired behavior. They would want to reap rewards for creating positive externality or they would want to avoid penalty for not contributing to its creation. Auditors monitor entrepreneurs externally and a competitive audit service fee, w is paid by entrepreneurs irrespective of their choice of reporting quality. Thus, the entrepreneurs will have the following payoffs:

$$\pi_h=y+\beta x-e+s-w \tag{3}$$

$$\pi_l=y+\beta x-\tau-w$$

It is assumed that the audit market is competitive. This implies that in a setting where auditors are completely honest, the optimality of their presence in the economy depends on the level of unit contribution to output that can be achieved through fair reporting. If the positive externality is higher than a certain threshold, then it is worth to impose external audits to create incentives towards the creation of that externality.

Now, consider that auditors may choose not to be honest and to use their superior information for extracting additional rents from the economy. There can be two circumstances, which enable these *dishonest auditors* to do that:

- They match with entrepreneurs choosing low quality financial reporting and take bribes to report that the quality is high.
- They match with entrepreneurs using high quality financial reporting and take bribes not to report that the quality is low or misreport so as to have an additional rent.

In both circumstances, the amount of rents that they receive can be no more than τ and s, because otherwise, it would not be feasible for entrepreneurs to bribe auditors. The auditors may accept only a share of τ or s. This share that they receive as bribes are denoted by $\sigma\in[0,1]$ and assumed to be homogenous across the auditors. However, dishonest behavior is not tolerated by the society since it undermines its intention to foster the use of high quality reporting. For this reason, auditors are inspected by the regulator with a probability of q. That is, with probability q, an inspector is matched with a dishonest auditor, resulting in him losing his service fee w, and paying a fine of ϕs, which is determined as a share of the subsidy. On the other hand, with probability $1-q$, the dishonest auditor receives both the service fee and a rent of $\sigma\tau$ or σs, without being detected by the regulator. Note that with an absolute detection with probability $q=1$, the risk-neutral auditors would not dare to demand rents and the optimum solution would be the same as that of the case with no dishonest behavior. However, inspection is costly and auditors are inspected with probability $q<1$.

This setting allows for a more realistic environment where independent auditors can be considered as self-interested payoff maximizers and analyze the appearance of corruption in the auditing sector as a strategic choice. It is possible for auditors to ask for rents from clients they found out to report inaccurate financial information, and in turn, giving assurance for this inaccurate information. However, they should do it taking the possibility of detection into account. That is, for instance, for a risk-neutral auditor, the expected value of taking an extra fee for giving assurance over a low quality reporting decreases when there is high probability of being punished for that action.

The competitive fee w plays an important role in this new framework since if the fee for the audit service is too low, it is understandable that auditors become prone to seek additional rents for compensation. Since these rents are at the expense of public interest, the society may interfere with the determination of these fees so that the auditors do not diverge from doing what is expected of them, and their inspection continue to be useful for the promotion of high quality reporting. In the auditor population, the proportion of dishonest auditors is denoted as p and the rest of the auditor population is honest.

The utility of an entrepreneur choosing low quality financial reporting and high quality financial reporting are π_l and π_h respectively and in this new setting with auditors involved, they are given by the following equations:

$$\pi_h = y + \beta x - e + s - p(1-q)\sigma s - w$$

$$\pi_l = y + \beta x - (1 - p(1-q))\tau + p(1-q)s(1-\sigma) - w \tag{4}$$

The utilities of the honest and dishonest auditors are given by π_H and π_D respectively and defined as:

$$\pi_H = w + \beta x$$

$$\pi_D = xw + (1-x)(1-q)w + \sigma s + \beta x - q(1-x)\phi s \tag{5}$$

It is clear that the probability of detection is important for determination of dishonest auditor's utility π_D. In order for an auditor to be punished for being dishonest, there should be an occasion, which will reveal that the auditor had been unable to detect a material misstatement. In addition to that, judicial authorities should gather sufficient evidence that this failure is related with auditor's negligent or fraudulent behavior, which requires substantial amount of physical and human resource. As the proportion of auditors choosing to pursue their self-interests increase in the economy, the government resources assignable for each case decreases. The less resource is assigned for detecting self-interested auditors, the less likely it is to spot and punish them. Thus, there is an inverse relationship between q and p. The authority finances the revenue to make these inspections by taxes levied from entrepreneurs who have chosen low quality financial reporting and who are not matched with dishonest auditors or otherwise who are matched with dishonest auditors and are caught through inspection. It is assumed here that the authority operates with a balanced budget. The following equation defines the budget balance:

$$p\left(-(1-q)s+q\left(-xs+(1-x)\tau+(1-x)\phi s\right)\right)+(1-p)\left(-xs+(1-x)\tau\right)=0 \tag{6}$$

Simplifying this equation, we get the following relationship:

$$q(p,x)=\frac{s\left(x+p(1-x)\right)-(1-x)(1-p)\tau}{p(1-x)(s+\tau+\phi s)} \tag{7}$$

For the society to be able to inspect with a non-negative probability, $s\left(x+p(1-x)\right)-(1-x)(1-p)\tau \geq 0$ so that $q(x^*,p^*)\geq 0$. The first term is the sum of subsidies distributed to high quality financial reporting entrepreneurs and low quality financial reporting entrepreneurs matched with self-interested auditors. The second term is the taxes collected by low quality financial reporting entrepreneurs matched with honest auditors. While $p(1-x)(s+\tau+\phi s)$, the term in the denominator is the welfare gain associated by catching dishonest auditors. When low quality financial reporting entrepreneurs matched with dishonest auditors, society loses the subsidy given to undeserved agents, the uncollected tax and if these are caught through inspection, the subsidy will be paid back together with the tax and penalty. $q(x^*,p^*)\leq 1$ when the welfare gain is greater

Replacing the value of $q(p, x)$ in the utility of the dishonest auditors, we obtain the following payoff for the dishonest auditors:

$$\pi_D = w+\sigma s+\beta xn-\frac{s-(1-x)(1-p)(s+\tau)}{p(1-x)(s+\tau+\phi s)}(w+\phi s) \tag{8}$$

This static model has an interior equilibrium where some of the entrepreneurs use good quality financial reporting and others use bad quality financial reporting and some others choose to be honest while others are dishonest if $\pi_l \leq \pi_h$ and $\pi_H \leq \pi_D$. Society can impact the distribution of behaviors under balanced budget by either determining a minimum fee for financial auditing services, choosing the frequency of visits to auditors and playing with the levels of subsidies and taxes. The mentioned conditions can be expressed by the following two inequalities:

$$p \geq \frac{s(w+\phi s)-(1-x)(s+\tau)(w+\phi s)}{\sigma s(s+\tau+\phi s)-(1-x)(s+\tau)(w+\phi s)} \tag{9}$$

$$1-x \geq \frac{s(s+\tau)}{p(s+\tau+\phi s)(s+\tau)+(1-p)(s+\tau)^2-(s+\tau+\phi s)(s+\tau)+e(s+\tau+\phi s)} \tag{10}$$

We can illustrate above conditions using Figure 1 and Figure 2. In Figure 1, the solid line represents the first and the dashed line represents the second curve and in the hatched zone, both constraints are satisfied. Figure 2 provides on the other hand with a point of intersection where both conditions are satisfied and both entrepreneurs and auditor populations are mixed i.e. in the economy both high and low quality financial reporting are used and we can see honest as well as dishonest auditors in the auditor population.

POPULATION DYNAMICS

Population Games and Revision Protocols

In above results, we see that an economy with both high and low quality financial reporting and honest as well as dishonest auditors is an equilibrium outcome of the static model. This equilibrium outcome results from a one-shot game setup where rational agents interact strategically under perfect information i.e. agents act under perfect information to maximize their self-interests. The concept of equilibrium, on the other hand, is based on the assumption that the equilibrium is known. In other words, agents are able to collectively locate the equilibrium outcome. In this context, the problems commonly faced are multiple equilibria, the realism of the concept of hyper-rationality and the lack of a dynamic aspect of the environment.

Figure 1. Mixed equilibrium conditions $(e,w,s,\tau,\sigma,\phi) = (1,2,0.5,0.4,0.4,0.1)$

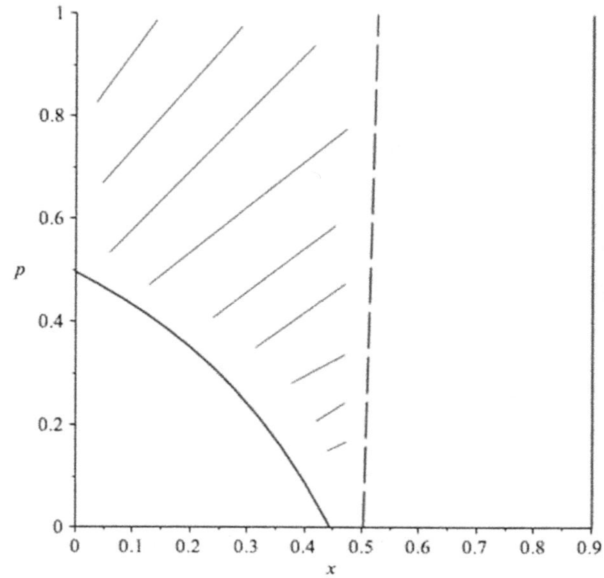

Figure 2. Mixed equilibrium conditions $(e,w,s,\tau,\sigma,\phi) = (0.8,1,0.5,0.4,0.3,0.4)$

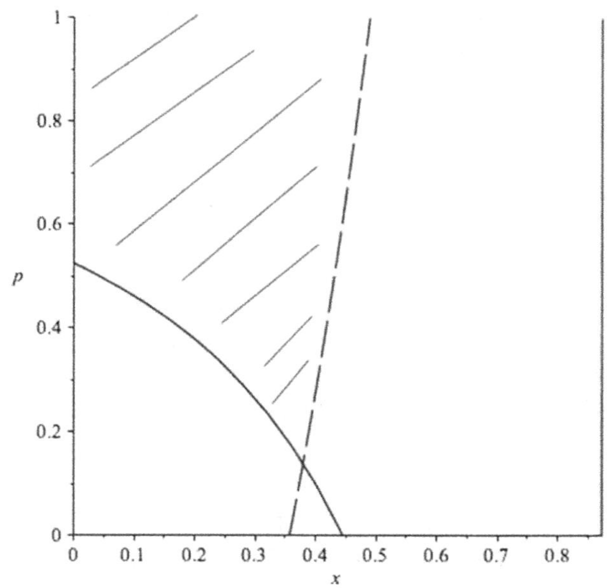

Strategic interaction can also be described in many situations such as externalities, macroeconomic spillovers, centralized markets, highway congestion, transportation mode choice, selfish routing; as population games. In these games, the individual is a small unit and the roles that he can get are limited. Each individual interacts anonymously and returns are continuous. Collective behavior models are derived from open micro foundations. In situations where there are large numbers of individuals, the above-mentioned assumption of equilibrium knowledge cannot be maintained. Hence, a more appropriate dynamic interpretation is needed to explain how individuals update their strategies or behaviors and come up with a theory of how the population evolves.

The choice procedure followed by individuals is called a revision protocol. When a revision protocol for players is defined in the context of a population game, one can generally derive dynamics that show how behavior changes (Sandholm, 2010). A revision process ensures that the better performing behaviors are selected. The imitation dynamics is a type of these revision processes and according to the *imitation dynamics*, individuals switch to strategies whose current payoffs are reasonably good; in our context whose current payoffs performs better than the average payoff in the population.

The above model can be considered as a normal form game as there is a strategic interaction between the agents in the economy. Acemoglu and Verdier (2000) propose a standard analysis of this framework using a fixed probability of detection of dishonest behavior. Fully rational agents play exactly once this game knowing all the details of the game including the preferences of the other agents. When this situation is interpreted as a population game, on the other hand, it allows for all behaviors or strategies in the population of agents, letting each agent to start playing the game initially with a specific strategy and assuming that the revision protocol operates over time on this initial distribution of behaviors. In other words, each population has a set of available strategies out of which their members are able to select. Thus, in this dynamic setting, agents' decision rules to choose a strategy does not depend on the expected utility evaluation based on probabilities, and rather depends on the strategy's expected utility assessment compared to the mean utility of the population, which is directly related to the fraction of population using it.

Entrepreneurs have two strategies: high quality financial reporting and low quality financial reporting. Each agent in the entrepreneur population is assigned initially one of these strategies. Auditors have two strategies: being honest and being dishonest. Each agent in the auditor population is assigned initially one of these strategies. The mean utility levels of entrepreneur and auditor populations are given by $\bar{\pi}_E$ and $\bar{\pi}_A$ respectively as follows:

$$\bar{\pi}_E = x\pi_h + (1-x)\pi_l \tag{11}$$

$$\bar{\pi}_A = p\pi_D + (1-p)\pi_H$$

However, agents of different occupations change their decisions at different paces. In particular, it is assumed in this model that an auditor becomes more easily corrupt than an entrepreneur choosing to change the information quality, since the latter is expected to involve certain structural investments that are costly to implement. The strategy revision pace of an auditor and an entrepreneur are denoted as v_A and v_E respectively, where $v_A > v_E$. Now, the imitation dynamics for each population can be defined as:

$$\frac{\dot{p}}{p} = v_A\left(\pi_D - \bar{\pi}_A\right) \tag{12}$$

$$\frac{\dot{x}}{x} = v_E\left(\pi_h - \bar{\pi}_E\right)$$

where $\dot{p}$ and $\dot{x}$ denotes the temporal changes in the proportions of dishonest auditors and high-quality reporting entrepreneurs, respectively. Thus, the left hand sides of the equations reflect the rate of changes from the initial proportions. Then the mean utilities are substituted into these equations to get:

$$\dot{p} = v_A p(1-p)\left(\pi_D - \pi_H\right) \tag{13}$$

$$\dot{x} = v_E x(1-x)\left(\pi_h - \pi_l\right) \tag{14}$$

It can be easily seen that there are trivial rest points for these equations, at which the economy would be in equilibrium. These are the points where all the members of a population choose one strategy or the other (i.e. p=0, p=1, x=0, x=1). We can refer to these states as *pure states*. In addition to these, it is possible to find interior population states that we can refer to *mixed states* where we can see different patters of behavior in the population. The roots of the following equations are mixed states of the model:

- Any x satisfying π_h-π_l=0 or

- Any p satisfying π_D-π_H=0 subject to the constraint $p,x\in(0,1)$

there is an interior rest point (x^*,p^*) for the system of dynamic equations. The first equation

$$\pi_h - \pi_l = -e + \left(1 - p\left(1 - q\right)\right)\left(s + \tau\right) = 0$$

has a root x^* and the second equation $\pi_D - \pi_H = -\left(1 - x\right)q\left(w + \phi s\right) + \sigma s = 0$ has the root p^* with

$$q\left(x^*, p^*\right) = \frac{\text{s} - \left(1 - x^*\right)\left(1 - p^*\right)\left(\text{s} + \tau\right)}{p^*\left(1 - x^*\right)\left(\text{s} + \tau + \phi s\right)}.$$

The values of p^* and x^* are given by the solution of the following set of non-linear equations:

$$p^*\left(1 - q\left(x^*, p^*\right)\right) = \frac{s + \tau - e}{s + \tau} \tag{15}$$

$$\left(1 - x^*\right)q\left(x^*, p^*\right) = \frac{\sigma s}{w + \phi s} \tag{16}$$

By replacing $q(x^*,p^*)$ in Equation 15 and 16, we get

$$p^* = \frac{\left(s + \tau - e\right)\left(\text{s} + \tau + \phi s\right)}{\phi s\left(s + \tau\right)} - \frac{\text{s} + \tau}{\phi s} + \frac{\text{s}}{\phi s\left(1 - x^*\right)}$$

and $s - \left(1 - x^*\right)\left(1 - p^*\right)\left(\text{s} + \tau\right) = p^* \dfrac{\sigma s\left(\text{s} + \tau + \phi s\right)}{w + \phi s}$ respectively.

Then by replacing p^* in the second equation, we get

$$\left(1 - x^*\right)\frac{\text{e}\left(w + \phi s\right)}{s} + \frac{1}{\left(1 - x^*\right)}\sigma s - \frac{\sigma\left(e - \phi s\right)\left(s + \tau\right) + \phi\sigma\text{es}}{\left(s + \tau\right)} - \frac{\left(\phi s + \text{s} + \tau\right)\left(w + \phi s\right)}{\left(\text{s} + \tau + \phi s\right)} = 0.$$

As $p,x \in (0,1)$, this equation can be rewritten as a second degree polynomial and for a certain range of the parameters of the model, there is solution to this polynomial[5]. An example of an interior rest point can be seen in Figure 1(b) where equation 15 and 16 are depicted and for the set of parameters $(e,w,s,\tau,\sigma,\phi) = (0.8,1,0.5,0.4,0.3,0.4)$ the mixed state is $(x^*,p^*) = (0.379,0.139)$.

Proposition 1. *There exists an interior rest point of the dynamic system (x^*,p^*) where some auditors are honest and some are rent seeking and some entrepreneurs use high quality financial reporting while some use low quality financial reporting for a certain range of exogenous variables: cost of the effort for higher quality financial auditing (e), competitive auditor fee (w), subsidies (s), taxes (t), rent rate (σ) asked over subsidy and penalty rate (ϕ) taken over subsidy.*

Note that the cost of choosing the higher financial reporting quality e affects only the interior rest point of entrepreneur population dynamics x*and the competitive auditor fee affects only the interior rest point of the auditor population dynamics p*. A population specific policy must aim either of these parameters.

Stability

The revision process determines how population shares corresponding to different pure strategies evolve over time. In this context, asymptotic stability of the rest points of the system of population dynamics given by equations 13 and 14 can be studied. We have pure states where all

Before starting with the stability of the pure states where agents choose one strategy or behavior over the other, note that the budget balance has to be satisfied. For $p=0$ where all auditors are honest, the budget balance is $\tau - x(s+t)$, where there is a surplus if $x=0$ and there is a deficit if $x=1$. For $p=1$, all auditors are dishonest, and the balanced budget condition is $-(1-q)s + q(\tau + \phi s) = 0$ and for a certain level of inspection probability q satisfying $-(1-q)s + q(\tau + \phi s) = 0$ the budget balance is satisfied if $x=0$ and there is a deficit if $x=1$. Consequently, the only possible pure state is $p=1$ and $x=0$ if the society aims at a balanced budget. This case refers to a situation where an entire entrepreneur population is using low quality financial reporting and all auditors are dishonest.

Proposition 2. *$(x^*,p^*)=(0,1)$ is stable if* $\sigma > \frac{(w+\phi s)}{(s+\tau+\phi s)}$ *and* $\frac{s+\tau}{s+\tau+\phi s} < e$.

To check the stability of the rest points of the dynamics, the eigenvalues of the Jacobian of this system of dynamic equations should be evaluated at (x^*,p^*). If all eigenvalues turns out to be negative at the rest points of the dynamic equations, these rest points are considered asymptotically stable (Medio & Lines, 2001). The rest point is unstable if at least one of the eigenvalues is positive. The Jacobian evaluated at (x^*,p^*) is as follows:

$$J\Big|_{(x,p)} = \begin{bmatrix} \begin{pmatrix} v_E(1-2x)\left(-e+(1-p(1-q))(s+\tau)\right) \\ +v_E x(1-x)p\dfrac{\partial q}{\partial x}(s+\tau) \end{pmatrix} & v_E x(1-x)(-(1-q)+p^*\dfrac{\partial q}{\partial p})(s+\tau) \\ v_A p(1-p)\left(q(w+\phi s)-(1-x)\dfrac{\partial q}{\partial x}(w+\phi s)\right) & \begin{pmatrix} v_A(1-2p)\left(-(1-x)q(w+\phi s)+\sigma s\right) \\ -v_A p(1-p)(1-x)\dfrac{\partial q}{\partial p}(w+\phi s) \end{pmatrix} \end{bmatrix}$$

Proof of Proposition 2 requires checking the eigenvalues of the Jacobian of the dynamic system. For $(x,p)=(0,1)$ the Jacobian is:

$$J\Big|_{(0,1)} = \begin{bmatrix} v_E\left(-e+\dfrac{s(s+\tau)}{(s+\tau+\phi s)}\right) & 0 \\ 0 & v_A\left(\dfrac{s(w+\phi s)}{(s+\tau+\phi s)}-\sigma s\right) \end{bmatrix}$$

For $(x,p)=(0,1)$, the eigenvalues of the Jacobian are $v_E\left(-e+\dfrac{s(s+\tau)}{(s+\tau+\phi s)}\right)$ and $v_A\left(\dfrac{s(w+\phi s)}{(s+\tau+\phi s)}-\sigma s\right)$. The second eigenvalue is negative if $\sigma > \dfrac{(w+\phi s)}{(s+\tau+\phi s)}$ and the first eigenvalue is negative if $\dfrac{s(s+\tau)}{(s+\tau+\phi s)} < e$ (in this case the trace of the Jacobian is negative and its determinant is positive). This means that the society will be trapped at an inefficient and corrupt state if high quality financial reporting is costly enough and self-interested auditors require a sufficiently high share of the subsidies. The only way to get out of this trap is to give entrepreneurs enough incentive to cover their efforts when transitioning to a higher quality of financial reporting.

Next, the stability of the mixed states, where some agents choose one strategy or behavior while some choose the other, will be analyzed. The rest point for the mixed states (x^*,p^*) are given above by equations 15 and 16. For $(x^*,p^*)\in(0,1)^2$ the Jacobian is

$$J|_{(x^*,p^*)} = \begin{bmatrix} v_E x^*(1-x^*)p^*\frac{\partial q}{\partial x}(s+\tau) & v_E x^*(1-x^*)(-(1-q)+p^*\frac{\partial q}{\partial p})(s+\tau) \\ v_A p^*(1-p^*)\left(-q(w+\phi s)+(1-x^*)\frac{\partial q}{\partial x}(w+\phi s)\right) & v_A p^*(1-p^*)(1-x^*)\frac{\partial q}{\partial x}(w+\phi s) \end{bmatrix}$$

Proposition 3. *$(x^*,p^*)\in(0,1)^2$ is stable if and only if*

1. $v_E x(1-x)p\frac{\partial q}{\partial x}(s+\tau)-v_A p(1-p)(1-x)\frac{\partial q}{\partial p}(w+\phi s)<0$ *and*
2. $-q(1-q)+qp\frac{\partial q}{\partial p}+(1-x)\frac{\partial q}{\partial x}(1-q)>0.$

Note that the first component of the first condition $v_E x(1-x)p\frac{\partial q}{\partial x}(s+\tau)$ relates to the match of a dishonest agent with an entrepreneur using low quality financial reporting and the inspection of the social loss of magnitude $s+\tau$. This social loss will be recaptured more (less) with the increase (decrease) in the probability of inspection that would be due to the increase (decrease) in the share of entrepreneurs changing to higher quality of financial reporting x. The second component of the first condition $-v_A p(1-p)(1-x)\frac{\partial q}{\partial p}(w+\phi s)$ relates to the dishonest auditor matched with an entrepreneur using low quality financial reporting and the resulting income if he is not inspected $w+\phi s$. This income will be earned if a honest auditor is inspected instead and this will be affected by the change in the probability of inspection created with the change in the population of dishonest auditors.

The eigenvalues of the Jacobian are the roots of a second degree equation. Instead of finding the roots explicitly, it is sufficient to look for the conditions ensuring that the eigenvalues are both negative. These conditions are that their sum shall be negative and their product shall be positive i.e. $J(1,1)+J(2,2)<0$ and $J(1,1)J(2,2)-J(1,2)J(2,1)>0$.These conditions are expressed in Proposition 3 and can be rewritten as

$$\frac{(s+\tau)sx}{(1-x)(s+\tau+\phi s)}+v_A\frac{(w+\phi s)(-\tau+x(s+\tau))(1-p)}{p(s+\tau+\phi s)}<0 \tag{17}$$

$$\frac{-s\left(-p\phi(s+\tau)(1-p)(1-x)^2+(\tau+s)x-\tau\right)}{\left(p(1-x)(s+\tau+\phi s)\right)^2}>0. \tag{18}$$

Note that the rates of adjustment v_A and v_E affect the stability of the rest points. We see that policy variables such as subsidies and penalties have an important impact on the stability of the mixed state as well as the level of rent seeking and punishment. The rates of adjustments of the dynamics also affect the stability of the mixed outcome.

Figure 3 provides an illustration of the rest points and their stability and is drawn based on the following values of the parameters $(e,w,s,\tau,\sigma,\phi)=(0.8,1,0.5,0.4,0.3,0.4)$. The mixed state for this set of parameters is $(x^*,p^*)=(0.379,0.139)$. This is a state where more than half of the entrepreneurs use low quality financial reporting and although a very high percentage is requested by dishonest auditors during external audits, the competitive audit fee and the punishment rate is high enough to create incentives to be honest. This mixed state where there is an inefficient allocation of capital but mostly honest auditors is a saddle. More than half of the auditors provide an honest external audit. The first stability condition is always satisfied ($0.250v_E$-$0.396v_A<0$) and the second stability condition is satisfied

$$(4.167>0)0.250v_E-0.396v_A<0v_E\frac{0.396}{0.250}v_Av_E1.584v_Av_Ev_A.$$

However the trace of the Jacobian is $0.250v_E$-$0.396v_A$ is negative if we satisfy above conditions for eigenvalues but the determinant is $-0.127v_Ev_A$ negative as well. Thus this interior rest point is a saddle. Figure 1provides with the simulation for another rest point of the dynamics: $(x^*,p^*)=(0,0.525)$ and $(x^*,p^*)=(0.357,0)$. For $(x^*,p^*)=0,0.525)$ the first stability condition is satisfied ($-0.396v_A<0$) and the second stability condition is satisfied (0.735>0). In Figure 3, we see that only $(x^*,p^*)=0,0.525)$ is stable. The velocity vectors on the graph give the direction of the motion along the trajectories. These directions are given by the signs of $\dot{x}$ and $\dot{p}$ and whether they are equal to 0. If both are zero, then we have an equilibrium point.

CONCLUSION

This study reinterprets the static corruption model of Acemoglu and Verdier (2000) in the context of the provision of incentives for entrepreneurs for higher quality financial reporting using external audits in order to analyze the impact of auditors' existence

Figure 3. Velocity plot for $(e,w,s,\tau,\sigma,\phi) = (0.8,1,0.5,0.4,0.3,0.4)$

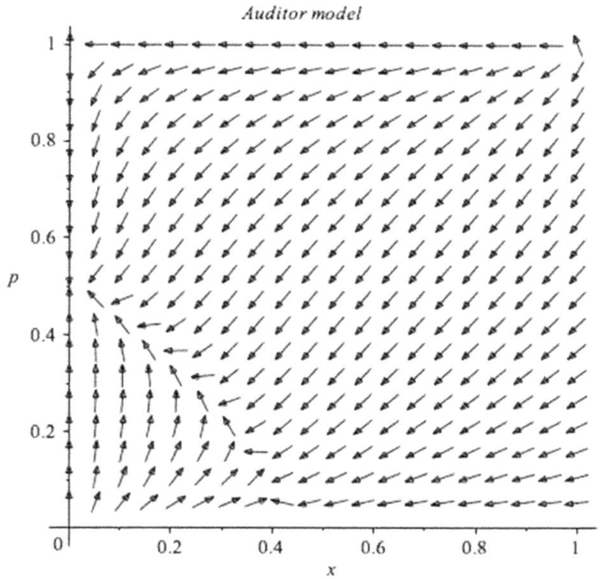

in the economy as monitoring agents. The results suggest that having auditors in the economy are feasible only if the positive externality being promoted is worth bearing the opportunity cost of allocating auditors in the production sector. Then the analysis is extended into a population game framework in which heterogeneous agents of different occupations are able to change their strategies or behaviors over time, based on their evaluations of their strategies' utility compared to the mean utility of their population.

Steady state equilibrium analysis of the dynamic equations shows that there is a pure state and a mixed state whose stability is affected by policy parameters such as subsidies, taxes, competitive auditor fee and rate of adjustment of different dynamics. The pure state is a bad equilibrium where the society is trapped at an inefficient and corrupt state. This happens only if high quality financial reporting is costly enough and self-interested auditors require a sufficiently high share of the subsidies. This result is in accordance with the results of Çule & Fulton (2009) where increase in penalties or inspection can have a perverse effect and create high cheating and corruption. In their study, when evasion and corruption are common, they become more acceptable and their cost is lowered and thus a perverse effect arises. In our model, the only way to get out of this trap is to give entrepreneurs enough incentive to cover their efforts when transitioning to a higher quality of financial reporting. The extent to which subsidies cover the cost of adopting a high quality financial reporting play an important role in avoiding a stable equilibrium where all entrepreneurs adopt a low-quality reporting system and all auditors are better off being dishonest. Practically, this implies that it is not enough for regulators on focusing solely on punishing auditors' misconduct, but also on keeping the benefits of high-quality financial reporting high enough at least to compensate the respective costs.

Finally, we have analyzed the stability of the mixed states, where some agents choose one strategy or behavior while some choose the other. At this point, we see that the rates of adjustment affect the stability of the rest points just like the policy variables such as subsidies and penalties and the level of rent seeking and punishment. Van Rijckeghem and Weder di Mauro (2001) study corruption in civil service and find using a data set on wages for low-income countries evidence of a statistically and economically significant relationship between relative civil-service pay and corruption where a rather large increase in wages is required to end corruption only using wages as a policy. In line with this empirical study, we also see that when enforcement is costly some dishonest rent seeking behavior may prevail in the economy. We also see that the behavioral aspects of the problem contribute to the complexity of the solution concept: when agents revise their behaviors at different rates, the enforcement of the honesty in the auditor sector may be difficult to achieve.

REFERENCES

Acemoglu, D., & Verdier, T. (2000). The choice between market failures and corruption. The American Economic Review, 90(1), 194–211. doi:10.1257/aer.90.1.194 doi:10.1257/aer.90.1.194

Agarwal, R., Echambadi, R., Franco, A. M., & Sarkar, M. B. (2004). Knowledge transfer through inheritance: Spin-out generation, development, and survival. Academy of Management Journal, 47(4), 501–522.

Ali, A. R. (2008). Does auditor industry specialization matter? Evidence from the bond market. Journal of Audit Practice, 5(3), 44–72.

Bonner, S. E., Palmrose, Z.-V., & Young, S. M. (1998). Fraud type and auditor litigation: An analysis of SEC accounting and auditing enforcement releases. The Accounting Review, 503–532.

Brickey, K. F. (2003). From Enron to WorldCom and beyond: Life and crime after Sarbanes-Oxley. Wash. ULQ, 81, 357.

Carruth, P. J. (2011). *Earnings Management: The Role Of Accounting Professionals*. Academic Press.

Çule, M., & Fulton, M. (2009). Business culture and tax evasion: Why corruption and the unofficial economy can persist. Journal of Economic Behavior & Organization, 72(3), 811–822. doi:10.1016/j.jebo.2009.08.005 doi:10.1016/j.jebo.2009.08.005

Dollery, B. E., & Wallis, J. L. (1997). Market failure, government failure, leadership and public policy. Journal of Interdisciplinary Economics, 8(2), 113–126. doi:10.1177/02601079X9700800202 doi:10.1177/02601079X9700800202

Doty, J. (2013). *The Role of the Audit in the Global Economy*. at the ICAS Aileen Beattie Memorial Lecture, London, UK.

Fairchild, R. (2008). *Does audit tenure lead to more fraud? A game theoretic approach*. Available at: https://papers.ssrn.com/sol3/papers.cfm?abstract_id=993400

Gul, F. A., Zhou, G., & Zhu, X. (2013). Investor protection, firm informational problems, big N auditors, and cost of debt around the world. Auditing, 32(3), 1–30. doi:10.2308/ajpt-50462 doi:10.2308/ajpt-50462

Infante, D., & Smirnova, J. (2009). Rent-seeking under a weak institutional environment. Economics Letters, 104(3), 118–121. doi:10.1016/j.econlet.2009.04.022 doi:10.1016/j.econlet.2009.04.022

Marriage, M. (2019). Auditors promote 'mythology' on fraud detection, claims regulator. *Financial Times*. Retrieved from https://www.ft.com/content/43ab3668-2959-11e9-88a4-c32129756dd8

Medio, A., & Lines, M. (2001). Nonlinear dynamics: A primer. Cambridge University Press. doi:10.1017/CBO9780511754050 doi:10.1017/CBO9780511754050

Miller, G. S. (2006). The press as a watchdog for accounting fraud. Journal of Accounting Research, 44(5), 1001–1033. doi:10.1111/j.1475-679X.2006.00224.x doi:10.1111/j.1475-679X.2006.00224.x

Moore, D. A., Tetlock, P. E., Tanlu, L., & Bazerman, M. H. (2006). Conflicts of interest and the case of auditor independence: Moral seduction and strategic issue cycling. Academy of Management Review, 31(1), 10–29. doi:10.5465/amr.2006.19379621 doi:10.5465/amr.2006.19379621

Sandholm, W. H. (2010). Population games and evolutionary dynamics. MIT Press.

Turner, L. E. (2001). *Quality, transparency, accountability*. Paper presented at the Speech to the Financial Executives Institute-AICPA's "Benchmarking the Quality of Earnings" Conference, New York, NY.

Van Rijckeghem, C., & Weder di Mauro, B. (2001). Bureaucratic corruption and the rate of temptation: Do wages in the civil service affect corruption, and by how much? Journal of Development Economics, 65(2), 307–331. doi:10.1016/S0304-3878(01)00139-0 doi:10.1016/S0304-3878(01)00139-0

Ye, M., & Simunic, D. A. (2016). *The impact of PCAOB-type regulations on auditors under different legal systems*. Rotman School of Management Working Paper (2697268).

Zimmerman, J. L. (2015). The role of accounting in the twenty-first century firm. Accounting and Business Research, 45(4), 485–509. doi:10.1080/00014788.2015.1035549 doi:10.1080/00014788.2015.1035549

ENDNOTES

1. Much of this evidence is investigated in prior literature such as Agarwal, Echambadi, Franco, and Sarkar (2004); Moore, Tetlock, Tanlu, and Bazerman (2006).
2. In fact, the model of Acemoglu and Verdier (2000) investigates the case where governments need bureaucrats to monitor entrepreneurs' choice of technology (good or bad) and implement policies accordingly. Our interpretation of it provides insight and basis to understand the dynamics of rent-seeking behavior of auditors in the economy.
3. In fact, those companies, statements of which have been detected to include inaccurate information are either asked to make the necessary corrections, often causing them to bear additional costs, or they end up in court trials being subject to high amounts of fine payments.
4. This can alternatively be seen as a compliance audit from which the companies are required to get a passing opinion to be able to receive a subsidy. This practice is especially common in tax immunity privileges conditionally granted to certain specialized startup companies.
5. $aX^2 - bX + \sigma s = 0$ where $X = (1 - x^*)$, $a = \frac{e(w+\phi s)}{s}$,

$$b = \left(\frac{\sigma(e-\phi s)(s+\tau)+\phi\sigma es}{(s+\tau)} + \frac{(\phi s+s+\tau)(w+\phi s)}{(s+\tau+\phi s)} \right),$$

$c = \sigma s$. For $b^2 - 4ac \geq 0$ there is a solution to this equation.

Chapter 4
Berezin Number Inequalities of an Invertible Operator and Some Slater Type Inequalities in Reproducing Kernel Hilbert Spaces

Ulaş Yamancı
Süleyman Demirel University, Turkey

Mehmet Gürdal
Süleyman Demirel University, Turkey

ABSTRACT

A reproducing kernel Hilbert space (shorty, RKHS) $H=H(\Omega)$ on some set Ω is a Hilbert space of complex valued functions on Ω such that for every $\lambda\in\Omega$ the linear functional (evaluation functional) $f\to f(\lambda)$ is bounded on H. If H is RKHS on a set Ω, then, by the classical Riesz representation theorem for every $\lambda\in\Omega$ there is a unique element $k_{H,\lambda}\in H$ such that $f(\lambda)=\langle f,k_{H,\lambda}\rangle$; for all $f\in H$. The family $\{k_{H,\lambda}:\lambda\in\Omega\}$ is called the reproducing kernel of the space H. The Berezin set and the Berezin number of the operator A was respectively given by Karaev in [26] as following $Ber(A)=\{\tilde{A}(\lambda):\lambda\in\Omega\}$ and $ber(A):=\sup_{\lambda\in\Omega}|\tilde{A}(\lambda)|$. In this chapter, the authors give the Berezin number inequalities for an invertible operator and some other related results are studied. Also, they obtain some inequalities of the slater type for convex functions of selfadjoint operators in reproducing kernel Hilbert spaces and examine related results.

DOI: 10.4018/978-1-7998-0134-4.ch004

INTRODUCTION

Firstly, we will mention about importance of the reproducing kernel as follows (Saitoh & Sawano, 2016; Saitoh, 1988).

The theory of reproducing kernels began with two papers of 1921 (Szego, 1921) and 1922 (Bergman, 1922) which is related with typical reproducing kernels of Szegö and Bergman, and since then the theory has been improved into a large and deep theory in complex analysis by many mathematicians. But, exactly, reproducing kernels appeared previously during the first decade of the twentieth century by S. Zaremba (1907) in his work on boundary value problems for harmonic and biharmonic functions. But he did not improve any further theory for the reproducing property. Furthermore, in fact, we know many concrete reproducing kernels for spaces of polynomials and trigonometric functions from much older days. On the other hand, the general theory of reproducing kernels was established in a complete form by N. Aronszajn (1950) in 1950. Furthermore, L. Schwartz (1964), who is a Fields medalist and founded distribution theory, improved the general theory remarkably in 1964 with a paper of over 140 pages (see Saitoh & Sawano, 2016; Saitoh, 1988).

When linear mappings in the framework of Hilbert spaces are considered, we will encounter in a natural way the notion of reproducing kernels; then the general theory is not restricted to Bergman and Szegö kernels, but the general theory is as important as the concept of Hilbert spaces. It is a main concept and important mathematics. The general theory of reproducing kernels is depending on elementary theorems on Hilbert spaces. The theory of Hilbert spaces is the minimum core of functional analysis. But, when the general theory is combined with linear mappings on Hilbert spaces, it will have many relations in various fields, and its fruitful applications will spread over to differential equations, integral equations, generalizations of the Pythagorean theorem, inverse problems, sampling theory, nonlinear transforms in connection with linear mappings, various operators among Hilbert spaces, and many other broad fields. Furthermore, when we apply the general theory of reproducing kernels to the Tikhonov regularization, it produces approximate solutions for equations on Hilbert spaces which contain bounded linear operators. Looking from the point of view of computer users at numerical solutions, we will see that they are fundamental and have practical applications (see Saitoh & Sawano, 2016; Saitoh, 1988).

Concrete reproducing kernels such as Bergman and Szegö kernels will produce many wide and broad results in complex analysis. They improved some important theory and lead to profound results in complex analysis containing several complex variables. On the other hand, the formal general theory given by Aronszajn also has favorable connections with various fields such as learning theory, support vector machines, stochastic theory, and operator theory on Hilbert spaces (see Saitoh & Sawano, 2016; Saitoh, 1988).

A reproducing kernel Hilbert space (shorty, RKHS) $\mathbf{H} = \mathbf{H}(\Omega)$ on some set Ω is a Hilbert space of complex valued functions on Ω such that for every $\lambda \in \Omega$ the linear functional (evaluation functional) $f \to f(\lambda)$ is bounded on $\mathbf{H}$. If $\mathbf{H}$ is RKHS on a set Ω then, by the classical Riesz representation theorem for every $\lambda \in \Omega$ there is a unique element $k_{\mathbf{H},\lambda} \in \mathbf{H}$ such that $f(\lambda) = \left\langle f, k_{\mathbf{H},\lambda} \right\rangle$ for all $f \in \mathbf{H}$. The family $\left\{k_{\mathbf{H},\lambda} : \lambda \in \Omega\right\}$ is called the reproducing kernel of the space $\mathbf{H}$. The reproducing kernel of the space $\mathbf{H}$ for any orthonormal basis $\{e_n(z)\}_{n\geq 0}$ of the space $\mathbf{H}$ can be represented by (see Aronzajn, 1950; Saitoh & Sawano, 2016)

$$k_{\mathbf{H},\lambda}(z) = \sum_{n=0}^{\infty} \overline{e_n(\lambda)} e_n(z).$$

The normalized reproducing kernel of the space $\mathbf{H}$ is denoted by $\hat{k}_{\mathbf{H},\lambda} = \frac{k_{\mathcal{H},\lambda}}{\|k_{\mathcal{H},\lambda}\|}$. For a bounded linear operator A on the RKHS $\mathbf{H}$, its Berezin symbol $\tilde{A}$ is defined by the formula (see Berezin, 1972)

$$\tilde{A}(\lambda) := \left\langle A\hat{k}_{\mathcal{H},\lambda}, \hat{k}_{\mathcal{H},\lambda} \right\rangle.$$

The Berezin symbol of the operator A is a bounded function because of $\left|\tilde{A}(\lambda)\right| \leq \|A\|$ for all $\lambda \in \Omega$. The behavior of the Berezin symbol of an operator provides important information about the operator. More information about reproducing kernels and Berezin symbols, can be found in Karaey (2013) and Karaey, Gurdal, and Yamanci (2013, 2014).

The Berezin symbol of an operator was first introduced by F. A. Berezin (1972) as an extension of Wick symbols on the Fock space. There are several branches of this topic from his original work. One branch uses the transform as an algebraic isomorphism to formulate function spaces with a non-commutative (non-pointwise) product which is useful in the quantization of physical systems (see Berezin, 1974). Another branch asks operator theoretic questions about how properties of the Berezin syrnbol are related to the properties of A. Among today's authors working in the fields of Toeplitz, Hankel and composition operators. The Berezin symbol has become another item of baggage carried by operators that is useful in the characterization of operator classes (see Potter, 2000).

Following Coburn (2004), note that since the Berezin map $A \xrightarrow{B} \tilde{A}$ is linear and in most familiar RKHSs it is one-to-one, it "encodes" operator-theoretic information into function theory in a striking but somewhat impenetrable way. In fact, since $k_{\lambda}^{\mathcal{H}} \to 0$ weakly as $\lambda \to \partial\Omega$ (of course, if the space $H(\Omega)$ is standard in sense of Nordgren and Rosenthal (1994), it is clear that B maps compact operators on these spaces into functions that vanish at the boundary $\partial\Omega$. Because of these properties, the mapping B has found useful applications in dealing with operators "of function-theoretic significance" such as Toeplitz and Hankel operators on the Hardy, Bergman and Fock spaces (for more information, see, for instance, Coburn (2004), Berger and Coburn (1987) and Engliš (1994, 1995, 1999).

The Berezin set and the Berezin number of the operator A was respectively given by Karaev (2006) as following

$$Ber(A) = \left\{\tilde{A}(\lambda) : \lambda \in \Omega\right\} \text{ and } ber(A) := \sup_{\lambda \in \Omega} \left|\tilde{A}(\lambda)\right|.$$

The Berezin number of the operator A has a relationship with its numerical radius as follows:

$$ber(A) \leq w(A) := \sup\left\{\left|\langle Af, f\rangle\right| : \|f\|_{\mathrm{H}} = 1\right\}.$$

The authors have obtained many results about numerical radius inequalities (see Abu-Omar & Kittaneh, 2015; Dragomir, 2006; Gustafson & Rao, 1994; Kittaneh, 2005; Kittaneh, Moslehian & Yamazaki, 2015; Sattari, Moslehian & Yamazaki, 2015). Recently, the concept of the Berezin number has attracted the attention of many authors; for example, using the Hardy-Hilbert type inequalities and some basic well-known inequalities, interesting results about the Berezin number inequalities were obtained (Bakherad, 2018; Bakherad & Garayev, 2019; Garayev, Gurdal & Okudan, 2016; Garayev, Gurdal & Saltabn, 2017; Garayev, Salton & Gundogdu, 2018; Hajmohamadi, Lashkaripour & Bakherad, n.d.; Yamanci, Gurdal & Garayev, 2017; Yamanci & Gurdal, 2017; Yamanci, Garayev & Celik, 2019).

INEQUALITIES FOR THE BEREZIN NUMBER OF AN INVERTIBLE OPERATOR

In this part, we suppose that $B : \mathbf{H} \to \mathbf{H}$ is an invertible bounded linear operator and $B^{-1} : \mathbf{H} \to \mathbf{H}$ is its inverse. Then, clearly,

$$\left\|B\hat{k}_{\lambda}\right\| \geq \frac{1}{\left\|B^{-1}\right\|}\left\|\hat{k}_{\lambda}\right\|$$

for any $\lambda \in \Omega$.

In the present paper, by using some ideas of papers (Dragomir, 2007a, 2007b), we give the Berezin number inequalities for an invertible operator and some other related results are studied.

Now we are ready to give our results.

Theorem 2.1. *Let* A, B be two bounded linear operators on reproducing kernel Hilbert space $\mathbf{H}$ *and let* B be invertible with the property that

$$\left\|A-B\right\| \leq \beta \tag{1}$$

for $\beta>0$. Then

$$\frac{\left\|A\right\|}{\left\|B^{-1}\right\|} \leq ber\left(B^{*}A\right)+\frac{1}{2}\beta^{2}.$$

Proof. From the condition (1), we obviously say that

$$\left\|A\hat{k}_{\lambda}\right\|^{2}+\left\|B\hat{k}_{\lambda}\right\|^{2} \leq 2\operatorname{Re}\left\langle\left(B^{*}A\right)\hat{k}_{\lambda},\hat{k}_{\lambda}\right\rangle+\beta^{2} \tag{2}$$

for any $\lambda \in \Omega$. Since B is an invertible operator, that is,

$$\left\|B\hat{k}_{\lambda}\right\|^{2} \geq \frac{1}{\left\|B^{-1}\right\|^{2}}\left\|\hat{k}_{\lambda}\right\|^{2}$$

and $\operatorname{Re}\left(\tilde{B^{*}A}\right)(\lambda) \leq \left|\left(\tilde{B^{*}A}\right)(\lambda)\right|$, the inequality (2) becomes as following:

$$\left\|A\hat{k}_{\lambda}\right\|^{2}+\frac{\left\|\hat{k}_{\lambda}\right\|^{2}}{\left\|B^{-1}\right\|^{2}} \leq 2\left|\left(\tilde{B^{*}A}\right)(\lambda)\right|+\beta^{2} \tag{3}$$

for any λ∈Ω.

Taking the supremum on λ∈Ω in above equation, we have

$$\|A\|^2 + \frac{1}{\|B^{-1}\|^2} \le 2ber\left(B^* A\right) + \beta^2. \tag{4}$$

From the elementary inequality

$$\frac{2\|A\|}{\|B^{-1}\|} \le \|A\|^2 + \frac{1}{\|B^{-1}\|^2}$$

and above inequality, we obtain

$$\frac{\|A\|}{\|B^{-1}\|} \le ber\left(B^* A\right) + \frac{1}{2}\beta^2.$$

Theorem 2.2. *Let A,B* and β be as in Theorem 2.1. Then

$$\|A\|\|B\| - \frac{\|B\|^2 \|B^{-1}\|^2 - 1}{\|B^{-1}\|^2} \le ber\left(B^* A\right) + \frac{1}{2}\beta^2.$$

Proof. Due to $\|A - B\| \le \beta$, we get that

$$\left\|A\hat{k}_\lambda\right\|^2 + \left\|B\hat{k}_\lambda\right\|^2 \le 2\operatorname{Re}\left\langle A\hat{k}_\lambda, B\hat{k}_\lambda\right\rangle + \beta^2$$

for any λ∈Ω, or equivalently

$$\|B\|^2 + \left\|A\hat{k}_\lambda\right\|^2 \le 2\operatorname{Re}\left\langle B^* A\hat{k}_\lambda, \hat{k}_\lambda\right\rangle - \left\|B\hat{k}_\lambda\right\|^2 + \|B\|^2 + \beta^2. \tag{5}$$

Due to

$$\left\| B\hat{k}_{\lambda} \right\|^2 \geq \frac{1}{\left\| B^{-1} \right\|^2} \left\| \hat{k}_{\lambda} \right\|^2, \quad \operatorname{Re}\left\langle \left(B^{*}A\right)\hat{k}_{\lambda}, \hat{k}_{\lambda} \right\rangle \leq \left| \left(\widetilde{B^{*}A}\right)(\lambda) \right|$$

and

$$\|B\|^2 + \left\| A\hat{k}_{\lambda} \right\|^2 \geq 2\left\| A\hat{k}_{\lambda} \right\| \|B\|$$

for any $\lambda \in \Omega$, therefore by (5) we have

$$2\left\| A\hat{k}_{\lambda} \right\| \|B\| \leq 2\left| \left(\widetilde{B^{*}A}\right)(\lambda) \right| + \beta^2 + \frac{\|B\|^2 \left\| B^{-1} \right\|^2 - 1}{\left\| B^{-1} \right\|^2}$$

$$\leq 2\sup_{\lambda \in \Omega} \left| \left(\widetilde{B^{*}A}\right)(\lambda) \right| + \beta^2 + \frac{\|B\|^2 \left\| B^{-1} \right\|^2 - 1}{\left\| B^{-1} \right\|^2}$$

for any $\lambda \in \Omega$. This implies that

$$\|A\| \|B\| - \frac{\|B\|^2 \left\| B^{-1} \right\|^2 - 1}{\left\| B^{-1} \right\|^2} \leq ber\left(B^{*}A\right) + \frac{1}{2}\beta^2,$$

which gives the desired result.

Theorem 2.3. *Let A, B and β be as in Theorem 2.1.* If B is invertible with the property that

$$\|A - B\| \leq \beta < \|B\|, \tag{6}$$

then

$$\sqrt{\|B\|^2 - \beta^2}\, \|A\| \leq ber\left(B^{*}A\right) + \frac{\|B\|^2 \left\| B^{-1} \right\|^2 - 1}{2\left\| B^{-1} \right\|^2}.$$

In particular, if $B=\alpha I$ with $|\alpha|>\beta$ and $B=\alpha A^*$ with $\|A\|\geq\frac{\beta}{|\alpha|}$ ($\alpha\neq 0$), then, respectively,

$$\sqrt{1-\left(\frac{\beta}{|\alpha|}\right)^2}\,\|A\|\leq ber(A) \tag{7}$$

and

$$\sqrt{\|A\|^2-\left(\frac{\beta}{|\alpha|}\right)^2}\,\|A\|\leq ber\left(A^2\right)+|\alpha|\frac{\|A\|^2\left\|A^{-1}\right\|^2-1}{2\left\|A^{-1}\right\|^2}. \tag{8}$$

Proof. Because of the first side of (6), we can write

$$\left\|A\hat{k}_\lambda\right\|^2+\left\|B\hat{k}_\lambda\right\|^2\leq 2\operatorname{Re}\left\langle A\hat{k}_\lambda, B\hat{k}_\lambda\right\rangle+\beta^2$$

for any $\lambda\in\Omega$, or equivalently

$$\|B\|^2+\left\|A\hat{k}_\lambda\right\|^2-\beta^2\leq 2\operatorname{Re}\left\langle B^*A\hat{k}_\lambda,\hat{k}_\lambda\right\rangle-\left\|B\hat{k}_\lambda\right\|^2+\|B\|^2. \tag{9}$$

Due to

$$\left\|B\hat{k}_\lambda\right\|^2\geq\frac{1}{\left\|B^{-1}\right\|^2}\left\|\hat{k}_\lambda\right\|^2,\quad \operatorname{Re}\left\langle\left(B^*A\right)\hat{k}_\lambda,\hat{k}_\lambda\right\rangle\leq\left|\left(\widetilde{B^*A}\right)(\lambda)\right|$$

and the second side of (6), we have

$$\|B\|^2+\left\|A\hat{k}_\lambda\right\|^2-\beta^2\leq 2\sqrt{\|B\|^2-\beta^2}\left\|A\hat{k}_\lambda\right\|$$

for any $\lambda\in\Omega$. Therefore, we obtain from (9)

$$2\sqrt{\|B\|^2-\beta^2}\,\|Ak_\lambda\|\le 2\left|\left(\tilde{B^*A}\right)(\lambda)\right|+\frac{\|B\|^2\left\|B^{-1}\right\|^2-1}{\left\|B^{-1}\right\|^2} \tag{10}$$

for any λ∈Ω.

Taking the supremum on λ∈Ω, we reach the desired result. In particular, putting the $B=\alpha I$ and $B=\alpha A^*$ in (10), we get the (7) and (8).

Theorem 2.4. Let A,B and β be as in Theorem 2.1. Also, if

$$\left\|B^{-1}\right\|<\frac{1}{\beta}, \tag{11}$$

then

$$\sqrt{1-\beta^2\left\|B^{-1}\right\|^2}\,\|A\|\le\left\|B^{-1}\right\|ber\left(B^*A\right).$$

Proof. From (4), we get

$$\|A\|^2+\frac{1-\beta^2\left\|B^{-1}\right\|^2}{\left\|B^{-1}\right\|^2}\le 2ber\left(B^*A\right). \tag{12}$$

Using the basic inequality and condition (11(, we have

$$2\frac{\|A\|}{\left\|B^{-1}\right\|}\sqrt{1-\beta^2\left\|B^{-1}\right\|^2}\le\|A\|^2+\frac{1-\beta^2\left\|B^{-1}\right\|^2}{\left\|B^{-1}\right\|^2}. \tag{13}$$

So, from (12) and (13), we obtain the desired result.

Theorem 2.5. *Let A,B be two bounded operators. If B is invertible such that* $\|A-B\|\le\beta$ and

$$\frac{1}{\sqrt{1+\beta^2}} \leq \left\|B^{-1}\right\| < \frac{1}{\beta}, \tag{14}$$

then

$$\|A\|^2 - ber^2\left(B^* A\right) \leq 2ber\left(B^* A\right) \frac{\left\|B^{-1}\right\| - \sqrt{1-\beta^2 \left\|B^{-1}\right\|^2}}{\left\|B^{-1}\right\|}.$$

Proof. We get from (4) that

$$\left\|A\hat{k}_{\lambda}\right\|^2 + \frac{1}{\left\|B^{-1}\right\|^2} \leq 2\left|\left(\tilde{B}^* A\right)(\lambda)\right| + \beta^2$$

for any λ∈Ω and since $\frac{1}{\left\|B^{-1}\right\|^2} > \beta^2$, we can say that $\left|\left(\tilde{B}^* A\right)(\lambda)\right| > 0$ for any λ∈Ω

If we divide in both sides of (5) with $\left|\left(\tilde{B}^* A\right)(\lambda)\right|$, we have

$$\frac{\left\|A\hat{k}_{\lambda}\right\|^2}{\left|\left(\tilde{B}^* A\right)(\lambda)\right|} + \frac{1}{\left\|B^{-1}\right\|^2 \left|\left(\tilde{B}^* A\right)(\lambda)\right|} \leq 2 + \frac{\beta^2}{\left|\left(\tilde{B}^* A\right)(\lambda)\right|}. \tag{15}$$

Subtracting $\left|\left(\tilde{B}^* A\right)(\lambda)\right|$ from both sides of above inequality, we have

$$\frac{\left\|A\hat{k}_{\lambda}\right\|^{2}}{\left|(B*A)(\lambda)\right|}-\left|\left(\tilde{B}^{*}A\right)(\lambda)\right|\leq 2-\left|\left(\tilde{B}^{*}A\right)(\lambda)\right|-\frac{1-\beta^{2}\left\|B^{-1}\right\|^{2}}{\left\|B^{-1}\right\|^{2}\left|(B*A)(\lambda)\right|}$$

$$=2-\frac{2\sqrt{1-\beta^{2}\left\|B^{-1}\right\|^{2}}}{\left\|B^{-1}\right\|}-\left(\left|\sqrt{\left(\tilde{B}^{*}A\right)(\lambda)}\right|-\frac{\sqrt{1-\beta^{2}\left\|B^{-1}\right\|^{2}}}{\left\|B^{-1}\right\|\sqrt{\left|\left(\tilde{B}^{*}A\right)(\lambda)\right|}}\right)$$

$$\leq 2\frac{\left\|B^{-1}\right\|-\sqrt{1-\beta^{2}\left\|B^{-1}\right\|^{2}}}{\left\|B^{-1}\right\|}$$

which shows that

$$\left\|A\right\|^{2}-\left|\left(\tilde{B}^{*}A\right)(\lambda)\right|^{2}\leq 2ber\left(B^{*}A\right)\frac{\left\|B^{-1}\right\|-\sqrt{1-\beta^{2}\left\|B^{-1}\right\|^{2}}}{\left\|B^{-1}\right\|}$$

for any $\lambda\in\Omega$.

Since, by (14), $\left\|B^{-1}\right\|-\sqrt{1-\beta^{2}\left\|B^{-1}\right\|^{2}}\geq 0$, taking the supremum in above inequality on $\lambda\in\Omega$, we get the desired inequality.

Theorem 2.6. *Let A,B* be two bounded operators. If *B* is invertible such that $\left\|A-B\right\|\leq\beta$ and $\left\|B^{-1}\right\|<\frac{1}{\beta}$, then

$$0\leq\left\|A\right\|^{2}\left\|B\right\|^{2}-ber^{2}\left(B^{*}A\right)\leq 2ber\left(B^{*}A\right).\frac{\left\|B\right\|}{\left\|B^{-1}\right\|}\left(\left\|B\right\|\left\|B^{-1}\right\|-\sqrt{1-\beta^{2}\left\|B^{-1}\right\|^{2}}\right).$$

Proof Subtracting $\dfrac{\left|\left(\tilde{B}^{*}A\right)(\lambda)\right|}{\left\|B\right\|^{2}}$ from both sides of (15), we obtain

$$
\begin{aligned}
0 &\le \frac{\left\|A\hat{k}_{\lambda}\right\|^{2}}{\left|\left(\widetilde{B^{*}A}\right)(\lambda)\right|}-\frac{\left|\left(\widetilde{B^{*}A}\right)(\lambda)\right|}{\|B\|^{2}} \le 2-\frac{\left|\left(\widetilde{B^{*}A}\right)(\lambda)\right|}{\|B\|^{2}}-\frac{1-\beta^{2}\left\|B^{-1}\right\|^{2}}{\left\|B^{-1}\right\|^{2}\left|\left(\widetilde{B^{*}A}\right)(\lambda)\right|} \\
&= 2-\frac{2\sqrt{1-\beta^{2}\left\|B^{-1}\right\|^{2}}}{\|B\|\left\|B^{-1}\right\|}-\left(\frac{\left|\sqrt{\left(\widetilde{B^{*}A}\right)(\lambda)}\right|}{\|B\|}-\frac{\sqrt{1-\beta^{2}\left\|B^{-1}\right\|^{2}}}{\left\|B^{-1}\right\|\sqrt{\left|\left(\widetilde{B^{*}A}\right)(\lambda)\right|}}\right)^{2} \\
&\le 2\frac{\|B\|\left\|B^{-1}\right\|-\sqrt{1-\beta^{2}\left\|B^{-1}\right\|^{2}}}{\|B\|\left\|B^{-1}\right\|},
\end{aligned}
$$

or equivalently

$$
\begin{aligned}
0 &\le \left\|A\hat{k}_{\lambda}\right\|^{2}\|B\|^{2}-\left|\left(\widetilde{B^{*}A}\right)(\lambda)\right|^{2} \\
&\le 2\frac{\|B\|}{\left\|B^{-1}\right\|}\left|\left(\widetilde{B^{*}A}\right)(\lambda)\right|\left(\|B\|\left\|B^{-1}\right\|-\sqrt{1-\beta^{2}\left\|B^{-1}\right\|^{2}}\right)
\end{aligned}
$$

for any $\lambda\in\Omega$. Since $\|B\|\left\|B^{-1}\right\|-\sqrt{1-\beta^{2}\left\|B^{-1}\right\|^{2}} \ge 0$ from above inequality, we obtain

$$
\sup_{\lambda\in\Omega}\left\|A\hat{k}_{\lambda}\right\|^{2}\|B\|^{2} \le \sup_{\lambda\in\Omega}\left|\left(\widetilde{B^{*}A}\right)(\lambda)\right|^{2}+2\frac{\|B\|}{\left\|B^{-1}\right\|}\sup_{\lambda\in\Omega}\left|\left(\widetilde{B^{*}A}\right)(\lambda)\right|\left(\|B\|\left\|B^{-1}\right\|-\sqrt{1-\beta^{2}\left\|B^{-1}\right\|^{2}}\right)
$$

for any $\lambda\in\Omega$. This implies the desired result.

SLATER TYPE INEQUALITIES IN REPRODUCING KERNEL HILBERT SPACES

Let I be an interval of real numbers with interior int(I) and $f : I \to \mathbb{R}$ be a convex function on I. Then f is continuous on int(I) and has finite right and left derivatives at every point of int(I). Also, if $x,y\in$ int(I) and $x<y$, then

$$
f'_{-}(x) \le f'_{+}(x) \le f'_{-}(y) \le f'_{+}(y)
$$

which gives that both $f_-^{'}$ and $f_+^{'}$ are nondecreasing function on int(I). A convex function must be differentiable except for at most countably many points.

The subdifferential of f for a convex function $f : I \to \mathbb{R}$, is the set of all functions ψ: $I \to [-\infty,\infty]$ with property that $\psi\left(\text{int}\left(I\right)\right) \subset \mathbb{R}$ and $f(y)+(x-y)\psi(y) \le f(x)$ for any $x,y \in I$ and it is denoted by ∂f.

Recall that if f is convex on I, then ∂f is nonempty, $f_-^{'}\ f'_+ \in \partial f$ and if $\psi \in \partial f$, then for any $x \in$ int(I).

$$f'_-\left(x\right) \le \psi\left(x\right) \le f'+(x).$$

Particularly, Ψ is a nondecreasing function.

If f is differentiable and convex on int(I), then $\partial f = \{f'\}$.

The following result is called as *the Slater inequality in the literature:*

Theorem 3.1 (Slater, 1981). If $f : I \to \mathbb{R}$ is a nondecreasing (nonincreasing) convex function, $p_i \ge 0$, $x_i \in I$, such that $P_n := \sum_{i=1}^n pi > 0$ and $\sum_{i=1}^n p_i \psi\left(x_i\right) \ne 0$ where $\psi \in \partial f$, then

$$\frac{1}{P_n}\sum_{i=1}^n p_i f\left(x_i\right) \le f\left(\frac{\sum_{i=1}^n p_i x_i \psi\left(x_i\right)}{\sum_{i=1}^n p_i \psi\left(x_i\right)}\right).$$

The monotonicity assumption for the derivative φ, as stated in Dragomir (2004), can be changed with the condition

$$\frac{\sum_{i=1}^n p_i x_i \psi\left(x_i\right)}{\sum_{i=1}^n p_i \psi\left(x_i\right)} \in I,$$

which is more general and can satisfy for convenient points in I and for not inevitably monotonic functions.

Let A be a selfadjoint linear operator on a complex Hilbert space **H**. The Gelfand map establishes a *-isometrically isomorphism Φ between the set $C(Sp(A))$ of all continuous functions defined on the spectrum of A, denoted by $Sp(A)$, and the C^*-algebra $C^*(A)$ generated by A and the identity operator $1_{\mathbf{H}}$ on **H** as follows (see for instance Furuta, et al., 2005).

For any $f,g \in C(Sp(A))$ and any $\alpha, \beta \in \mathbb{C}$, we get:

1. $\Phi(\alpha f+\beta g) = \alpha\Phi(f) = \beta\Phi(g)$;
2. $\Phi(fg)=\Phi(f)\Phi(g)$ and $\Phi(\overline{f}) = \Phi(f)^*$;
3. $\|\Phi(f)\| = \|f\| := \sup_{t\in Sp(A)} |f(t)|$;
4. $\Phi(f_0)=I$ and $\Phi(f_1)=A$, where $f_0(t)=1$ and $f_1(t)=t$, for $t\in Sp(A)$.

With this concept we define $f(A):=\Phi(f)$ for all $f\in C(Sp(A))$ and it is called the continuous functional calculus for the selfadjoint operator A.

If A is a selfadjoint operator and f is a real valued continuous function on $Sp(A)$, then $f(t)\geq 0$ for any $t\in Sp(A)$ implies that $f(A)\geq 0$ on $\mathbf{H}$. Therefore, if f and g are real valued functions on $Sp(A)$ then the following basic property holds:

$$f(t)\geq g(t) \text{ for any } t\in Sp(A) \text{ implies that } f(A)\geq g(A) \tag{16}$$

in the operator order of $B(\mathbf{H})$.

A real valued continuous function f on an interval J is said to be operator convex (operator concave) if

$$f(((1-\lambda)A+\lambda B)\leq(\geq)(1-\lambda)f(A)+\lambda f(B)$$

in the operator order, for all $\lambda\in[0,1]$ and for any self-adjoint operators A and B on $\mathbf{H}$ whose spectra are contained in J. Obviously, a function f is operator concave if $-f$ is operator convex.

A real valued continuous function f on an interval J is said to be operator monotone if it is monotone with respect to the operator order, i.e., $A\leq B$ with $Sp(A)$, $Sp(B)\subset J$ imply $f(A)\leq f(B)$. (For more facts on operator convex (operator concave) and operator monotone functions, the reader can be consult in Furuta et al., 2005 and its references).

In this section, we obtain some inequalities of the slater type for convex functions of selfadjoint operators in reproducing kernel Hilbert spaces and examine related results.

Before giving our results, we need some well-known results.

The following result is due to Mond & Pečarić (1933):

Theorem 3.2 (Mond & Pečarić, 1993). *Let A be a selfadjoint operator on the Hilbert space H and assume that $Sp(A)\subseteq[m,M]$ for some scalars m,M with $m<M$. If f is a convex function on $[m,M]$, then for each $x\in H$ with $\|x\|=1$,*

$$f(\langle Ax,x\rangle)\leq\langle f(A)x,x\rangle.$$

The next result is a special case of Theorem 3.2.

Theorem 3.3 (McCarthy, 1967). *Let A be a selfadjoint positive operator on a Hilbert space* $\mathbf{H}$ *and let* $x \in \mathbf{H}$ be any unit vector. Then(i) $\langle A^r x, x\rangle \geq \langle Ax, x\rangle^r$ for all $r>1$;(ii) $\langle A^r x, x\rangle \leq \langle Ax, x\rangle^r$ for all $0<r<1$;(iii) If A is invertible, then $\langle A^r x, x\rangle \geq \langle Ax, x\rangle^r$ for all $r<0$.

Our aim in this paper is to give some Slater's type inequalities for the convex functions of selfadjoint operators in reproducing kernel Hilbert space.

Theorem 3.4. Let $f : I \to \mathbb{R}$ be a convex and differentiable function on int(I) (the interior of I) whose derivative f' is continuous on int(I). Then

$$0 \leq f\left(\frac{\widetilde{Af'}(A)(\lambda)}{\left(\tilde{f}'(A)\right)(\lambda)}\right) - \tilde{f}(A)(\lambda) \leq f'\left(\frac{\widetilde{Af'}(A)(\lambda)}{\tilde{f}'(A)(\lambda)}\right)\left[\frac{\widetilde{Af'}(A)(\lambda) - \tilde{A}(\lambda)\widetilde{Af'}(A)(\lambda)}{\tilde{f}'(A)(\lambda)}\right] \tag{17}$$

for any selfadjoint operator A on the reproducing kernel Hilbert space $\mathbf{H}(\Omega)$ with $Sp(A) \subseteq [m, M] \subset \text{int}(I)$ and $f'(A)$ is a positive definite operator on $\mathbf{H}$ and *any* $\lambda \in \Omega$.

Proof. Since F is differentiable and convex on int(I), then we get that

$$f'(s).(t-s) \leq f(t) - f(s) \leq f'(t).(t-s)$$

for any $t, s \in [m, M]$.

Let us fix $t \in [m, M]$. By applying the property (16) for the self adjoint operator A, then we get

$$\left\langle f'(A).(t-1_H - A)\hat{k}_\lambda, \hat{k}_\lambda\right\rangle \leq \left\langle \left[f(t).1_H - f(A)\right]\hat{k}_\lambda, \hat{k}_\lambda\right\rangle \leq \left\langle f'(t).(t.1_H - A)\hat{k}_\lambda, \hat{k}_\lambda\right\rangle$$

for any $\lambda \in \Omega$ and $t \in [m, M]$ or equvialently

$$t\tilde{f}'(A)(\lambda) - \tilde{f}'(A)\tilde{A}(\lambda) \leq f(t) - \tilde{f}(A)(\lambda) \leq f'(t)t - f'(t)\tilde{A}(\lambda) \tag{18}$$

for any $\lambda \in \Omega$ and $t \in [m,M]$.

Because A is selfadjoint with $mI \leq A \leq MI$ and $f'(A)$ is positive definite, then

$$m\tilde{f}'(A)(\lambda) \leq A\tilde{f}'(A)(\lambda) \leq M\tilde{f}'(A)(\lambda)$$

for any $\lambda \in \Omega$, which gives that

$$t_0 := \frac{A\tilde{f}'(A)(\lambda)}{\tilde{f}'(A)(\lambda)} \in [m, M] \text{ for any } \lambda \in \Omega.$$

As a result, putting $t=t_0$ in the equation (18), then we have the desired result (17).

Remark 3.1. *The assumption that* $f'(A)$ is a positive definite operator on $\mathbf{H}(\Omega)$ can be changed with the more general condition that

$$\frac{A\tilde{f}'(A)(\lambda)}{\tilde{f}'(A)(\lambda)} \in \operatorname{int}(I) \text{ for any } \lambda \in \Omega, \tag{19}$$

which might be confirmed for special convex functions f.

Remark 3.2. *Let the function be concave on* int(I). If the condition (19) satisfies, then we get the following inequality

$$0 \leq \tilde{f}(A)(\lambda) - f\left(\frac{A\tilde{f}'(A)(\lambda)}{\tilde{f}'(A)(\lambda)}\right) \leq f'\left(\frac{A\tilde{f}'(A)(\lambda)}{\tilde{f}'(A)(\lambda)}\right).\left[\frac{\tilde{A}(\lambda)\tilde{f}'(A)(\lambda) - A\tilde{f}'(A)(\lambda)}{\tilde{f}'(A)(\lambda)}\right]$$

for any $\lambda \in \Omega$.

Let us define

$$\tilde{B}(f', A; \lambda) := \frac{1}{\tilde{f}'(A)(\lambda)}.f'\left(\frac{A\tilde{f}'(A)(\lambda)}{\tilde{f}'(A)(\lambda)}\right).$$

The following results provide more useful upper bounds for the nonnegative quantity

$$f\left(\frac{A\tilde{f}'(A)(\lambda)}{\tilde{f}'(A)(\lambda)}\right)-\tilde{f}'(A)(\lambda) \text{ for } \lambda\in\Omega.$$

Theorem 3.5. Let $f: I \to \mathbb{R}$ be a convex and differentiable function on int(I) (the interior of I) whose derivative f' is continuous on int(I). Then

$$\begin{aligned}
&(0\leq) f\left(\frac{A\tilde{f}'(A)(\lambda)}{\tilde{f}'(A)(\lambda)}\right)-\tilde{f}(A)(\lambda) \\
&\leq \tilde{B}(f',A;\lambda)\times\begin{cases} \frac{1}{2}.(K-k)\left[\left\|f'(A)\hat{k}_{\lambda}\right\|^2-\tilde{f}'(A)(\lambda)^2\right]^{1/2} \\ \frac{1}{2}.\left(f'(K)-f'(k)\right)\left[\left\|A\hat{k}_{\lambda}\right\|^2-\tilde{A}(\lambda)^2\right]^{1/2} \end{cases} \\
&\leq \frac{1}{4}(K-k)\left(f'(K)-f'(k)\right)\tilde{B}(f',A;\lambda)
\end{aligned} \tag{20}$$

and

$$\begin{aligned}
&(0\leq) f\left(\frac{A\tilde{f}'(A)(\lambda)}{\tilde{f}'(A)(\lambda)}\right)-\tilde{f}(A)(\lambda) \\
&\leq \tilde{B}(f',A;\lambda)\times\left[\frac{1}{4}.(K-k)\left(f'(K)-f'(k)\right)\right. \\
&\left.-\begin{cases}\left[\left\langle (K-A)\hat{k}_{\lambda},(A-k)\hat{k}_{\lambda}\right\rangle\left\langle \left(f'(K)-f'(A)\right)\hat{k}_{\lambda},\left(f'(A)-f'(k)\right)\hat{k}_{\lambda}\right\rangle\right]^{1/2} \\ \left|\tilde{A}(\lambda)-\frac{K+k}{2}\right|\left|\tilde{f}'(A)(\lambda)-\frac{f'(K)+f'(k)}{2}\right| \end{cases}\right] \\
&\leq \frac{1}{4}(K-k)\left(f'(K)-f'(k)\right)\tilde{B}(f',A;\lambda)
\end{aligned} \tag{21}$$

for any selfadjoint operator A on the reproducing kernel Hilbert space $\mathbf{H}(\Omega)$ with $Sp(A)\subseteq[m,M]\subset \operatorname{int}(I)$ and $f'(A)$ is a positive definite operator on $\mathbf{H}$ and any $\lambda\in\Omega$, respectively.

Furthermore, if A is a positive definite operator, then

$$
\begin{aligned}
&(0\leq)\, f\left(\frac{\widetilde{Af'(A)}(\lambda)}{\widetilde{f'(A)}(\lambda)}\right)-\tilde{f}(A)(\lambda) \\
&\leq \tilde{B}(f',A;\lambda)\times\begin{cases}\dfrac{1}{4}\cdot\dfrac{(K-k)(f'(K)-f'(k))}{\sqrt{Kkf'(K)f'(k)}}\tilde{A}(\lambda)\widetilde{f'(A)}(\lambda)\\ \left(\sqrt{K}-\sqrt{k}\right)\left(\sqrt{f'(K)}-\sqrt{f'(k)}\right)\left[\tilde{A}(\lambda)\widetilde{f'(A)}(\lambda)\right]^{1/2}\end{cases} \\
&\leq \tilde{B}(f',A;\lambda)\times\begin{cases}\dfrac{1}{4}\cdot\sqrt{\dfrac{Kf'(K)}{kf'(k)}}(K-k)(f'(K)-f'(k))\\ \left(\sqrt{K}-\sqrt{k}\right)\left(\sqrt{f'(K)}-\sqrt{f'(k)}\right)\sqrt{Kf'(K)}\end{cases}
\end{aligned}
\tag{22}
$$

for any $\lambda\in\Omega$.

Proof. From Corollary 1 in [11, 13], we have that

$$
\widetilde{Af'(A)}(\lambda)-\tilde{A}(\lambda)\widetilde{f'(A)}(\lambda)\leq\frac{1}{2}.(K-k)\left[\left\|f'(A)\hat{k}_{\lambda}\right\|^{2}-\widetilde{f'(A)}(\lambda)^{2}\right]^{1/2}
$$
$$
\leq\frac{1}{4}(K-k)(f'(K)-f'(k))
$$

and

$$
\widetilde{Af'(A)}(\lambda)-\tilde{A}(\lambda)\widetilde{f'(A)}(\lambda)\leq\frac{1}{2}.(f'(K)-f'(k))\left[\left\|A\hat{k}_{\lambda}\right\|^{2}-\tilde{A}(\lambda)^{2}\right]^{1/2}
$$
$$
\leq\frac{1}{4}(K-k)(f'(K)-f'(k)),
$$

for each $\lambda\in\Omega$, which together with (17) give the desired result (20).

By using Lemma 3 in Dragomir (2008, 2011), we have that

$$\widetilde{Af'(A)}(\lambda)-\tilde{A}(\lambda)\widetilde{f'(A)}(\lambda)\leq\frac{1}{4}(K-k)(f'(K)-f'(k))$$
$$-\begin{cases}\left[\left\langle (K-A)\hat{k}_{\lambda},(A-k)\hat{k}_{\lambda}\right\rangle\left\langle (f'(K)-f'(A))\hat{k}_{\lambda},(f'(A)-f'(k))\hat{k}_{\lambda}\right\rangle\right]^{1/2}, \\ \left|\tilde{A}(\lambda)-\frac{K+k}{2}\right|\left|\widetilde{f'(A)}(\lambda)-\frac{f'(K)+f'(k)}{2}\right|,\end{cases}$$

for each $\lambda\in\Omega$. This inequality together with (17) gives the desired result (21).

Lastly, taking advantage of Lemma 3 in [7, 8], we get that

$$\widetilde{Af'(A)}(\lambda)-\tilde{A}(\lambda)\widetilde{f'(A)}(\lambda)$$
$$\leq\begin{cases}\frac{1}{4}\cdot\frac{(K-k)(f'(K)-f'(k))}{\sqrt{Kkf'(K)f'(k)}}\tilde{A}(\lambda)\widetilde{f'(A)}(\lambda), \\ \left(\sqrt{K}-\sqrt{k}\right)\left(\sqrt{f'(K)}-\sqrt{f'(k)}\right)\left[\tilde{A}(\lambda)\widetilde{f'(A)}(\lambda)\right]^{1/2},\end{cases} \tag{23}$$

for each $\lambda\in\Omega$. So, above inequality together with (17) gives the desired result (22).

Remark 3.3. *We get from the first inequality in* (23) that

$$(1\leq)\frac{\widetilde{Af'(A)}(\lambda)}{\tilde{A}(\lambda)\widetilde{f'(A)}(\lambda)}\leq\frac{1}{4}\cdot\frac{(K-k)(f'(K)-f'(k))}{\sqrt{Kkf'(K)f'(k)}}+1$$

which means that

$$f'\left(\frac{\widetilde{Af'(A)}(\lambda)}{\widetilde{f'(A)}(\lambda)}\right)\leq f'\left(\left[\frac{1}{4}\cdot\frac{(K-k)(f'(K)-f'(k))}{\sqrt{Kkf'(K)f'(k)}}+1\right]\tilde{A}(\lambda)\right),$$

for all $\lambda\in\Omega$, since f' is monotonic nondecreasing and A is positive definite.

The first inequality in (22) implies the following result

$$(0\le) f\left(\frac{A\tilde{f}'(A)(\lambda)}{\tilde{f}'(A)(\lambda)}\right)-\tilde{f}'(A)(\lambda)\le\frac{1}{4}.\frac{(K-k)(f'(K)-f'(k))}{\sqrt{Kkf'(K)f'(k)}}$$

$$\times f'\left(\left[\frac{1}{4}.\frac{(K-k)(f'(K)-f'(k))}{\sqrt{Kkf'(K)f'(k)}}+1\right]\tilde{A}(\lambda)\right)\tilde{A}(\lambda)$$

$$\le\frac{1}{4}.\frac{(K-k)(f'(K)-f'(k))}{\sqrt{Kkf'(K)f'(k)}}f'\left(\left[\frac{1}{4}.\frac{(K-k)(f'(K)-f'(k))}{\sqrt{Kkf'(K)f'(k)}}+1\right]M\right)M,$$

for all $\lambda\in\Omega$.

We also have from the second inequality in (22)

$$(0\le) f\left(\frac{A\tilde{f}'(A)(\lambda)}{\tilde{f}'(A)(\lambda)}\right)-\tilde{f}'(A)(\lambda)\le\left(\sqrt{K}-\sqrt{k}\right)\left(\sqrt{f'(K)}-\sqrt{f'(k)}\right)$$

$$\times f'\left(\left[\frac{1}{4}.\frac{(K-k)(f'(K)-f'(k))}{\sqrt{Kkf'(K)f'(k)}}+1\right]\tilde{A}(\lambda)\right)\left[\frac{\tilde{A}(\lambda)}{\tilde{f}'(A)(\lambda)}\right]^{\frac{1}{2}}$$

$$\times f'\left(\left[\frac{1}{4}.\frac{(K-k)(f'(K)-f'(k))}{\sqrt{Kkf'(K)f'(k)}}+1\right]K\right)\sqrt{\frac{K}{f'(k)}},$$

for all $\lambda\in\Omega$.

REFERENCES

Abu-Omar, A., & Kittaneh, F. (2015). Notes on some spectral radius and numerical radius inequalities. *Studia Mathematica*, *272*(2), 97–109. doi:10.4064m227-2-1

Aronzajn, N. (1950). Theory of reproducing kernels. *Transactions of the American Mathematical Society*, *68*(3), 337–404. doi:10.1090/S0002-9947-1950-0051437-7

Bakherad, M. (2018). Some Berezin Number Inequalities for Operator Matrices. *Czechoslovak Mathematical Journal*, *68*(143), 997–1009. doi:10.21136/CMJ.2018.0048-17

Bakherad, M., & Garayev, M. T. (2019). Berezin number inequalities for operators. *Concr. Oper.*, *6*(1), 33–43. doi:10.1515/conop-2019-0003

Berezin, F. A. (1972). Covariant and contravariant symbols for operators, Math. *USSR-Izv.*, *6*(5), 1117–1151. doi:10.1070/IM1972v006n05ABEH001913

Berezin, F. A. (1974). Quantization. *Math. USSR Izvestija*, *8*(5), 1109–1165. doi:10.1070/IM1974v008n05ABEH002140

Berger, C. A., & Coburn, L. A. (1987). Toeplitz operators on the Segal-Bargmann space. *Transactions of the American Mathematical Society*, *301*(2), 813–829. doi:10.1090/S0002-9947-1987-0882716-4

Bergman, S. (1922). Über die Entwicklung der harmischen Funktionen der Ebene und des Rümes nach Orthogonal Functionen. *Mathematische Annalen*, *86*(3-4), 238–271. doi:10.1007/BF01457987

Coburn, L. A. (2004). A Lipschitz estimate for Berezin's operator calculus. *Proceedings of the American Mathematical Society*, *133*(01), 127–131. doi:10.1090/S0002-9939-04-07476-3

Dragomir, S. S. (2004). *Discrete Inequalities of the Cauchy-Bunyakovsky-Schwarz Type*. NY: Nova Science Publishers.

Dragomir, S. S. (2006). Reverse inequalities for the numerical radius of linear operators in Hilbert spaces. *Bulletin of the Australian Mathematical Society*, *73*(2), 255–262. doi:10.1017/S0004972700038831

Dragomir, S. S. (2007). A survey of some recent inequalities for the norm and numerical radius of operators in Hilbert spaces. *Banach Journal of Mathematical Analysis*, *1*(2), 154–175. doi:10.15352/bjma/1240336213

Dragomir, S. S. (2007). Inequalities for the norm and the numerical radius of linear operators in Hilbert spaces. *Demonstratio Math.*, *40*(2), 411–417.

Dragomir, S. S. (2011). Some Slater type inequalities for convex functions of selfadjoint operators in Hilbert spaces. *Revista de la Unión Matemática Argentina*, *52*(1), 109–120.

Dragomir, S.S. (2008a). Grüss' type inequalities for functions of selfadjoint operators in Hilbert spaces. Preprint. *RGMIA Res. Rep. Coll.*, *11*(e), Art. 11.

Dragomir, S.S. (2008b). Some new Grüss' type Inequalities for functions of selfadjoint operators in Hilbert spaces. Preprint. *RGMIA Res. Rep. Coll.*, *11*(e), Art. 12.

Engliš, M. (1994). Functions invariant under the Berezin transform. *Journal of Functional Analysis*, *121*(1), 233–254. doi:10.1006/jfan.1994.1048

Engliš, M. (1995). Toeplitz operators and the Berezin transform on H^2. *Linear Algebra and Its Applications*, *223/224*, 171–204. doi:10.1016/0024-3795(94)00056-J

Engliš, M. (1999). Compact Toeplitz operators via the Berezin transform on bounded symmetric domains. *Integral Equations and Operator Theory*, *33*(4), 426–455. doi:10.1007/BF01291836

Furuta, T., Mićić Hot, J., Pečarić, J., & Seo, Y. (2005). *Mond-Pečarić Method in Operator Inequalities. Inequalities for Bounded Selfadjoint Operators on a Hilbert Space*. Zagreb: Element.

Garayev, M., Saltan, S., & Gundogdu, D. (2018). On the Inverse Power Inequality for the Berezin Number of Operators. *J. Math. Inequal.*, *12*(4), 997–1003. doi:10.7153/jmi-2018-12-76

Garayev, M. T., Gürdal, M., & Okudan, A. (2016). Hardy-Hilbert's inequality and a power inequality for Berezin numbers for operators. *Mathematical Inequalities & Applications*, *3*(3), 883–891. doi:10.7153/mia-19-64

Garayev, M. T., Gürdal, M., & Saltan, S. (2017). Hardy Type Inequality For Reproducing Kernel Hilbert Space Operators and Related Problems. *Positivity*, *21*(4), 1615–1623. doi:10.100711117-017-0489-6

Gustafson, K. E., & Rao, D. K. M. (1997). *Numerical Range*. Springer-Verlag.

Hajmohamadi, M., Lashkaripour, R., & Bakherad, M. (n.d.). Improvements of Berezin number inequalities. *Linear and Multilinear Algebra*. doi:10.1080/03081087.2018.1538310

Karaev, M. T. (2006). Berezin symbol and invertibility of operators on the functional Hilbert spaces. *Journal of Functional Analysis*, *238*(1), 181–192. doi:10.1016/j.jfa.2006.04.030

Karaev, M. T. (2013). Reproducing Kernels and Berezin symbols Techniques in Various Questions of Operator Theory. *Complex Analysis and Operator Theory*, *7*(4), 983–1018. doi:10.100711785-012-0232-z

Karaev, M. T., Gürdal, M., & Yamancı, U. (2013). Special operator classes and their properties. *Banach Journal of Mathematical Analysis*, *7*(2), 74–85. doi:10.15352/bjma/1363784224

Karaev, M. T., Gürdal, M., & Yamancı, U. (2014). Some results related with Berezin symbols and Toeplitz operators. *Mathematical Inequalities & Applications, 17*(3), 1031–1045. doi:10.7153/mia-17-76

Kittaneh, F. (2005). Numerical radius inequalities for Hilbert space operators. *Studia Mathematica, 168*(1), 73–80. doi:10.4064m168-1-5

Kittaneh, F., Moslehian, M. S., & Yamazaki, T. (2015). Cartesian decomposition and numerical radius inequalities. *Linear Algebra and Its Applications, 471*, 46–53. doi:10.1016/j.laa.2014.12.016

McCarthy, C. A. (1967). Cp. *Israel Journal of Mathematics, 5*(4), 249–271. doi:10.1007/BF02771613

Mond, B., & Pečarić, J. (1993). Convex inequalities in Hilbert space. *Houston Journal of Mathematics, 19*, 405–420.

Nordgren, E., & Rosenthal, P. (1994). Boundary values of Berezin symbols, Nonselfadjoint operators and related topics. Oper. Theory Adv. Appl., 73.

Potter, M. J. A. (2000). *Berezin symbols and operator theory*. Thesis.

Saitoh, S. (1988). *Theory of reproducing kernels and its applications*. Pitman Research Notes in Mathematics Series 189.

Saitoh, S., & Sawano, Y. (2016). *Theory of reproducing kernels and applications*. Springer. doi:10.1007/978-981-10-0530-5

Sattari, M., Moslehian, M. S., & Yamazaki, T. (2015). Some generalized numerical radius inequalities for Hilbert space operators. *Linear Algebra and Its Applications, 470*, 216–227. doi:10.1016/j.laa.2014.08.003

Schwartz, L. (1964). Sous-espaces hilbertiens d'espaces vectoriels topologiques et noyaux associés (noyaux reproduisants). *Journal d'Analyse Mathématique, 13*(1), 115–256. doi:10.1007/BF02786620

Slater, M. S. (1981). A companion inequality to Jensen's inequality. *Journal of Approximation Theory, 32*(2), 160–166. doi:10.1016/0021-9045(81)90112-X

Szegö, S. (1921). Über orthogonale Polynome, die zu einer gegebenen Kurve der Komplexen Ebenegehören. *Mathematische Zeitschrift, 9*(3-4), 218–270. doi:10.1007/BF01279030

Yamancı, U., Garayev, M. T., & Çelik, C. (2019). Hardy-Hilbert type inequality in reproducing kernel Hilbert space: Its applications and related results. *Linear and Multilinear Algebra, 67*(4), 830–842. doi:10.1080/03081087.2018.1490688

Yamancı, U., & Gürdal, M. (2017). On numerical radius and Berezin number inequalities for reproducing kernel Hilbert space. *New York Journal of Mathematics*, *23*, 1531–1537.

Yamancı, U., Gürdal, M., & Garayev, M. T. (2017). Berezin Number Inequality for Convex Function in Reproducing Kernel Hilbert Space. *Filomat*, *31*(18), 5711–5717. doi:10.2298/FIL1718711Y

Zaremba, S. (1907). L'equation biharminique et une class remarquable de fonctions foundamentals harmoniques. Bull. *Int. de l'Academie des Sci. de Cracovie*, *39*, 147–196.

Chapter 5
COGARCH Models:
An Explicit Solution to the Stochastic Differential Equation for Variance

Yakup Arı
https://orcid.org/0000-0002-5666-5365
Alanya Alaaddin Keykubat University, Turkey

ABSTRACT

In this chapter, the features of a continuous time GARCH (COGARCH) process is discussed since the process can be applied as an explicit solution for the stochastic differential equation which is defined for the volatility of unequally spaced time series. COGARCH process driven by a Lévy process is an analogue of discrete time GARCH process and is further generalized to solutions of Lévy driven stochastic differential equations. The Compound Poisson and Variance Gamma processes are defined and used to derive the increments for the COGARCH process. Although there are various parameter estimation methods introduced for COGARCH, this study is focused on two methods which are Pseudo Maximum Likelihood Method and General Methods of Moments. Furthermore, an example is given to illustrate the findings.

INTRODUCTION

Mandelbrot (1963) pointed out that the financial time series are not usually stationary, and the increments of these series have no autocorrelation but their squares present out a correlation. Furthermore, he showed the volatility of financial series which exist in volatility cluster is not constant and the distribution of the series is not normal since leptokurtic shape. Black (1976) showed leverage-effect is another stylized

DOI: 10.4018/978-1-7998-0134-4.ch005

fact of financial series. The leverage-effect occurs when the negative fluctuations on stock price have a greater influence on volatility than positive fluctuations do where negative and positive fluctuations have the same magnitude. The long-term persistence is another important subject to discuss for financial series. Fractal mathematics helps us to identify the long-term persistence via models which are the generalization of short-term memory models such as Autoregressive Integrated Moving Average (ARIMA) models with Chaos theory. The randomness generated by deterministic systems, as in how the sensitivity of chaotic systems to their initial conditions can be assumed as a definition of Chaos theory. After the study of Hosking (1981) that is the introduction of the fractional differentiation operator makes these models especially important. The chaotic systems, which have complex forms, are becoming indistinctive with respect to the randomness when the time is going up. Thus, the term pseudo-randomness is sometimes used to define the behaviour of chaotic systems. On the other hand, the scope of non-linear models is the non-stationary of increments assuming the presence of heteroskedasticity in the model. The Autoregressive Conditional Heteroscedastic (ARCH), Generalized ARCH (GARCH) and Continuous GARCH (COGARCH) models and their extensions are the part of this approach.

Engle (1982), The Nobel Prize laureates, has introduced the ARCH models that are proper to represent the typical empirical findings of financial time series. Although the ARCH model can yield volatility clusters, it also has some weaknesses in describing the stylized fact of financial series. It can be concluded that a mechanical way to describe the behaviour of the conditional variance. But it gives no indication of the behaviour of conditional variance because it assumes that positive and negative shocks have the same effects on volatility since it depends on the square of the previous shocks (Tsay, 2013).

After Engle, Bollerslev (1986) whose study was another milestone for financial time series analysis was GARCH model. The discrete-time GARCH models, which are improper models to deal with heteroscedasticity in time series, capture some of the most prominent features in financial data, particularly in the volatility process. In ordinary discrete time GARCH models, time series are assumed to be equally spaced. But time series have often irregular space between two observations. Tick-by-tick data and daily data are examples of this situation. To accommodate the irregularity of time spaces, Klüppelberg et al. (2004) proposed a continuous time GARCH (COGARCH) process driven by a Lévy process, which is an analogue of discrete time GARCH process and is further generalized to solutions of Lévy driven stochastic differential equations. Like in the discrete time case, the COGARCH model cannot model this stylized fact of time series. Therefore, Haug (2006) have developed the exponential COGARCH model as the first extension of the COGARCH model that is based on the discrete time EGARCH model.

The aim of this chapter is to define and to discuss the properties of COGARCH models with examples. In this context, the outline of the chapter is designed as follows;

The general form of the stochastic differential equations (SDEs) is given in Section 2. Then, the properties of the discrete time GARCH model is discussed in section 3. COGARCH process is a Lévy-driven process, therefore the basis of this process is constructed on Compound Poisson and Variance Gamma which are also stochastic processes like Brownian Motion, are defined in Section 4. In the subsection 4.1, derivation of Continuous time GARCH Process and second-order properties of the volatility process will be defined and discussed, depending on the approach of Klüppelberg et al. (2004) and Klüppelberg et al. (2011). Moreover, some classical and Bayesian estimation methods for COGARCH models will be mentioned in Section 5. Compound Poisson COGARCH(1, 1) and Variance Gamma COGARCH(1, 1) simulations will be given as examples and furthermore, the model will be applied to real-life data.

FROM DETERMINISTIC TO STOCHASTIC

It is useful to begin with an example of ordinary differential equation (ODE) to understand the structure of stochastic differential equations. An ODE has a general form as following;

$$\frac{dx(t)}{dt} = f(t,x), \quad dx(t) = f(t,x)d \tag{1}$$

with initial condition $x(t)=x_0$By taking the integral of above Equation 1, the integral form can be written as

$$x(t) = x_0 + \int_0^t f(s, x(s))ds \tag{2}$$

where $x(t)=x(t,x_0,t_0)$ is the solution of ODE depending on initial condition. Let

$$\frac{dx(t)}{dt} = a(t)x(t), \quad x(0) = x_0 \tag{3}$$

be an example, where a(t) is a determistic parameter. When it is assumed that a(t) is a deterministic parameter but rather a stochastic parameter for ODE in Equation 3, the stochastic parameter a(t) can be defined as

$$a(t) = f(t) + h(t)\varepsilon(t) \tag{4}$$

where ε(t) is awhite noise process. Thus, the following stochastic differential equation (SDE) is obtained;

$$\frac{dX(t)}{dt} = f(t)X(t) + h(t)X(t)\mu(t) \tag{5}$$

If Equation 5 is rewirtten in the differential form using dW(t)=ε(t) dt, where dW(t) denotes the differential form of the Brownian motion, the following is obtained

$$dX(t) = f(t)X(t)dt + h(t)X(t)dW(t) \tag{6}$$

The general form of an SDE is given as

$$dX(t,\omega) = f(t,X(t,\omega))dt + g(t,X(t,\omega))dW(t,\omega) \tag{7}$$

where ω denotes that X=X(t,ω) is a random variable and having an initial condition $X(0,\omega)=X_0$ with probability one. The Equation 7 can be written in the integral equation as

$$X(t,\omega) = X_0 + \int_0^t f(s,X(s,\omega))ds + \int_0^t g(s,X(s,\omega))dW(s,\omega) \tag{8}$$

where $f(t,X(t,\omega)) \in \mathbb{R}$, $g(t,X(t,\omega)) \in \mathbb{R}$ and $W(t,\omega) \in \mathbb{R}$.

THE DIFFERENCE EQUATIONS OF VOLATILITY: ARCH AND GARCH MODELS

Engle (1982) has shown that the ARCH models are proper models to represent the typical empirical findings of financial time series, e.g. the conditional heteroscedasticity. The amplitude of the returns varies over time is described as "volatility clustering." The ARCH model is designed to deal with just this set of issues.

They have become common implements for dealing with time series heteroskedastic models. Such models that can be used in financial decisions concerning risk analysis of holding an asset or the value of an option, portfolio selection and derivative pricing (Engle, 2001) provide a volatility measure, like a standard deviation. The basic concept of the ARCH model is that the series of errors are serially uncorrelated, though dependent on its p squared lag values.

Definition 1: ARCH(q) Process

Given that the volatilities σ_t^2 are stationary and independent random variables and, assuming that ϵ_t is a sequence of random variables, independently and identically distributed, the process a_t is an ARCH(q) model if $a_t=\sigma_t\epsilon_t$, $\epsilon_t \sim f_v(0,1)$ where $\omega_t^2 = \alpha_0 + \sum_{i=1}^{q} \alpha_i a_{t-i}^2$, $t \in \mathbb{N}$ where α_0 and α_i are the parameters and they are satisfying $\alpha_0>0$ and $\sum_{i=1}^{q} \alpha_i < 1$ to guarantee that the variance is positive and is a stationary process.

Although the ARCH model can yield volatility clusters, it also has some weaknesses; first of all, it assumes that positive and negative shocks have the same effects on volatility because it depends on the square of the previous shocks. But, the financial asset responds differently to positive and negative shocks. The ARCH model does not provide any new insight for understanding the source of variations of a financial time series. We only conclude a mechanical way to describe the behaviour of the conditional variance. It gives no indication of the behaviour of conditional variance (Tsay, 2012).

Bollerslev (1986) extended the ARCH model to the Generalized Autoregressive Conditional Heteroscedastic (GARCH) model, which assumes that the conditional variance depends on its own p past values and q past values of the squared error terms. This model is denoted as GARCH(p,q). In most applications, the GARCH (1,1) model captures the volatility of financial data and has been utilized widely by practitioners and academicians and therefore has been studied extensively, see for example Tsay (2012).

Definition 2: GARCH(p,q) Process

Assuming that volatilities σ_t^2 are random variables and ϵ_t is a sequence of random variables, independently and identically distributed, then a_t follow a general GARCH(p, q) model if $a_t=\sigma_t\epsilon_t$ where $\epsilon_t \sim f_v(0,1)$. The variance equation of the GARCH (p,q) model can be expressed as

$$\sigma_t^2 = \alpha_0 + \sum_{i=1}^{q} \alpha_i a_{t-i}^2 + \sum_{i=1}^{q} \beta_i \sigma_{t-i}^2 \tag{9}$$

where the innovations follow the probability density function $f_v(0,1)$ with zero mean and unit variance. In non-normal case, v is used as additional distributional parameters for the scale and the shape of the distribution.

Bollerslev (1986) has shown that the GARCH(p,q) process is covariance stationary with $E(a_t) = 0$, $var(a_t) = \alpha_0 / \left(1-\left(\sum_{i=1}^{q}\alpha_i + \sum_{i=1}^{p}\beta_i\right)\right)$ and $cov(a_t, a_s) = 0$ for $t \neq s$ if and only if $\sum_{i=1}^{q}\alpha_i + \sum_{i=1}^{p}\beta_i < 1$. He used the MLE method by maximizing the given log-likelihood function

$$L(\vartheta) = \ln \prod_t f_v\left(a_t, E\left(a_t \mid I_{t-1}\right), \sigma_t\right)$$

where f_v is the conditional distribution function. The second argument of f_v denotes the mean, and the third argument the standard deviation. The full set of parameters ϑ includes the parameters from the variance equation $\vartheta=(\alpha_0,\alpha_1,\ldots,\alpha_p, \beta_1,\ldots,\beta_q)$ and the distributional parameters (v) in the case of a non-normal distribution function.

COGARCH MODELS

In this section, the definitions and properties of Brownian motion and Lévy processes will be given. Then Compound Poisson process, Variance Gamma process and The Lévy -Ito decomposition will be defined based on the study of Applebaum (2009).

Definition 3: Brownian Motion

A Brownian motion is a stochastic process B_t for t>0 which satisfies the following conditions;

1. The starting value of Brownian motion is $B_0=0$
2. The increments of Brownian motion are independent, which means that $B_{t+s} - B_t$ is independent of $\sigma(B_t)$ for all $0 \leq s, t \leq \infty$.
3. $B_{t+s} - B_t \sim N(0, s)$ for all $0 \leq s, t \leq \infty$ means that the increments are normally distirbuted.

4. Brownian motion has continuous trajectories.

Some basic properties of the Brownian motion are

- The random variables B_t-B_s and B_{t-s} follow a normal distribution with zero mean and (t-s) variance where t<s. Larger fluctuations are observed where the intervals are larger in Brownian motion since the variance is the length of the interval.
- The finite-dimensional distributions of the Brownian motion are multivariate Gaussian distributions since Brownian motion is a Gaussian process.
- $B_{t_1} - B_{t_0}, B_{t_2} - B_{t_1}, \dots, B_{t_n} - B_{t_{n-1}}$ are independent since the process has independent increments where $0 \le t_0 < t_1 < \dots < t_n < \infty$ and $n \ge 0$.
- $Cov(B_t, B_s = \min(s,t)$ for $0 \le t < s < \infty$

Definition 4: Lévy Processes

A Lévy process is a stochastic process $\left(L_t\right)_{t\ge0}$ that satisfies the following properties:

1. The starting value of the Lévy process is L_0=0
2. It has independent and stationary increments.
3. It is stochastically continuous.

Having continuous sample paths with probability one is a characteristic of the Brownian motion. But, this assumption is required that the process should be a càdlàg process which is

- $\lim_{s\to t} L_s = L_t$ means that the process is right-continuous.
- $L_{t^-} = \lim_{s\to t} L_s$ means that the process has limits from the left with probability one.

Thus, it can be concluded that jumps of the Lévy process $\Delta L_t = L_t - L_{t^-}$ are also a càdlàg process and these jumps are observed often in financial time series.

The Lévy process L_t is infinitely divisible for $\forall t \ge 0$. So, the process characterized by its characteristic exponent that satisfies the Lévy -Khintchine formula. Then, its characteristic function can be defined as $\varnothing_t(u) = E\left(e^{iuL_t}\right)$ and it can be written in terms of the characteristic function of L_1, $\varnothing(u) = E\left(e^{iuL_1}\right)$.

The characteristic function $\varnothing_L(u)$ can be used to characterize the distribution of a Lévy process where

$$\varnothing_L(u) = E\left(e^{iuL_t}\right) = exp\left\{t\psi(u)\right\}$$

where $\Psi(u)$ is the characteristic exponent of L_1 and $\Psi(u)$ is given by

$$\psi(u) = \mathrm{iau} - \frac{\sigma^2 u^2}{2} + \int_{\mathbb{R}\setminus\{0\}} \left(e^{iux} - 1 - \mathrm{iux}1_{\{|x|<1\}}\right)v(dx)$$

where $a \in \mathbb{R}$, σ>0 and v is the Lévy measure of L_t that satisfies $v(\{0\}=0$ and $\int \min\{x^2, 1\}v(dx) < \infty$. (See Applebaum, 2009, Theorem 1.2.14 Lévy–Khintchine)

Definition 5: Compound Poisson Process

A Poisson process N for t≥0 and parameter λ>0, If a Poisson process is independent of an i.i.d. (independent and identically distributed) sequence of random variables $\left(\mathrm{Y}_i\right)_{i\in\mathbb{N}}$, then a compound Poisson process L is defined as

$$\mathrm{L}_t = \sum_{i=1}^{\mathrm{N}_t} \mathrm{Y}_i, \quad t \geq 0 \tag{9}$$

The compound Poisson (CP) process has jumps with random size instead of the constant jumps of size 1 of a Poisson process.

Definition 6: Variance Gamma Process

A Variance Gamma (VG) process is obtained by evaluating Brownian motion with drift at a random time given by a Gamma process. Let B(t) be a standard Brownian motion then Brownian motion with drift can be defined as

$$b(t,\theta,\sigma) = \theta t + \sigma B(t) \; t \geq 0 \tag{10}$$

where $\theta \in \mathbb{R}$ is the drift term and σ is variance? The time change of the Brownian motion is done with respect to a Gamma process $\left(\mathrm{H}_t\right)_{t\geq 0}$ with parameters a, b>0, such that each of the i.i.d increments is Gamma distributed with density

$$f_{H_t}(x) = \frac{b^{at}}{\Gamma(at)} e^{at-1} e^{-bx} \quad \text{for } x \geq 0$$

where $\Gamma(.)$ is Gamma function. Thus, the VG process $(V_t)_{t\geq0}$ can be obtained by

$$V_t = \theta H_t + \sigma B_{H_t} \quad t \geq 0 \tag{11}$$

Continuous Time GARCH Process

In this subsection, derivation of Continuous time GARCH process and second-order properties of the volatility process will be defined and discussed, depending on the approach of Klüppelberg et al. (2004) and Klüppelberg et al. (2011).

The approach of Klüppelberg et al. (2004) includes only one source of uncertainty (two of the same kind). The increments of a Lévy process replace the innovations of a discrete time GARCH model. Additionally, it contains the autoregressive property, which can be found in the discrete time case. The idea of Klüppelberg et al. (2004) for the construction of a continuous time GARCH model is to preserve the structure and the main characteristics of a discrete time GARCH model.

Klüppelberg et al. (2004) used the discrete time GARCH (1,1) process to build continuous time analogue. Let $a_n = \sigma_n \epsilon_n$ where $\epsilon_n \sim f_v(0,1)$ and $\sigma_n^2 = \alpha_0 + \alpha_1 a_{n-1}^2 + \beta_1 \sigma_{n-1}^2$ be a discrete time GARCH(1,1) process where α_0, α_1 and β_1 are constant parameters. The idea is to replace the white noise sequence of the discrete equation with the increments of a Lévy process. So, the continuous time GARCH process $(G_t)_{t\geq0}$ can be defined as

$$(G_t)_{t\geq0} = \int_0^t \sigma_{s-} dL_s \tag{12}$$

The volatility process is defined as a random recurrence equation and iterating the recurrence:

$$\sigma_n^2 = \alpha_0 \sum_{t=0}^{n-1} \prod_{j=i+1}^{n-1} \left(\beta_1 + \alpha_1 a_j^2\right) + \sigma_0^2 \prod_{j=0}^{n-1} \left(\beta_1 + \alpha_1 a_j^2\right) \tag{13}$$

Replacing the sum in the Equation 13 by integration the volatility process becomes

$$\sigma_n^2 = \left[\alpha_0 \int_0^n \exp\left(-\sum_{j=0}^{|s|} \log\left(\beta_1 + \alpha_1 a_j^2\right) ds\right) + \sigma_0^2\right] \exp\left(\sum_{j=0}^{n-1} \log\left(\beta_1 + \alpha_1 a_j^2\right)\right) \tag{14}$$

Replacing the $\beta = \alpha_0$, $\eta = -\log\beta_1$ and $\varphi = \frac{\alpha_1}{\beta_1}$ in the Equations 14 the volatility process is obtained by

$$\sigma_n^2 = \left(\beta \int_0^n e^{X_s} ds + \sigma_0^2\right) e^{-X_t} \tag{15}$$

with the auxiliary process $X_t = t\eta - \sum_{0 \le s \le t} \log\left(1 + \varphi\left(\Delta L_s\right)^2\right)$ where $t \ge 0$ and L_t is a Lévy process.

Then, the COGARCH(1,1) can be defined as the càdlàg process that satisfies the following stochastic differential equation

$$dG_t = \sigma_{t-} dL_t, \text{ for } t \ge 0 \text{ and } G_0 = 0 \tag{16}$$

The process $\left(\sigma_t^2\right)_{t \ge 0}$ satisfies the following stochastic differential equation

$$d\sigma_t^2 = \left(\beta - \eta\sigma_{t-}\right) dt + \varphi\sigma_{t-} d\left[L, L\right]_t^d \tag{17}$$

where $[L,L]_t$ is the quadratic variation of the Lévy process

$$\left[L, L\right]_t = \sigma^2 t + \sum_{0 \le s \le t} \left(\Delta L_s\right)^2 = \sigma^2 t + \left[L, L\right]_t^d$$

where $\left[L, L\right]_t^d = \sum_{0 \le s \le t} \left(\Delta L_s\right)^2$ is the discrete part of the quadratic variation of L_t.

Definition 7: Continuous time GARCH(1,1) Process

Let $L = \left(L_t\right)_{t \ge 0}$ be a Lévy process with triplet (a, σ^2, ν) where ν is a Lévy measure. Given a finite random variable independent of L_t, the COGARCH process $\left(G_t\right)_{t \ge 0}$ and the variance process $\sigma^2 = \left(\sigma_t^2\right)_{t \ge 0}$ are defined by the stochastic differential equations

$$dG_t = \sigma_{t-} dL_t$$

$$d\sigma_t^2 = (\beta - \eta\sigma_{t-})dt + \varphi\sigma_{t-} d[L,L]_t^d$$

for $t \geq 0$ and $G_0=0$, $\beta>0$, $\eta>0$, $\varphi \geq 0$ and $[L,L]_t^d$ is the discrete part of the quadratic variation Lévy process.

It can be inferred from the Definition 7 that the process G jumps at the same times as L does, and the size of its jumps is $\Delta G_t = \sigma_t \Delta L_t$ for $t \geq 0$. So, ΔL_t can be assumed as innovations in case of discrete GARCH models.

THE ESTIMATION METHODS FOR COGARCH MODELS

The several frequentists and Bayesian estimation methods have been introduced to estimate the parameters of COGARCH and its extensions models in the last decade. The method of moments (MM) procedure is used by Haug et al. (2007). They estimate the model parameters from the log-returns by matching the empirical autocorrelation function and moments to their theoretical counterparts. However, their methodology is only valid if the driving Lévy process has finite variance, e.g., is a Compound Poisson process with normally distributed jumps or a Variance-Gamma process. For MM, there is no need to specify the driving Lévy process, however, it can only be applied to equally spaced observations and is, therefore, not applicable in many interesting situations. Moreover, it has been shown that, as for many other models, this moment method - although being consistent - is not very efficient.

Müller (2010) proposed that it is also possible to use Bayesian methods to estimate COGARCH(1,1) parameters by MCMC simulations. This method is also restricted by the condition that the driving Lévy process is Compound Poisson.

Maller et al. (2008) used an approximation to fit the model to unequally spaced time data, by deriving a pseudo-maximum likelihood (PML) function and numerically maximizing it in order to estimate the corresponding parameters. In this way, given a COGARCH model, there exists a sequence of GARCH models, which converges to it. They proved that the COGARCH model occurs as a continuous-time limit of GARCH in a strong sense and construct a sequence of discrete time stochastic processes that converge in probability and in the Skorokhod metric to a COGARCH model. The result is useful for the estimation of the continuous model defined for irregularly spaced time series data. The estimation procedure is extended and is implemented in the R package "yuima" by Iacus et al. (2015). A simulation study demonstrates that if a compound Poisson with normal jumps, using PML the mean squared errors is reduced with respect to the MM drives the model, but relevant biases can be found. Although being not consistent, PML method gives only slightly better estimates for realistic sample sizes. Behme et al. (2014) used PML method to estimate the parameters of Asymmetric Power COGARCH (APCOGARCH) and its reduced form Glosten, Jagannathan and Runkle COGARCH (GJR- COGARCH).

Bayraci et al. (2014) introduced a simulation-based indirect inference approach in order to estimate the COGARCH(1,1) model. The basic idea of the indirect inference is that when a model leads to a complicated structural or reduced form and therefore to intractable likelihood functions, estimation of the original model can be indirectly achieved by estimating an auxiliary model which is constructed as an approximation of the original one.

Marín et al. (2015) use the data cloning methodology, which is another Bayesian approach; one can obtain approximate maximum likelihood estimators of COGARCH models avoiding numerically maximization of the pseudo-likelihood function. They take the pseudo-likelihood function to build the joint posterior distribution of the parameters, as a previous step to deal with the data cloning methodology. But, in this case, as the procedure is based on an MCMC algorithm, it is not necessary to numerically maximize the pseudo-likelihood function as in Maller et al. (2008).

Marín et al. (2016) proposed two Bayesian methods which are Hamiltonian Monte-Carlo (HMC) methodology and Approximate Bayesian Computation (ABC) methodology. HMC methodology can be applied using the following steps; the first generate the initial values for model parameters and for momentum variables; then the second step is to implement the called Leapfrog algorithm, which depends on the derivative function of the logarithm of the posterior density function and on a scale factor and the last step to include the accept-reject Metropolis-Hasting method for the previously obtained values. In ABC method, they generate the parameter values from the prior distributions, simulate with them a sample path and consider the proposed values as good if the obtained trajectory is similar enough to the original data, considering several goodness of fit statistics. They complete the ABC algorithm using accept-reject Metropolis-Hasting algorithm.

The estimation is essential since it enables us to make the following possible interpretations using the parameters of the model. The values of the parameters in COGARCH model demonstrate the speed of the decline of a volatility burst which appears due to the arrival of new information to markets. Moreover, those values show the level of volatility and the magnitude of the volatility jumps that may be considered as a measure of how information affects volatility and how fast market assumes new events. Hence; the estimation helps us understanding how large-scale financial activities will develop.

Estimation of COGARCH(1,1) by Pseudo-Maximum Likelihood (PML) Method

Maller et al. (2008) developed PML method based on the GARCH approximation to COGARCH for irregularly observed time series data from the COGARCH(1,1) model. They suppose that the observations are $G(t_i)$, $0=t_0<t_1<\ldots<t_N=T$, on the integrated COGARCH $\left(G_t\right)_{t\geq 0}=\int_0^t \sigma_{s-}dL_s$ and assumed to be stationary. The $\{t_i\}$ are assumed fixed time points. Let $a_i = Y_i = G(t_i) - G(t_{i-1})$ denote the observed returns and $\Delta t_i := t_i - t_{i-1}$. So, the observed return can be written as following $Y_i = \int_0^t \sigma_{s-}dL_s$ where L is a Lévy process with $E[L(1)]=0$ and $E[L^2(1)]=1$.

The purpose is to estimate (β,η,φ) from the observed $Y_1,Y_2,\ldots,Y_N$ using pseudo-maximum likelihood (PML) method. Y_i is conditionally independent of $Y_{i-1}, Y_{i-2},\ldots$ given information set F_{t-1} since σ is Markovian. So, $E(Y_i \mid F_{t-1}) = 0$ for the conditional expectation of Y_i, and, for the conditional variance,

$$\rho_i^2 := E(Y_i^2 \mid F_{t_{i-1}}) = \left(\sigma^2(t_{i-1}) - \frac{\beta}{\eta-\phi}\right)\left(\frac{e^{(\eta-\phi)\Delta t_i}-1}{\eta-\phi}\right) + \frac{\beta \Delta t_i}{\eta-\phi}$$

$E\left(\sigma^2(0)\right) = \frac{\beta}{\eta-\phi}$, with η>φ and $E[L^2(1)]=1$ satisfy the stationarity of the model.

The pseudo-maximum likelihood function for $Y_1,Y_2,\ldots,Y_N$ can be written as following with the assumption of Y_i, are conditionally $N(0,\rho_i^2)$

$$L = L(\beta,\eta,\phi) = \sum_{i=1}^{N}\left(-\frac{1}{2}\log(2\pi) - \frac{1}{2}\ln\rho_i^2 - \frac{1}{2}\frac{Y_i^2}{\rho_i^2}\right)$$

Above equation needs a calculable quantity for ρ_i^2. Hence $\sigma^2(t_{i-1})$ should be substituted by $\sigma_i^2 = \beta\Delta t_i + e^{-\eta\Delta t_i}\sigma_{i-1}^2 + \phi e^{-\eta\Delta t_i}Y_i^2$. After substituting σ_i^2 for $\sigma^2(t_{i-1})$ and resulting modified ρ_i^2, pseudo-maximum likelihood function can be found for fitting a GARCH model to the unequally spaced series. The recursion of σ_i^2 can be easily done taking $\sigma^2(0)=\beta/(\eta-\phi)$ as an initial value. The maximization of $L=L(\beta,\eta,\varphi)$ gives PMLEs of (β,η,φ).

General Method of Moment (GMM) Estimation in the COGARCH(1,1) Model

In this part, the method of Haug et. al. (2007) is followed. The auxiliary process X_t is a spectrally negative Lévy process, with drift η, no Gaussian part and with Lévy measure v_X.

$$v_X[0,\infty)=0, v_X(-\infty,-x]=v_L\left(\left\{y\in\mathbb{R}:|y|\geq\sqrt{e^x/\varphi}\right\}\right),\quad x>0 \tag{18}$$

The Laplace transform $E\left[e^{-sX_t}\right]=e^{t\Psi(s)}$ with Laplace exponent is also used in this method where

$$\Psi(s)=-\eta s+\int_{\mathbb{R}}\left(\left(1+\varphi x^2\right)^s-1\right)v_L dx,\quad s\geq 0.$$

Proposition 1

Suppose that the Lévy process $\left(\mathrm{L}_t\right)_{t\geq 0}$ has finite variance and zero mean, and that $\Psi(1)<0$. Let $\left(\sigma_t^2\right)_{t\geq 0}$ be the stationary volatility process, so that $\left(\mathrm{G}_t\right)_{t\geq 0}$ has stationary increments. Then $E\left(\mathrm{G}_t^2\right)<\infty$ for all $t\geq 0$, and for every $t,h\geq r>0$ it holds

$$E\left(\mathrm{G}_t^{(r)}\right)=0,\quad E\left(\mathrm{G}_t^{(r)}\right)^2=\frac{\beta r}{|\Psi(1)|}E\left(\mathrm{L}_1^2\right),\quad Cov\left(\mathrm{G}_t^{(r)},\mathrm{G}_{t+h}^{(r)}\right)=0$$

If further $E\left(\mathrm{L}_1^2\right)<\infty$ and $\Psi(1)<0$, then $E\left(\mathrm{G}_t^4\right)<\infty$ for all $t\geq 0$ and, if additionally the Lévy measure v_L of L is such that $\int_{\mathbb{R}}x^3 v_L(dx)=0$, then it holds for every $t,h\geq r>0$

$$E\left(\mathrm{G}_t^{(r)}\right)^2=6E\left(\mathrm{L}_1^2\right)\frac{\beta^2}{\Psi(1)^2}\left(\frac{2\eta}{\varphi}+2\tau_{\mathrm{L}}^2-E\left(\mathrm{L}_1^2\right)\right)\left(\frac{2}{|\Psi(2)|}-\frac{1}{|\Psi(1)|}\right)\left(r-\frac{1-e^{r|\Psi(1)|}}{|\Psi(1)|}\right)$$
$$+\frac{2\beta^2}{\varphi^2}\left(\frac{2}{|\Psi(2)|}-\frac{1}{|\Psi(1)|}\right)r+3\frac{\beta^2}{\Psi(1)^2}\left(E\left(\mathrm{L}_1^2\right)\right)^2 r^2$$

and

$$Cov\left(\left(G_t^{(r)}\right)^2, \left(G_{t+h}^{(r)}\right)^2\right)$$

$$= E\left(L_1^2\right)\frac{\beta^3}{\Psi(1)^3}\left(\frac{2\eta}{\varphi} + 2\tau_L^2 - E\left(L_1^2\right)\right)\left(\frac{2}{|\Psi(2)|} - \frac{1}{|\Psi(1)|}\right)\left(1 - e^{r|\Psi(1)|}\right)\left(e^{r|\Psi(1)|} - 1\right)e^{-h|\Psi(1)|} < 0$$

where $\tau_L^2 \geq 0$ the variance of the Brownian motion component of *L*.

Let $\gamma(h) = Cov\left(\left(G_{ri}^{(r)}\right)^2, \left(G_{r(i+h)}^{(r)}\right)^2\right)$, $h \in \mathbb{N}$ the autocovariance function then $\rho(h) = Cov\left(\left(G_{ri}^{(r)}\right)^2, \left(G_{r(i+h)}^{(r)}\right)^2\right)$, $h \in \mathbb{N}$ is the autocorrelation function of the discrete time process $\left(\left(G_{ri}^{(r)}\right)^2\right)_{i \in \mathbb{N}}$ where $G_t^{(r)} := G_t - G_{t-r}$. Then,

$$\frac{\rho(h)}{\rho(1)} = \frac{\gamma(h)}{\gamma(1)} = e^{-(h-1)r|\Psi(1)|}, \quad h \geq 1$$

Let $\mu, \gamma(0)$, $k, p > 0$ be constants such that

$$E\left(\left(G_i^{(1)}\right)^2\right) = \mu$$

$$Var\left(\left(G_i^{(1)}\right)^2\right) = \gamma(0)$$

$$\rho(h) = Cov\left(\left(G_{ri}^{(r)}\right)^2, \left(G_{r(i+h)}^{(r)}\right)^2\right) = ke^{hp}, \quad h \in \mathbb{N}$$

Then the first and the second moments of COGARCH(1,1) process for equally spaced returns are

$$M_1 := \gamma(0) - 2\mu^2 - 6\frac{1 - p - e^{-p}}{\left(1 - e^{p}\right)\left(1 - e^{-p}\right)}k\gamma(0) \tag{19}$$

$$M_2 := \frac{2k\gamma(0)p}{M_1\left(1-e^{p}\right)\left(1-e^{-p}\right)} \tag{20}$$

The estimations of the parameters (β,η,φ) using the Equations 19 and 20 are obtained by

$$\beta=\mu \tag{21}$$

$$\varphi = \mathrm{p}\sqrt{1+M_2} - p \tag{22}$$

$$\eta = \mathrm{p}\sqrt{1+M_2}\left(1-\tau_{\mathrm{L}}^2\right) + p\tau_{\mathrm{L}}^2 = p + \varphi\left(1-\tau_{\mathrm{L}}^2\right) \tag{23}$$

where the variance τ_{L}^2 of the Brownian motion component of L is known with $0 < \tau_{\mathrm{L}}^2 < 1$.

EXAMPLE of COGARCH(1, 1) PROCESS

In this section, CP COGARCH(1, 1) and VG COGARCH(1, 1) simulations are given as examples. The simulation study is done using the R package "yuima" by Iacus et al. (2015). The estimation results for CP COGARCH(1, 1) and VG COGARCH(1, 1) are given in Table 1 with standard errors of estimates. The estimations are done by the general method of moments. The standard error for the parameter β is not provided in the summary since its value is obtained once the parameters φ and η are estimated and then the variance-covariance matrix refers only to these parameters. According to the estimations, stationarity conditions are satisfied and the variance process is positive. The summary statistics for the increments are given in Table 2.

Table 1. The estimation results compound poisson and variance gamma COGARCH(1, 1)

Param	Real Param	GMM-CP	Std. Error	GMM-VG	Std. Error
β	0.755	1.032	NA	0.595	NA
φ	0.038	0.034	0.029	0.0248	0.016
η	0.053	0.068	0.069	0.0326	0.026

Table 2. The summary statistics for the increments

	Compound Poisson	Variance Gamma
Number of increments	24000	24000
Average of increments	-0.00211	0.000186
Standard Dev. of increments	0.25657	0.255978
Minimum of increments	-2.839877	-3.468155
Maximum of increments	3.685463	3.694253
Log.obj. Function	-3.52325	-3.146079

Figure 1. a) Simulated sample paths of CP COGARCH(1, 1) and VG COGARCH(1, 1) b) Estimated increments of CP COGARCH(1, 1) and VG COGARCH(1, 1) c) Estimated sample paths of CP COGARCH(1, 1) and VG COGARCH(1, 1)

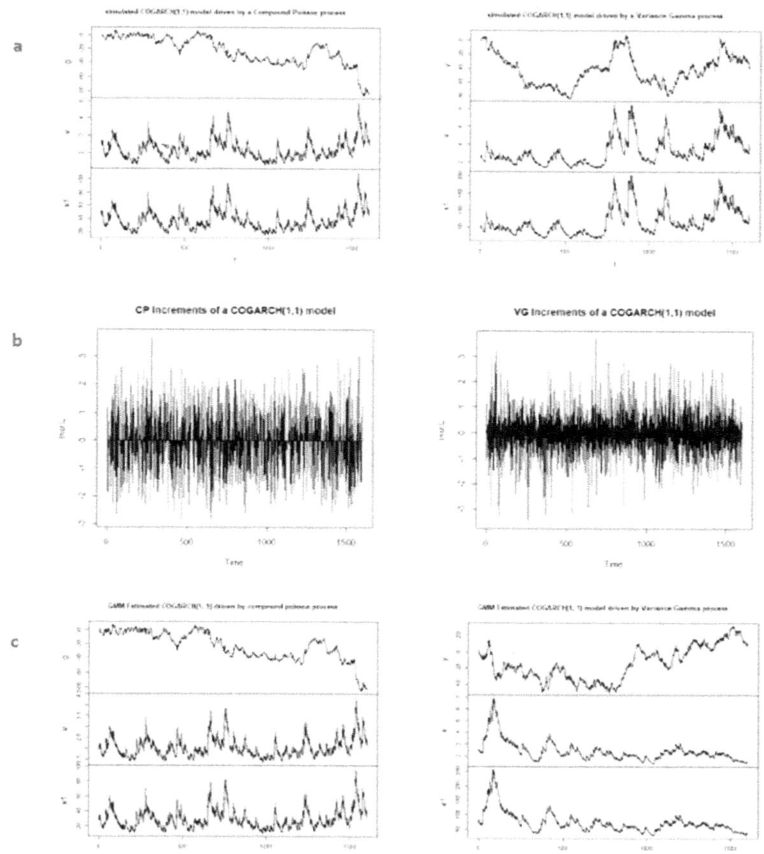

The plots of the simulations of COGARCH(1, 1) driven by Compound Poisson Process and Variance Gamma Process are given in Figure 1. Moreover, the estimated increments and the estimated sample paths of CP COGARCH(1, 1) and VG COGARCH(1, 1) models with estimated increments are given in the same figure.

REFERENCES

Applebaum, D. (2009). *Lévy Processes and Stochastic Calculus* (2nd ed.). Cambridge, UK: Cambridge University Press.

Bayracı, S., & Ünal, G. (2014). Stochastic interest rate volatility modeling with a continuous-time GARCH(1, 1) model. *Journal of Computational and Applied Mathematics, 259*, 464–473. doi:10.1016/j.cam.2013.10.017

Black, F. (1976). Studies of Stock Price Volatility Changes. *Proceedings from the American Statistical Association, Business and Economic Statistics Section*, 177-181.

Behme, A., Klüppelberg, C., & Mayr, K. (2014). Asymmetric COGARCH processes. *Journal of Applied Probability, 51A*(ASpec.), 161–173. doi:10.1239/jap/1417528473

Engle, R. F. (1982). Autoregressive conditional heteroscedasticity with estimates of the variance of United Kingdom inflation. *Econometrica, 50*(4), 987–1007. doi:10.2307/1912773

Haug, S. (2006). *Exponential COGARCH and other continuous time models with applications to high frequency data* (Dissertation). Technische Universitat München.

Haug, S., Klüppelberg, C., Lindner, A., & Zapp, M. (2007). Moment Estimating the COGARCH (1,1) model - a first go. *The Econometrics Journal, 10*(2), 320–341. doi:10.1111/j.1368-423X.2007.00210.x

Hosking, J. R. (1981). Fractional Differencing. *Biometrika, 68*(1), 165–176. doi:10.1093/biomet/68.1.165

Iacus, S. M., Mercuri, L., & Rroji, E. (2015a). *Discrete time approximation of a COGARCH(p,q) model and its estimation.* arXiv preprint arXiv:1511.00253

Iacus, S. M., Mercuri, L., & Rroji, E. (2015b). Estimation and Simulation of a COGARCH(p,q) model in the YUIMA project. arXiv preprint arXiv:1505.03914

Klüppelberg, C., Lindner, A., & Maller, R. A. (2004). Continous time GARCH process driven by a Lévy process: Stationarity and second order behavior. *Journal of Applied Probability, 41*(3), 601–622. doi:10.1239/jap/1091543413

Klüppelberg, C., Maller, R., & Szimayer, A. (2011). The COGARCH: A Review, with News on Option Pricing and Statistical Inference. In *Surveys in Stochastic Processes. Proc. Of the 33rd SPA Conference in Berlin* (pp. 29–58). EMS Publishing House.

Maller, R. A., Mueller, G., & Szimayer, A. (2008). GARCH modelling in continuous time for irregularly spaced time series data. *Bernoulli*, *14*(2), 519–542. doi:10.3150/07-BEJ6189

Mandelbrot, B. (1963). The Variation of Certain Speculative Prices. *The Journal of Business*, *36*(4), 394–419. doi:10.1086/294632

Marín, J. M., Rodríguez-Bernal, M. T., & Romero, E. (2015). Data cloning estimation of GARCH and COGARCH models. *Journal of Statistical Computation and Simulation*, *85*(9), 1818–1831. doi:10.1080/00949655.2014.903948

Marín, J. M., Rodríguez-Bernal, M. T., & Romero, E. (2016). *ABC and Hamiltonian Monte-Carlo methods in COGARCH models*. UC3M Working Papers, Statistics and Econometrics, 16-01.

Müller, G. (2010). MCMC estimation of the COGARCH (1, 1) model. *Journal of Financial Econometrics*, *8*(4), 481–510. doi:10.1093/jjfinec/nbq029

Platen, E., & Bruti-Liberati, N. (2010). *Numerical Solution of Stochastic Differential Equations with Jumps in Finance*. Springer-Verlag Berlin Heidelberg. doi:10.1007/978-3-642-13694-8

Tsay, R. S. (2012). *Analysis of Financial Time Series*. Wiley and Sons.

Chapter 6
Cooperative Grey Games:
An Application on Transportation Situations

Sirma Zeynep Alparslan-Gök
Süleyman Demirel University, Turkey

Emad Qasım
Süleyman Demirel University, Turkey

Osman Palancı
Süleyman Demirel University, Turkey

Mehmet-Onur Olgun
Süleyman Demirel University, Turkey

ABSTRACT

In this chapter, the authors extend transportation situations under uncertainty by using grey numbers. Further, they try in this research building models for grey game problems on transportation situations proposing the ideas of grey solutions and their corresponding structures. They introduce cooperative grey games and grey solutions. They focus on the grey Shapley value and the grey core of the modeled game arising from transportation situations. Moreover, they prove the nonemptiness of the grey core for the transportation grey games, and some results on the relationship between the grey core.

DOI: 10.4018/978-1-7998-0134-4.ch006

INTRODUCTION

Game theory is a modern section of decision theory, having various applications in socio-economic, political, organizational, ecological processes. The subject of this study is conflict and cooperation. The situations in which the interests of participants collide. Essentially all aspects of human activities affect to some extent the interests of different parties and therefore belong to the field of game theory. However, at present methods theory of games in real control procedures (primarily in the construction of organizational systems, the formation of economic mechanism and procedures political negotiations, socio-economic planning and forecasting) are not widely used. This is due to both the lack of theoretical and game training of management experts, and the fact that classic game models are too abstract and difficult to adapt to real processes of management and decision-making. Currently, various sections of the theory of games are included in the programs compulsory and special courses of many higher educational institutions. The study and teaching of this discipline entail serious difficulties, associated with a lack of necessary literature. This report is an exposition of transportation situations under grey games. By and large, makers and retailers are going to minimize their costs or maximizing their benefits. Makers and retailers can shape coalitions to get however much as possible. Constitutionally, a transportation situation comprises two sets of agents called makers and retailers which deliver/request merchandises. The transport of the merchandise from the makers to the retailers must be beneficial. Thusly, the primary goal is to transport the products from the makers to the retailers at greatest benefit (Aparicio et al. 2010). Such a participation can happen in transportation situations (Aziz et al. 2014; Frisk et al. 2010; Zener and Ergun 2008; Snchez-Soriano 2006; Snchez-Soriano et al. 2002, 2001; Soons 2011; Theys et al. 2008). Be that as it may, when the agents included concurring on a coalition, the subject of conveying the acquired benefit or expenses among the specialists emerges. The cooperative game theory is broadly utilized on intriguing sharing cost/benefit issues in numerous regions of Operational Research, for example, association, steering, planning, creation, stock, transportation, and so forth. (See Borm et al. 2001 for a survey on Operational Research Games). Transportation games are inspected in Sanchez-Soriano et al. (2001). Our paper studies the core of the transportation games and illustrates the non-emptiness of the core of transportation games. Also, Sanchez-Soriano et al. (2001) give a few outcomes about the connection between the core and dual optimal solutions of the transportation issue. The paper Sanchez-Soriano (2003) presents a specially appointed solution idea for transportation games called the pairwise egalitarian solution. In the continuation, the article Sanchez-Soriano (2006) looks at the relationship between the pairwise solutions and the core of transportation games. Besides, Sanchez-Soriano (2006) demonstrates that each core component of a transportation game is contained in a pairwise solution with a particular weight framework. In the traditional way to deal with the issue, the parameters are precisely known. In this case, the issue is completely understood utilizing the results of Sanchez-Soriano et al. (2001). However, in real-life transportation situations, issue parameters are not known precisely. Agents considering cooperation can rather figure lower and upper limits on the result of their cooperation. In this manner, we have a transportation interval situation and to solve the related sharing benefit issues, we require suitable sets of solutions. To deal with transportation situations with interval information, the theory of cooperative interval games is appropriate (Alparslan Gok et al. 2008, 2009a, b). The per-user is alluded to Branzei et al. (2010a, b) for a short study on cooperative solution concepts and for a guide for utilizing interval solutions when uncertainty about information is deleted Alparslan Gok et al. 2011). This paper broadens the examination of two-sided transportation situations (Sanchez-Soriano et al. 2001). Furthermore, their related cooperative games to a setting with interval information, i.e., the benefit b_{ij} of goods j by a maker i, the generation p_i of products of the maker i, and the request q_i of goods retailer j, in the transportation model now lie in intervals of genuine numbers acquired by forecasting their values from the viewpoint of a specialist see. This report consists of five sections which as follows: In the first section, we will review basic concepts and definitions from grey numbers and game theory. In the second section, we consider special classes of cooperative grey games, classification and mathematical models of decision-making problems. The third section, we study transportation interval situations, where we present transportation interval situations and related games, we will give some properties of the grey Shapley value on transportation interval situations and we use transportation situations dealing with the application of scientific methods for making a decision, and especially to the allocation of connections and routing games. Subsequent sections show our new results. In the fourth section, we introduce the transportation grey situations and related games, we will calculate both of the grey Shapley value and the core of the transportation grey game. The main purpose of transportation situations is to provide a rational basis for decisions making in the absence of complete information, because the systems composed of human, machine, and procedures may do not have complete information.

BASIC CONCEPTS

Grey Numbers

Grey theory (Deng, 1982), originally developed by Professor Deng in 1982, has become a very effective method of solving uncertainty problems under discrete data and incomplete information. Grey theory has now been applied to various areas such as forecasting, system control, decision making and computer graphics. Here, we give some basic definitions regarding relevant mathematical background of grey system, grey set and grey number in grey theory:

Definition 2.1.1. A grey number takes an unknown distribution between fixed lower and upper bounds, denoted as $\otimes \in [\underline{a}, \overline{a}]$, where $\underline{a}$ and $\overline{a}$ are respectively, the lower and upper bounds for $\otimes$.

Operations on interval grey numbers: Let $\otimes_1 \in [\underline{a}, \overline{a}]$, $\otimes_2 \in [\underline{b}, \overline{b}]$ and α is a positive real number, then

$$\otimes_1 \in [\underline{a}, \overline{a}] + \otimes_2 \in [\underline{b}, \overline{b}] \Leftrightarrow \otimes_1 + \otimes_2 \in [\underline{a} + \underline{b}, \overline{a} + \overline{b}]$$

1. The scalar multiplication of α and $\otimes$ is defined as follows:

$$\alpha \otimes \in [\alpha \underline{a}, \alpha \overline{a}].$$

We denote by $\mathcal{G}(\mathbb{R})$ the set of interval grey numbers in R. Let $\otimes_1, \otimes_2 \in \mathcal{G}(\mathbb{R})$ with $\otimes_1 \in [\underline{a}, \overline{a}]$, $\otimes_2 \in [\underline{b}, \overline{b}]$,

$|\otimes_1| = \underline{a} - \overline{a}$ and $\alpha \in \mathbb{R}_+$. Then by parts 1 and 2 we see that $\mathcal{G}(\mathbb{R})$ has a cone structure.

2. In general, the difference of $\otimes_1$ and $\otimes_2$ is defined as follows:

$$\otimes_1 \ominus \otimes_2 = \otimes_1 + (-\otimes_2) \in [\underline{a} - \underline{b}, \overline{a} - \overline{b}].$$

Different from the above subtraction we use a partial subtraction operator. We define $\otimes_1 \ominus \otimes_2$, only if $|\underline{a} - \overline{a}| \geq |\underline{b} - \overline{b}|$, by $\otimes_1 - \otimes_2 = [\underline{a} - \underline{b}, \overline{a} - \overline{b}]$. Note that

$\underline{a}-\underline{b}\leq\overline{a}-\overline{b}$. We recall that $\left[\underline{a},\overline{a}\right]$ is weakly better than $\left[\underline{b},\overline{b}\right]$, which we denote by $\left[\underline{a},\overline{a}\right]\succeq\left[\underline{b},\overline{b}\right]$, if and only if $\underline{a}\geq\underline{b}$ and $\overline{a}\geq\overline{b}$. We also use the reverse notation $\left[\underline{a},\overline{a}\right]\preceq\left[\underline{b},\overline{b}\right]$, if and only if $\underline{a}\leq\underline{b}$ and $\overline{a}\leq\overline{b}$ (Alparslan Gok et al., 2009).

Cooperative Interval Games

In this section we give some basic facts and definitions from cooperative interval games.

Definition 2.2.1. (Cooperative interval game) A cooperative game is an ordered pair (N,w), where $N=\{1,2,\ldots,n\}$ is a set of players and $w:2^N\rightarrow\mathbb{R}$ is a characteristic function of cooperative game. We further assume that $w(\phi)=[0,0]$. Set of all interval cooperative games on player set N is denoted by IG^N. We often write $w(i)$ instead of $w(\{i\})$.

It will be useful to name the following important games associated with each element of IG^N.

Definition 2.2.2. (Border games) For every $(N,w)\in IG^N$, border games $\left(N,\underline{w}\right)$ (lower border game) and $\left(N,\overline{w}\right)$ (upper border game) are given by $\underline{w}\left(S\right)=\underline{w\left(S\right)}$ and $\overline{w}\left(S\right)=\overline{w\left(S\right)}$ for every $S\in 2^N$.

Definition 2.2.3. Length game of $(N,w)\in IG^N$, is a games $(N,|w|)\in IG^N$, where

$$|w|\left(S\right)=\overline{w}\left(S\right)-\underline{w}\left(S\right),\ \forall S\in 2^N$$

Definition 2.2.4. (Interval imputation)) Set of interval imputations of $(N,w)\in G^N$, is defined as

$$I\left(w\right)=\left\{\left(I_1,I_2,\ldots,I_n\right)\in IR^N\mid\sum_{i\in N}I_i=w\left(N\right),\ I_i\succeq w\left(i\right),\ \forall i\in N\right\}$$

Definition 2.2.5. (Interval core) Set of interval core of $(N,w)\in G^N$, is defined as

$$C\left(w\right)=\left\{\left(I_1,I_2,\ldots,I_n\right)\in I\left(w\right)\mid\sum_{i\in S}I_i\succeq w\left(S\right),\ \forall S\in 2^N\setminus\{\varnothing\}\right\}$$

Important difference between these four definitions is that selection concepts yield a payoff vectors from $\mathbb{R}^N$, while $I(w)$ and $C(w)$ yield vectors from $I\mathbb{R}^N$. That means that the selection-based approach gives us payoffs with no additional uncertainty. However, this approach was never systematically studied and very little is known. This thesis is trying to fix this and we concentrate almost purely on selections.

Existing Classes of Interval Games

This subsection aims on presenting existing classes of interval games which have been studied earlier see Alparslan Gok et al. (2009). This is necessary later when we recall selection-based classes and show their relations with the existing ones.

Definition 2.3.1. (Size monotonicity) A game $(N,w)\in IG^N$ is size monotonic if holds

$$|w||(T)\leq|w|(S) \text{ for every } T\subseteq S\subseteq N.$$

That is when the length game of w is monotonic. Class of size monotonic games on player set N is denoted by $SMIG^N$. As we can see, size monotonic games capture situations in which an interval uncertainty grows with the size of coalition.

Definition 2.3.2. (Convex interval game) An interval game (N,w) is convex interval if its border games and length game are convex.

We write CIG^N for a set of convex interval games on player set N.

Convex interval game is supermodular as well but converse does not hold in general.

See Alparslan Gok at al. (2009) for characterizations of convex interval games and discussion on their properties.

Proposition 2.3.3. An interval game (N,w) is selection convex if and only if for every $S,T\in 2^N$ such that $S\subsetneq T$, $T\subsetneq S$, $S\neq\emptyset$, $T\neq\emptyset$ holds

$$w(S)+w(T)\leq\underline{w}(S\cup T)+\underline{w}(S\cap T)$$

Cooperative Grey Games

In this section, we introduce the notion of cooperative grey games. A cooperative grey game is an ordered pair N, w' with the player set $N=\{1,\ldots,n\}$ in which $w' = \otimes : 2^N \to \mathcal{G}(\mathbb{R})$ is the grey payoff characteristic function such that $w'(\varnothing) = \otimes_{\varnothing} \in [0,0]$, grey payoff function $w'(S) = \otimes_S \in \left[\underline{A_S}, \overline{A_S}\right]$ refers to the valuing area of the grey expectation benefit which is belonged to a coalition $S \in 2^N$, where $\underline{A_S}$ and $\overline{A_S}$ represent the maximum and minimum possible profits of the coalition S. So, a cooperative grey game can be considered as a classical cooperative game with grey profits $\otimes$. Grey solutions are useful to solve reward/cost sharing problems with grey data using cooperative grey games as a tool. Building blocks for grey solutions are grey payoff vectors, i.e. vectors whose components belong to $\mathcal{G}(\mathbb{R})$. We denote by $\mathcal{G}(\mathbb{R})^N$ the set of all such grey payoff vectors. We denote by $\mathcal{G}G^N$ the family of all coopvvdyrioiwmbndbso9erative grey games.

The following example illustrates a grey game.

Example 2.4.1. (Grey glove game). Let $N=\{1,\ldots,n\}$ be the set of players consisting of two disjoint subsets L and R. The members of L possess each one left-hand glove, the members of R one right-hand glove. A single glove is worth nothing, a right-left pair of gloves is worth between 10 and 20 Euros. In case $L=\{1,2\}$ and $R=\{3\}$, this situation can be modelled as a three-person grey game, where the coalitions formed by players 1 and 3, players 2 and 3, and the grand coalition obtain an element of the worth [10,20]. The worth gained in other cases is [0,0], i.e.

$$\otimes_{13} = w'(1,3) = \otimes_{23} = w'(2,3) = \otimes_N = w'(N) \in [10,20]$$

and $\otimes_S = w'(S) \in [0,0]$, otherwise (Palanci et al., 2015b).

The grey game in Example 4.1 has a nice interpretation with the Technology-Emissions-Means model (in short: TEM model) which was prepared within the occasion of the Kyoto Contract (1997) and Kyoto Protocol (1997c). The TEM model allows a simulation of technical and financial parameters and describes the economical interactions between several players (countries, companies) which intend to minimize their emissions or, in that process of emission reduction, to optimize their payoff function, by means of cooperative game theory. In fact, the game theoretical character of the TEM model was early recognized and approached from different perspectives (Pickl at al., 2000 and Pickl, 2002). Let us consider two complementary countries in technological sense, which do not have enough money to reduce their greenhouse gas emissions. In a sense we can consider these countries like one producing left-hand and the other right-hand gloves (e.g. Australia and New Zealand have different roles compared with Indonesia). These countries have to cooperate to reduce their emissions which shows the necessity of constructing a cooperative game.

The next example models a sequencing production situation with grey data (Branzei at al., 2011).

Example 2.4.2. Consider a production situation with 3 departments involved in the working process of a raw material. Each department assures one stage of processing and there is a hierarchy between them: the material is processed at stage i only after its processes in stages $1,\ldots,i$-1. At any stage i, i=1,2,3, there is a fixed cost necessary to process the material. However, the cost at stage 2 may increase with an additional amount, for example due to a machinery accident and related maintenance. Suppose that the cost at stages 1 and 3 are 7 and 12, respectively, whereas the cost of stage 2 is in between 5 and 10. The uncertainty due to department 2 affects the departments that are not its superiors. This situation is modelled as the cooperative grey N, w' with $N=\{1,\ldots,n\}$ and

$$\otimes_1 = w'(1) = \otimes_{13} = w'(1,3) \in [7+5, 7+10] = [12,17],$$

$$\otimes_{123} = w'(1,2,3) \in [7+5+12, 7+10+12] = [24,29]$$

and $\otimes_S = w'(S) \in [0,0]$ in any other case (Palanci et al., 2015b).

Now, we introduce some theoretical notions from the theory of cooperative grey games. For $w, w_1, w_2 \in IG^N$ and $w', w_1', w_2' \in \mathcal{G}G^N$ we say that $w_1' \in w_1 \leq w_2' \in w_2$ if $w_1'(S) \leq w_2(S)$, where $w_1'(S) \in w_1(S)$ and $w_2'(S) \in w_2(S)$, for each $S \in 2^N$. For $w_1', w_2' \in \mathcal{G}G^N$ and $\alpha \in \mathbb{R}_+$ we define $N, w_1' + w_2'$ and $\langle N, \alpha w' \rangle$ by $(w_1' + w_2')(S) = w_1'(S) + w_2'(S)$ and $(\alpha w')(S) = \alpha w'(S)$ for each $S \in 2^N$. So, we conclude that $\mathcal{G}G^N$ endowed with "$\leq$" has a cone structure with respect to addition and multiplication with non-negative scalars above. For $w_1', w_2' \in \mathcal{G}G^N$ where $w_1' \in w_1, w_2' \in w_2$ with $|w_1(S)| \geq |w_2(S)|$ for each $S \in 2^N$, $N, w_1' - w_2'$ is defined by $(w_1' - w_2')(S) = w_1'(S) - w_2'(S) \in w_1(S) - w_2(S)$. We call a game N, w' grey size monotonic if $N,|w|$ is monotonic, i.e. $|w|(S) \leq |w|(T)$ for all $S,T \in 2^N$ with

$S \subset T$. For further use we denote by $SM\mathcal{G}G^N$ the class of grey size monotonic games with player set *N*. The grey marginal operators and the grey Shapley value are defined on $SM\mathcal{G}G^N$.

Grey Shapley Value

Denote by $\Pi(N)$ the set of permutations $\sigma:N\rightarrow N$ of *N*. The grey marginal operator $m^{\sigma}: SM\mathcal{G}G^N \rightarrow \mathcal{G}(\mathbb{R})^N$ corresponding to σ, associates with each $w' \in SM\mathcal{G}G^N$ the grey marginal vector $m^{\sigma}(w')$ of w' with respect to σ defined by

$$m_i^{\sigma}(w') := w'(P^{\sigma}(i) \cup \{i\}) - w'(P^{\sigma}(i)) \in \left[\underline{A_{P^{\sigma}(i)\cup\{i\}}} - \underline{A_{P^{\sigma}(i)}}, \overline{A_{P^{\sigma}(i)\cup\{i\}}} - \overline{A_{P^{\sigma}(i)}}\right],$$

for each $i \in N$, where $P^{\sigma}(i) = \{r \in N | \sigma^{-1}(r) < \sigma^{-1}(i)\}$, and $\sigma^{-1}(i)$ denotes the entrance number of player *i*. For grey size monotonic games N, w', $w'(T) - w'(S) \in w(T) - w(S)$ is defined for all $S, T \in 2^N$ with $S \subset T$ since $|w(T)| = |w|(T) \geq |w|(S) = |w(S)|$. We notice that for each $w' \in SM\mathcal{G}G^N$ the grey marginal vectors $m^{\sigma}(w')$ are defined for each $\sigma \in \Pi(N)$, because the monotonicity of $|w|$ implies $\overline{A_{S\cup\{i\}}} - \underline{A_{S\cup\{i\}}} \geq \overline{A_S} - \underline{A_S}$, which can be rewritten as $\overline{A_{S\cup\{i\}}} - \overline{A_S} \geq \underline{A_{S\cup\{i\}}} - \underline{A_S}$. So, $w'(S \cup \{i\}) - w'(S) \in w(S \cup \{i\}) - w(S)$ is defined for each $S \subset N$ and $i \notin S$. Next, we notice that all the grey marginal vectors of a grey size monotonic game are efficient grey payoff vectors. The grey Shapley value $\Phi': SM\mathcal{G}G^N \rightarrow \mathcal{G}(\mathbb{R})^N$ is defined by

$$\Phi'(w') := \frac{1}{n!} \sum_{\sigma \in \Pi(N)} m^{\sigma}(w') \in \left[\frac{1}{n!} \sum_{\sigma \in \Pi(N)} m^{\sigma}(\underline{A}), \frac{1}{n!} \sum_{\sigma \in \Pi(N)} m^{\sigma}(\overline{A})\right],$$

for each $w' \in SM\mathcal{G}G^N$. (1)

We can write the last equation as follows:

$$\Phi_i'(w') := \frac{1}{n!}\sum_{\sigma\in\Pi(N)}\left[w'\left(P^{\sigma}(i)\cup\{i\}\right) - w'\left(P^{\sigma}(i)\right)\right]$$

$$\in\left[\frac{1}{n!}\sum_{\sigma\in\Pi(N)}\underline{A_{P^{\sigma}(i)\cup\{i\}}} - \underline{A_{P^{\sigma}(i)}}, \frac{1}{n!}\sum_{\sigma\in\Pi(N)}\overline{A_{P^{\sigma}(i)\cup\{i\}}} - \overline{A_{P^{\sigma}(i)}}\right] \tag{2}$$

The terms after the summation sign in Equation (2) are of the form $w'\left(S\cup\{i\}\right) - w'\left(S\right)$, where S is a subset of N not containing i. Note that there are exactly $|S|!(n\text{-}1\text{-}|S|)!$ orderings for which one has $P^{\sigma}(\{i\})=S$. The first factor, $|S|!$, corresponds to the number of orderings of S and the second factor, $(n\text{-}1\text{-}|S|)!$, is just the number of orderings of $N/\left(S\cup\{i\}\right)$. Using this, we can rewrite Equation (2) as

$$\Phi_i'(w') = \sum_{S:i\notin S}\frac{|S|!\left(n-1-|S|\right)!}{n!}\left(w'\left(S\cup\{i\}\right) - w'\left(S\right)\right),$$

Let us note that

$$\sum_{S:i\notin S}\frac{|S|!\left(n-1-|S|\right)!}{n!} = 1$$

The following example illustrates the calculation of the grey Shapley value.

Example 2.4.3. (Palanci et al., 2015b) Let N, w', be a cooperative grey game with $N=\{1,2,3\}$ and

$$\otimes_1 = w'(1) = \otimes_{13} = w'(1,3) \in [7,7], \quad \otimes_{12} = w'(12) \in [12,17],$$

$\otimes_N = w'(123) \in [24,29]$ and $\otimes_S = w'(S) \in [0,0]$ otherwise. Then the grey marginal vectors are given in the following table, where $\sigma:N\rightarrow N$ is identified with $(\sigma(1),\sigma(2),\sigma(3))$. Firstly, for $\sigma_1=(1,2,3)$, we calculate the grey marginal vectors. Then,

$$m_1^{\sigma_1}(w') = w'(1) \in [7,7],$$

$$m_2^{\sigma_1}(w') = w'(12) - w'(1) \in [12,17] - [7,7] = [5,10],$$

$$m_3^{\sigma_1}(w') = w'(123) - w'(12) \in [24,29] - [12,17] = [12,12].$$

The others can be calculated similarly, which is shown in Table 1.

The average of the six grey marginal vectors is the grey Shapley value of this game which can be calculated as:

$$\Phi'(w') \in \left(\left[\frac{27}{2},16\right],\left[\frac{13}{2},9\right],[4,4]\right).$$

TRANSPORTATION INTERVAL SITUATIONS AND GAMES

In this section, we present the transportation interval situations propelled by Sánchez-Soriano et al. (2001). In a transportation interval situation, the set of players is parceled into two disjoint subsets *P* and *Q*, containing *n* and *m* players, separately. The individuals from *P* will be called makers, while the individuals from *Q* will be the retailers. Every inception player $i \in P$ has a positive integer interval number of units of a certain indivisible good, p_i', and every goal player $j \in Q$ requests a positive integer interval number of units of this good, q_j'. The delivery of one unit from

Table 1. Grey marginal vectors of the cooperative grey game

σ	$m_1^{\sigma}(w')$	$m_2^{\sigma}(w')$	$m_3^{\sigma}(w')$
σ_1=(1,2,3)	$m_1^{\sigma_1}(w') \in [7,7]$	$m_2^{\sigma_1}(w') \in [5,10]$	$m_3^{\sigma_1}(w') \in [12,12]$
σ_2=(1,3,2)	$m_1^{\sigma_2}(w') \in [7,7]$	$m_2^{\sigma_2}(w') \in [17,22]$	$m_3^{\sigma_2}(w') \in [0,0]$
σ_3=(2,13)	$m_1^{\sigma_3}(w') \in [12,17]$	$m_2^{\sigma_3}(w') \in [0,0]$	$m_3^{\sigma_3}(w') \in [12,12]$
σ_4=(2,3,1)	$m_1^{\sigma_4}(w') \in [24,29]$	$m_2^{\sigma_4}(w') \in [0,0]$	$m_3^{\sigma_4}(w') \in [0,0]$
σ_5=(3,1,2)	$m_1^{\sigma_5}(w') \in [7,7]$	$m_2^{\sigma_5}(w') \in [17,22]$	$m_3^{\sigma_5}(w') \in [0,0]$
σ_6=(3,2,1)	$m_1^{\sigma_6}(w') \in [24,29]$	$m_2^{\sigma_6}(w') \in [0,0]$	$m_3^{\sigma_6}(w') \in [0,0]$

source player i to goal player j creates a nonnegative interval real benefit b'_{ij}. Here we have

$$p'_i := \left[\underline{p'_i}, \overline{p'_i}\right], q'_i := \left[\underline{q'_i}, \overline{q'_i}\right] \text{ and } b'_i := \left[\underline{b'_i}, \overline{b'_i}\right] \in I(\mathbb{R}).$$

A transportation interval situation like this is described by a 5-tuple (P, Q, B', p', q'), where B' is the $n \times m$ grid of interval benefits, p' is the n-dimensional vector of accessible interim units at the starting points, and q' is the m-dimensional vector of interval requests.

For each transportation interval situation (P, Q, B', p', q') and each coalition $S \subset N := P \cup Q$, with makers $S_P := S \cap P$ and retailers $S_Q := S \cap Q$, and accepting that these sets are both nonempty, we can characterize the maximization issue of the pessimistic scenario by

$$\begin{aligned} \underline{T}: \quad & \text{maximize} \quad \sum_{i \in S_P} \sum_{j \in S_Q} \underline{b'_{ij}} x_{ij} \\ & \text{such that} \quad \sum_{j \in S_Q} x_{ij} \le \underline{p'_i}, \; i \in S_P, \\ & \qquad \sum_{i \in S_P} x_{ij} \le \underline{q'_j}, \; j \in S_Q, \\ & \qquad x_{ij} \ge 0, \; (i, j) \in S_P \times S_Q, \end{aligned} \tag{5}$$

and the maximization issue of the optimistic scenario give as

$$\begin{aligned} \overline{T}: \quad & \text{maximize} \quad \sum_{i \in S_P} \sum_{j \in S_Q} \overline{b'_{ij}} x_{ij} \\ & \text{such that} \quad \sum_{j \in S_Q} x_{ij} \le \overline{p'_i}, \; i \in S_P, \\ & \qquad \sum_{i \in S_P} x_{ij} \le \overline{q'_j}, \; j \in S_Q, \\ & \qquad x_{ij} \ge 0, \; (i, j) \in S_P \times S_Q, \end{aligned} \tag{6}$$

Let's take $\vartheta(T(S))$ which means the optimal interval estimation of the issue $T(S)$ such that $\vartheta(T(S)) = \left[\vartheta(\underline{T}(S)), \vartheta(\overline{T}(S))\right] \in I(R)$ and here $\vartheta(\underline{T}(S))$ is the optimal value of the maximization issue of the pessimistic scenario, $\vartheta(\overline{T}(S))]$ is the optimal value of the maximization issue of the optimistic scenario. So, we can define a cooperative interval game associated with every transportation interval situation (P,Q,B',p',q') in the going with way:

- The set of players is $N = P \cup Q$;
- The characteristic function $\langle N, v' \rangle$ is given by

$$v'(S) := \begin{cases} [0,0], & \text{if } S = \phi \text{ or } S \text{ is contained in } P \text{ or } Q, \\ , & \text{otherwise,} \end{cases}$$

where $v'(S)$ satisfies the condition

$$\vartheta(\underline{T}(S)) + \vartheta(\overline{T}(T)) \geq \vartheta(\underline{T}(T)) + \vartheta(\overline{T}(S)) \text{ for all } S \subset T.$$

Presently, we display definition of a transportation interval game.

Definition 3.1. A transportation interval game is a cooperative interval game $v' \in IG^N$ arising from a transportation situation (P,Q,B',p',q'). Often, we identify a transportation interval situation (P,Q,B',p',q') with its associated transportation game v'.

Now, we examine the properties of transportation interval games. We begin by taking note of that, in view of its definition, a transportation interval games is interim zero-standardized. Plainly a transportation interval game is superadditive, measure monotonic but not necessarily convex. In the following section, we show that transportation interval games are size monotonic.

Example 3.2. Consider the 3-person transportation interval situation (P,Q,B',p',q') which has one maker and two retailers:

$$P = \{1\}, Q = \{2,3\}, B' = ([3,5],[5,6]), p' = [3,5], q' = ([2,4],[1,3]).$$

Now, we characterize a transportation interval game related with a transportation interval situation (P,Q,B',p',q'). Here, N={1,2,3} is the set of players and the trademark elements of the transportation interval game are as the following:

$$v'(1)=v'(2)=v'(3)=v'(23)=[0,0], v'(12)=[6,20], v'(13)=[5,18], v'(123)=[11,28].$$

Along these lines, we acquire the transportation interval game $\langle N,v'\rangle$ comparing to a transportation interval game. We take note of that this game is interval zero-normalized since for all $i \in N$ we have $v'(i)=[0,0]$. Obviously this game is superadditive be that as it may, not raised. For the coalitions S=(12) and T=(13), this game does not satisfy the state of convexity:

$$|v'|(12)+|v'|(13)\leq |v'|(1)+|v'|(123)\rightarrow 27\otimes 19.$$

The Interval Shapley Value of the Transportation Interval Game

In this area, we calculate the interval Shapley value of the transportation interval game. To begin with, we recall the definition of the interval Shapley value. For this, we require reviewing some notions from the theory of cooperative interval games (Alparslan Gök et al. 2009a).

Interval solutions are helpful to understand reward/cost sharing problems with interval information utilizing cooperative interval games as an instrument.The interval payoff vectors, which are the building blocks for interval solutions, are the vectors whose components belong to $(\mathbb{R})$. We indicate by $I(\mathbb{R})^N$ the set of all such interval payoff vectors. We call a game $\langle N,v'\rangle$ size monotonic if $\langle N,|v'|\rangle$ is monotonic, i.e., $|v'|(S)\leq|v'|(T)$ for all $S,T \in 2^N$ with $S\subset T$. For further use we mean by $SMIG^N$ the class of size monotonic interval games with player set N.

The following theorem demonstrates that the transportation interval games are size monotonic.

Theorem 3.1.1. The transportation interval game $|N,v'|$ belongs to class of $SMIG^N$.
***Proof*.** We have

$$|v'|(S)\leq|v'|(T) \text{ for all } S,T\in 2^N \text{ with } S\subset T$$

Take $S,T \in 2^N$ with $S\subset T$. If $S=\phi$ or S is contained in P or Q, then $v'(S)=[0,0]=v'(T)$. Now, it is obvious that

$$|v'|(S) \leq |v'|(T)$$

If $S \neq \phi$ or S is not contained in P or Q, then $v'(S) = \left[\vartheta(\underline{T}(S)), \vartheta(\overline{T}(S))\right]$ with $\vartheta(\underline{T}(S)) \leq \vartheta(\overline{T}(S))$. For $S \subset T$, using the definition of a transportation interval game we obtain

$$\vartheta(\underline{T}(S)) \leq \vartheta(\underline{T}(T))$$

$$\vartheta(\overline{T}(S)) \leq \vartheta(\overline{T}(T))$$

Then,

$$|v'|(S) \leq |v'|(T),$$

$$\vartheta(\overline{T}(S)) - \vartheta(\underline{T}(S)) \leq \vartheta(\overline{T}(T)) - \vartheta(\underline{T}(T)),$$

$$\vartheta(\underline{T}(T)) + \vartheta(\overline{T}(S)) \leq \vartheta(\underline{T}(S)) + \vartheta(\overline{T}(T)),$$

for all $S,T \in 2^N$ with $S \subset T$. Since v' satisfies the condition

$$\vartheta(\underline{T}(T)) + \vartheta(\overline{T}(S)) \leq \vartheta(\underline{T}(S)) + \vartheta(\overline{T}(T)),$$

$\langle N, v' \rangle$ belongs to the class of $SMIG^N$.

We realize that if an interval game is belonging to $SMIG^N$, then the interval Shapley value is constantly given (Alparslan Gök et al. 2009a).

The interval marginal operators and the interval Shapley value were defined on $SMIG^N$ in Alparslan Gök et al. (2009a) as follows:

Mean by $\Pi(N)$ the set of permutations $\sigma: N \rightarrow N$ of $N=\{1,2,3\}$. The interval marginal operator $m^\sigma : SMIG^N \rightarrow I(\mathbb{R})^N$ relating to σ, partners with each $v' \in SMIG^N$ the interim minor vector $m^\sigma(v')$ of v' as for σ, characterized by $m_i^\sigma(v') = v'(P^\sigma(i) \cup \{i\}) - v'(P^\sigma(i))$ for every $i \in N$, where $P^\sigma(i) := \{r \in N \mid \sigma^{-1}(r) < \sigma^{-1}(i)\}$. Here, $\sigma^{-1}(i)$ indicates the passageway number of player i. For size monotonic games $\langle N, v' \rangle$, $v'(T) - v'(S)$ is defined for all

$S,T \in 2^N$ with $S \subset T$, since $\left|v'(T)\right| = \left|v'\right|(T) \geq \left|v'\right|(S) = \left|v'(S)\right|$. Now, we see that for all $v' \in SMIG^N$ the interval marginal vectors $m^\sigma(v')$ are defined for all $\sigma \in \Pi(N)$, because the monotonicity of $|v'|$ implies $\overline{v'}(S \cup \{i\}) - \underline{v'}(S \cup \{i\}) \geq \overline{v'}(S) - \underline{v'}(S)$, which can be reworked as $\overline{v'}(S \cup \{i\}) - \overline{v'}(S) \geq \underline{v'}(S \cup \{i\}) - \underline{v'}(S)$. So, $v'(S \cup \{i\}) - v'(S)$ is defined for all $S \subset N$ and $i \notin S$. We take note of that all the interval marginal vectors of a size monotonic game are efficient interval payoff vectors.

The interval Shapley value relegates to every cooperative interval game a payoff vector whose segments are minimized intervals of real numbers. Cooperative games in the added substance cone on which we utilize the interval Shapley value emerge from some Operations Researches and economic situations with interval information.

We can define the interval Shapley value as following:

$$\Phi : SMIG^N \to I(\mathbb{R})^N,$$

such that

$$\Phi(v') := \frac{1}{n!} \sum_{\sigma \in \Pi(N)} m^\sigma(v'),$$

for each $v' \in SMIG^N$.

Now, we would computation be able to of the interval Shapley value in the transportation interval game from the last example as follows:

Example 3.1.2. Consider $\langle N, v' \rangle$ shown in the last example as the transportation interval game. We have that, $N\{1,2,3\}$ and the characteristic function v' is given by

$$v'(1) = v'(2) = v'(3) = v'(23) = [0,0], v'(12) = [6,20], v'(13) = [5,18], v'(123) = [11,28].$$

Here, the set of permutations of N is

$$\Pi(N)$$
$$= \{\sigma_1 = (1,2,3), \sigma_2 = (1,3,2), \sigma_3 = (2,1,3), \sigma_4 = (2,3,1), \sigma_5 = (3,1,2), \sigma_6 = (3,2,1)\}$$

In the first, we compute the interval marginal vectors for σ$_2$=(1,2,3), that is

$$m_1^{\sigma_1}(v') = v'(1) = [0,0],$$

$$m_2^{\sigma_1}(v') = v'(12) - v'(1) = [6,20] - [0,0] = [6,20],$$

$$m_3^{\sigma_1}(v') = v'(123) - v'(12) = [11,28] - [6,20] = [5,8].$$

Similarly, we can compute the other values, as shown in Table 2.

Table 2 is shows the interval marginal vectors of the cooperative transportation interval game in Example 3.1. So, the interval Shapley value of this game is the average of the six interval marginal vectors, which can be written as

$$\Phi(v') = \left(\left[\frac{11}{2}, \frac{47}{3}\right], \left[3, \frac{20}{3}\right], \left[\frac{5}{2}, \frac{17}{3}\right]\right)$$

The paper Pulido et al. (2002) analyzes a contention circumstance in college administration. The creators contemplate how to apportion cash among the offices to purchase gear for teaching laboratories. They present a broadened bankruptcy issue in which two methods for estimating the requests of the administrative entities exist: the " objective privileges" and the " claims". They examine and propose reasonable bankruptcy rules for the new kind of issue.

Table 2. Interval marginal vectors

σ	$m_1^{\sigma}(v')$	$m_2^{\sigma}(v')$	$m_3^{\sigma}(v')$
σ$_1$=(1,2,3)	$m_1^{\sigma_1}(v') = [0,0]$	$m_2^{\sigma_1}(v') = [6,20]$	$m_3^{\sigma_1}(v') = [5,8]$
σ$_2$=(1,3,2)	$m_1^{\sigma_2}(v') = [0,0]$	$m_2^{\sigma_2}(v') = [6,20]$	$m_3^{\sigma_2}(v') = [5,18]$
σ$_3$=(2,1,3)	$m_1^{\sigma_3}(v') = [6,20]$	$m_2^{\sigma_3}(v') = [0,0]$	$m_3^{\sigma_3}(v') = [5,8]$
σ$_4$=(2,3,1)	$m_1^{\sigma_4}(v') = [11,28]$	$m_2^{\sigma_4}(v') = [0,0]$	$m_3^{\sigma_4}(v') = [0,0]$
σ$_5$=(3,1,2)	$m_1^{\sigma_5}(v') = [5,18]$	$m_2^{\sigma_5}(v') = [6,10]$	$m_3^{\sigma_5}(v') = [0,0]$
σ$_6$=(3,2,1)	$m_1^{\sigma_6}(v') = [11,28]$	$m_2^{\sigma_6}(v') = [0,0]$	$m_3^{\sigma_6}(v') = [0,0]$

The work Pulido et al. (2008) studies new bankruptcy situations where what's more to claims an exogenously given reference point for the allocation of the estate is available. The authors present and investigate two sorts of bargain solutions and demonstrate that they coincide with the τ-value of two corresponding transferable utility games. Also, they explain a bankruptcy situation with references by implies of a compromise solution. They decide for every player which mix of references and claims prompts the most astounding result for him/her and which to the least result. This leads to an upper and lower bound for the allocation estate. The compromise solution is then basically defined as the unique efficent convex combination of these two vectors. We use a method which was presented in Branzei et al. (2010b) that changes an interval allocation into a payoff vector, under the presumption that exclusive the uncertainty as to the value of the stupendous coalition has been resolved. The research question addressed here is as follows: How to determine for all players their payoff generated by cooperation within the grand coalition in the promised range of payoffs to establish such a cooperation after the uncertainty in the payoff for the grand coalition has been resolved? This question is an important one that deserves attention both in the literature and in game practice.

Now, we utilize the basic (one-stage) procedure, presented by Branzei et al. (2010b).

Let N be a set of players that consider cooperation underneath uncertainty of coalition values, i.e., understanding what each group S of players (coalition) can obtain between two bounds, $\underline{v'(s)}$ and $\overline{v'(s)}$, through cooperation. If the players use cooperative game theory as a tool, they are able to pick out an interval solution concept, say the value-kind Solution Ψ that associates with the related cooperative interval game $\langle N, v' \rangle$, the interval allocation $\Psi(v') = (J_1, J_2, \ldots, J_n)$ which ensures for every player $i \in N$ a final payoff within the interval $J_i = \left[\underline{J}_i, \overline{J}_i\right]$, when the value of the grand coalition is known. Surely, $\underline{v'}(N) = \sum_{i \in N} \underline{J}_i$ and $\overline{v'}(N) = \sum_{i \in N} \overline{J}_i$. For each $i \in N$ the interval $\left[\underline{J}_i, \overline{J}_i\right]$ can be viewed as the interval claim of i on the realization R of the payoff for the grand coalition $N\left(\underline{v'}(N) \leq R \leq \overline{v'}(N)\right)$. One should determine payoffs $x_i \in \left[\underline{J}_i, \overline{J}_i\right]$, $i \in N$ (the feasibility condition) such that $\sum_{i \in N} x_i = R$ (the efficiency condition). We take note of that for the case $R = \underline{v'}(N)$ the payoff vector x is $\left(\underline{J}_1, \underline{J}_2, \ldots, \underline{J}_n\right)$, in the case $R = \overline{v'}(N)$ we get $x = \left(\overline{J}_1, \overline{J}_2, \ldots, \overline{J}_n\right)$, but in the case $\underline{v'}(N) < R < \overline{v'}(N)$ there are endlessly numerous approaches to determine allocations $(x_1, \ldots, x_n)$ fulfilling both the efficiency and the feasibility conditions. In the last case, we require appropriate distribution tenets to determine fair allocations

$(x_1,\ldots,x_n)$ of R fulfilling the above conditions. As players incline toward as expansive payoffs as possible and as the sum R to be divided between them is littler than $\sum_{i\in N}\overline{J}_i$, the players are facing a bankruptcy-like situation, suggesting that bankruptcy rules are great possibility for changing an interval allocation $(J_1,J_2,\ldots,J_n)$ into a payoff vector $(x_1,x_2,\ldots,x_n)$.

Since R appears as a realization of $v'(N)$, one can with the exception of that

$$\underline{v}'(N)\le R\le \overline{v}'(N).$$

One basic idea is to determine $\lambda\in[0,1]$ such that

$$R=\lambda\underline{v}'(N)+(1-\lambda)\overline{v}'(N), \tag{7}$$

and for each $i\in N$ the payoff is

$$x_i=\lambda\underline{J}_i+(1-\lambda)\overline{J}_i.$$

We note that $\underline{J}_i\le x_i\le(1-\lambda)\overline{J}_i$ and

$$\sum_{i\in N}x_i=\lambda\sum_{i\in N}\underline{J}_i+(1-\lambda)\sum_{i\in N}\overline{J}_i=\lambda\underline{v}'(N)+(1-\lambda)\overline{v}'(N)=R.$$

So, x is an efficient payoff vector corresponding to R.

We review that a bankruptcy situation with set of claimants N is a pair (E,d), where $E\ge 0$ is the estate to be divided and $d\in\mathbb{R}_+^N$ is the vector of claims such that $\sum_{i\in N}d_i\ge E$. Let BR^N is the set of bankruptcy situations with player set N. So, we can define a function $f:BR^N\to\mathbb{R}^N$ which allocates to every bankruptcy situation $(E,d)\in BR^N$ apayoff vector $f(E,d)\in\mathbb{R}^N$ such that $0\le f(E,d)\le d$ (reasonability) and $\sum_{i\in N}f_i(E,d)=E$ (efficiency). In this paper, we use three bankruptcy rules: the proportional rule (PROP), the constrained equal awards rule (CEA) and the constrained equal losses (CEL) rule. Now, we recall those three bankrupcty rules. The bankrupcty rule PROP is defined by $PROP_i(E,d)=(d_i/\sum_{j\in N}d_j.E$ for each bankruptcy problem (E,d) and all $i\in N$. This rule gives allocations in the core of the (pessimistic) bankruptcy game related with a bankruptcy situation and are generally used in applications.

The bankrupcty rule CEA is defined by $CEA_i(E,d) = \min\{d_i,\alpha\}$, where α is determined by $\sum_{i\in N} CEA_i(E,d) = E$ for each bankruptcy problem (E,d) and all $i\in N$, while the bankrupcty rule CEL is defined by $CEL_i(E,d) = \min\{d_i\text{-}\beta,0\}$, where β is determined by $\sum_{i\in N} CEL_i(E,d) = E$ for each bankruptcy problem (E,d) and all $i\in N$.

For additionally use, we present the notation $\mathcal{F} = \{PROP, CEA, CEL\}$ and let $f \in \mathcal{F}$. Then, we can partition the sum R accomplished by N through handing out the sum $\underline{J}_i + f_i(E,d)$ to each player $i\in N$, where $E = R - \sum_{i\in N} \underline{J}_i$ and $d_i = \overline{J}_i - \underline{J}_i$ for each $i\in N$.

Now, we introduce the one-stage procedures where some bankruptcy rule $f \in \mathcal{F}$ is utilized.

Example 3.1.3. Let $\langle N, v'\rangle$ be the three-person transpotation interval game with

$$v'(1) = v'(2) = v'(3) = v'(23) = [0,0],$$

$$v'(12) = [6,20], v'(13) = [5,18], v'(123) = [11,28].$$

Let the realization of $v'(N)$ is R=20 and think about that cooperation inside the grand coalition depended on the utilization of interval Shapley value. Then,

$$\Phi(v') = \left(\left[\frac{11}{2},\frac{47}{3}\right],\left[3,\frac{20}{3}\right],\left[\frac{5}{2},\frac{17}{3}\right]\right)$$

In the beginning, we note that condition $\underline{v'}(N) \le R \le \overline{v'}(N)$ is satisfied. Then, from (7) we have 20 = λ(11) + (1-λ)(28), we get $\lambda = \frac{8}{17}$, so the payoff vector is $x = \left(\frac{122}{17},\frac{84}{17},\frac{71}{17}\right)$.

Let us decide individual uncertainty-free shares by using PROP, CEA and CEL to distribute the sum R- $(J_1+J_2+J_3)$ =9 among the three agents. We note that here we focus on a classical bankruptcy problem (E,d) with E=9 and $d = \left(\frac{61}{6},\frac{11}{3},\frac{19}{6}\right)$.

Table 3. The bankrupcty rules

PROP(E,d)	*CEA(E,d)*	*CEL(E,d)*
$\left(5\frac{13}{34},1\frac{16}{17},4\frac{23}{34}\right)$	$\left(4\frac{3}{8},4\frac{3}{8},3\frac{1}{6}\right)$	$\left(7\frac{3}{4},1\frac{1}{4},0\right)$

Table 4. A payoff vecrors

PROP(E,d)	*CEA(E,d)*	*CEL(E,d)*
$\left(10\frac{15}{17},4\frac{16}{17},4\frac{3}{17}\right)$	$\left(9\frac{7}{8},7\frac{3}{8},5\frac{2}{3}\right)$	$\left(13\frac{1}{4},4\frac{1}{4},2\frac{1}{2}\right)$

Using the one-stage producere with PROP, CEA and CEL in the role of *f*, this is shown in Table 3. Then, we get $x=\left(\frac{11}{2},3,\frac{5}{2}\right)+f\left(6,\left(\frac{61}{6},\frac{11}{3},\frac{19}{6}\right)\right)$, see Table 4.

A correlation of the payoff vectors got using PROP, CEA, and CEL can be valuable by and by to help the decision of the favored bankrupcty rule to be implemented.

The Interval Core of the Transportation Interval Game

In this segment, we exhibit some fascinating outcomes concerning the interval core of a transportation interval game. It is outstanding that the transportation games have nonempty cores (Branzei et al. 2010b). Now, we check if this property can be expanded to transportation interval situations.

The dual problem of the maximization issue for the critical situation (1) is given by the minimization issue:

$$
\begin{aligned}
T^D(S):\quad &\text{minimize} && \sum_{i\in S_P}\underline{p_i}u_i+\sum_{j\in S_Q}\underline{q_j}v_j\\
&\text{such that} && u_i+v_j\geq \underline{b_{ij}},\ (i,j)\in S_P\times S_Q,\\
& && u_i,v_j\geq 0,\ i\in S_P,\ j\in S_Q,
\end{aligned}
$$

Also, the dual issue of the maximization issue for the hopeful situation (2) is given by the minimization issue:

$$T^D(S): \text{ minimize } \sum_{i\in S_P}\overline{p_i}u_i + \sum_{j\in S_Q}\overline{q_j}v_j$$

$$\text{such that } u_i + v_j \geq \overline{b_{ij}},\ (i,j)\in S_P\times S_Q,$$
$$u_i, v_j \geq 0,\ i\in S_P,\ j\in S_Q,$$

Remark 3.2.1. An optimal solution for a linear program is an achievable solution with the biggest objective function value. The value of the objective function for the optimal solution is said to be the value of the linear program. A linear program may have multiple optimal solutions but only one optimal solution value. For our situation, there are optimal interval solutions. But some of them are not all around characterized. We need to pick well-defined characterized optimal interval solutions. From that point forward, we just use the optimal interval solutions with the well-characterized one.

Now, we give an example for this situation.

Example 3.2.2. Consider the three-person transportation interval situation (P,Q,B',p',q') such that

$$P=\{1\},Q=\{2,3\},B'=([1,3],[2,4]),p'=[3,5],q'=([2,4],[1,3]).$$

The dual issue of the maximization issue for the pessimistic situation $T^D(\{1,2,3\})$ is

$$\begin{aligned} \text{minimize } & 3u_1 + 2v_2 + 1v_1 \\ \text{such that } & u_1 + v_2 \geq 1, \\ & u_1 + v_3 \geq 2, \\ & u_1, v_2, v_3 \geq 0. \end{aligned} \tag{8}$$

When we solve the linear programming problem in Eq. (7), we see that there are numerous optimal solutions. One of them is (0,1,2) and the other solution is (0,1,0).

The dual issue of the maximization issue for the optimistic situation $T^D(\{1,2,3\})$ is

$$\begin{aligned} \text{minimize} \quad & 5u_1 + 4v_2 + 3v_1 \\ \text{such that} \quad & u_1 + v_2 \geq 3, \\ & u_1 + v_3 \geq 4, \\ & u_1, v_2, v_3 \geq 0. \end{aligned} \tag{9}$$

Similarly, when we solve the linear programming problem in Eq. (8), (3,0,1) is the optimal solution of this linear programming problem. If we take (0,1,2) as an optimal solution of the linear programming problem in Eq. (7) and (3,0,1) as an optimal solution of the linear programming problem in Eq. (8), then the interval core imputation ([0,15], [1,0], [1,3]) is not well defined. Then, we could not choose from the optimal solutions which are not well defined. Moreover, we can select (1,0,1) as an optimal solution of the linear programming in Eq. (7) and (3,0,1) as an optimal solution of the linear programming problem in Eq. (8). Finally, if we take (1,0,1) as an optimal solution in Eq. (7) and (3,0,1) as an optimal solution in Eq. (8), then the interval core imputation ([3,15], [0,0], [1,3]) is well defined. As we can see, there are optimal interval solutions but some of them are well defined. Then, we choose optimal interval solutions as well-defined optimal interval solutions.

The optimal interval value (well defined the optimal interval solution) of $T^D(S)$ will be denoted by $\vartheta(T^D(S))$. Here, such that $\vartheta(\underline{T}D(S))$ is the optimal value of the minimization problem of the pessimistic situation. Here, $\vartheta\left(\overline{T^D}(S)\right)$ is the optimal value of the minimization problem of the optimistic situation. In the next theorem, we make use of the well-known Duality. Theorem from linear programming to show that every optimal solution of $T^D(S)$ determines an imputation in the core of the corresponding transportation game. It is clear that this theorem can be extended for transportation interval games. As it is well known, the Duality Theorem asserts that the objective functions of a linear programming problem and its dual attain the same optimal value, provided that both problems have finite optimal values, as it happens in our case (Schrijver 1986). Thus, we have $\vartheta(\underline{T}(S)) = \vartheta(\underline{T}D(S))$ and $\vartheta\left(\bar{T}(S)\right) = \vartheta\left(\overline{T^D}(S)\right)$, for every $S \subset N$ such that $S \cap P$ and $S \cap Q$ are nonempty.

We would like to recall the idea of Owen set on transportation games (Llorca et al. 2004). Thus, for an arbitrary transportation interval game we can always select a interval core element, but with great difficulty. Owen (1975) introduced the class of linear production games and presented a method to find a nonempty subset of the core of these games. Since a transportation game may seem as a special case of linear production games, we can use this method to derive interval core elements.

This set is the so-called interval Owen set of the transportation interval situation, which is defined by

$$OS_d(v') = x_i = [\underline{x_i}, \overline{x_i}] = [\underline{p_i u_i}, \forall i \in P, (x, y) \in I(\mathbb{R})^P \times I(\mathbb{R})^Q \mid y_j$$
$$= \left[\underline{y_j}, \overline{y_j}\right] = \left[\underline{q_j v_j}, \overline{q_j v_j}\right], \forall j \in Q, \text{with}\left(u^*; v^*\right) \in OS_d(v'),$$

where $OS_d(v')$ is the set of optimal solutions of the dual program of the transportation game for the grand coalition.

Corollary 3.2.3. Let (P, Q, B', p', q') be a transportation interval situation and consider its corresponding transportation interval game $\langle N, v' \rangle$. Then,

1. the interval Owen set of transportation interval game is nonempty,
2. $OS_d(v') \subset C(v')$,
3. transportation interval games are *I*-balanced.

TRANSPORTATION GREY SITUATIONS AND GAMES

In this section, we introduce the transportation grey situations inspired by Sánchez-Soriano J et al. (2001) and PalancıO et al. (2016). In a transportation grey situation the set of players is partitioned into two disjoint subsets *P* and *Q*, containing *n* and *m* players respectively. The members of *P* will be called producers, whereas the members of *Q* will be the retailers. Each origin player *i*∈*P* has a positive integer grey number of units of a certain indivisible good, $p_i^{'}$, and each destination player *j*∈*Q* demands a positive integer grey number of units of this good, $q_j^{'}$. The shipping of one unit from origin player *i* to destination player *j* produces a nonnegative grey real profit $b_{ij}^{'}$. Here, $p_i^{'} \in \left[\underline{p_i^{'}}, \overline{p_i^{'}}\right], q_j^{'} \in \left[\underline{q_j^{'}}, \overline{q_j^{'}}\right]$ and $b_{ij}^{'} \in \left[\underline{b_{ij}^{'}}, \overline{b_{ij}^{'}}\right] \in \mathcal{G}(\mathbb{R})$.

A *transportation grey situation* like this is characterized by a 5-tuple (P, Q, B', p', q'), where B' is the n×m matrix of grey profits, p' is the n-dimensional vector of available grey units at the origins, and q' is the m-dimensional vector of grey demands.

For every transportation grey situation (P, Q, B', p', q') and every coalition $S \subset N := P \cup Q$, with producers $S_P := S \cap P$ and retailers $S_Q := S \cap Q$, and assuming that these sets are both non-empty, we can define the maximization problem of the pessimistic scenario by:

$$\underline{T}(S): \quad \text{maximize} \quad \sum_{i \in S_P} \sum_{j \in S_Q} \underline{b_{ij}'} x_{ij}$$
$$\text{such that} \quad \sum_{j \in S_Q} x_{ij} \leq \underline{p_i'},\ i \in S_P,$$
$$\sum_{i \in S_P} x_{ij} \leq \underline{q_j'},\ j \in S_Q, \tag{12}$$
$$x_{ij} \geq 0,\ (i,j) \in S_P \times S_Q,$$

and the maximization problem of the optimistic scenario is stated as:

$$\overline{T}(S): \quad \text{maximize} \quad \sum_{i \in S_P} \sum_{j \in S_Q} \overline{b_{ij}'} x_{ij}$$
$$\text{such that} \quad \sum_{j \in S_Q} x_{ij} \leq \overline{p_i'},\ i \in S_P,$$
$$\sum_{i \in S_P} x_{ij} \leq \overline{q_j'},\ j \in S_Q, \tag{13}$$
$$x_{ij} \geq 0,\ (i,j) \in S_P \times S_Q.$$

We denote by $\vartheta(T(S))$ the optimal grey value of the problem $T(S)$ Here, $\vartheta(T(S)) \in \left[\vartheta(\underline{T}(S)), \vartheta(\overline{T}(S))\right]$ such that $\vartheta(\underline{T}(S))$ is the optimal value of the maximization problem of the pessimistic scenario, $\vartheta(\overline{T}(S))$ is the optimal value of the maximization problem of the optimistic scenario. Then, we can define a *cooperative grey game* associated with every transportation grey situation (P, Q, B', p', q') in the following way:

- The set of players is $N = P \cup Q$;
- The characteristic function $\langle N, w' \rangle$ is given by:

$$w'(S) \in \begin{cases} [0,0], & \text{if } S = \varnothing \text{ or } S \text{ is contained in } P \text{ or in } Q, \\ \left[\vartheta(\underline{T}(S)), \vartheta(\overline{T}(S))\right], & \text{in any other case,} \end{cases}$$

and $w'(S)$ satisfies the condition

$$\vartheta(\underline{T}(S)) + \vartheta(\overline{T}(T)) \geq \vartheta(\underline{T}(T)) + \vartheta(\overline{T}(S)) \text{forall} S \subset T.$$

Now, we present the definition of a transportation grey game.

Definition 4.1 A transportation grey game is a cooperative grey game $w' \in \mathcal{G}G^N$ arising from a transportation situation (P,Q,B',p',q'). Often, we identify a transportation grey situation (P,Q,B',p',q') with its associated transportation game v'.

Now, we study the properties of transportation grey games. We start by noting that, in view of its definition, a transportation grey game is grey zero-normalized. It is clear that a transportation grey game is superadditive, size monotonic but not necessarily convex. In the following section, we show that transpotation grey games are size monotonic.

Finally, we present an example of a 3-person transportation grey game.

Example 4.2. Consider the 3-person transportation grey situation (P,Q,B',p',q') which has one producer and two retaliers:

$$P=\{1\},Q=\{2,3\},B'=([3,5] \quad [5,6]),p'\in[3,5],q'\in([2,4],[1,3]).$$

Now, we define a transportation grey game associated with a transportation grey situation (P,Q,B',p',q'). Here, $N=\{1,2,3\}$ is the set of players and the characteristic functions of the transportation grey game are as follows:

$$w'(1)=w'(2)=w'(3)=w'(23)\in[0,0],$$

$$w'(12)\in[6,20],w'(13)\in[5,18],w'(123)\in[11,28].$$

So, we obtain the transportation grey game $\langle N,w'\rangle$ corresponding to a transportation grey situation.

The Grey Shapley Value of the Transportation Grey Game

In this section, we calculate the grey Shapley value of the transportation grey game. Firstly, we recall the definition of the grey Shapley value. For this, we need to recall some notions from the theory of cooperative grey games (Alparslan Gök SZ et al., 2009a).

Grey solutions are useful to solve reward/cost sharing problems with grey data using cooperative grey games as a tool. The grey payoff vectors, which are the building blocks for grey solutions, are the vectors whose components belong to $\mathcal{G}(\mathbb{R})$. We denote by $\mathcal{G}(\mathbb{R})^N$ the set of all such grey payoff vectors.

We call a game $\langle N, w' \rangle$ *grey size monotonic* if $\langle N, |w| \rangle$ is monotonic, i.e., $|w|(S) \leq |w|(T)$ for all $S,T \in 2^N$ with $S \subset T$. For further use we denote by $SM\mathcal{G}G^N$ the class of grey size monotonic games with player set *N*.

The following theorem shows that the transportation grey games are size monotonic.

Theorem 4.1.1. The transportation grey game $\langle N, w' \rangle$ belongs the class of $SM\mathcal{G}G^N$.

Proof. We show that the transportation grey game v' belongs to the class of $SM\mathcal{G}G^N$. For this,

$$|w'|(S) \leq |w'|(T) \text{ for all } S,\ T \in 2^N \text{ with } S \subset T.$$

Take $S,T \in 2^N$ with $S \subset T$. If $S = \varnothing$ or *S* is contained in *P* or in *Q*, then $w'(S) = w'(T) \in [0,0]$. Now, it is obvious that

$$|w'|(S) \leq |w'|(T).$$

If $S \neq \varnothing$ or *S* is not contained in *P* or in *Q*, then $w'(S) \in \left[\vartheta\left(\underline{T}(S)\right), \vartheta\left(\overline{T}(S)\right)\right]$ with $\vartheta\left(\underline{T}(S)\right) \leq \vartheta\left(\overline{T}(S)\right)$. For $S \subset T$, by using the definition of a transportation grey game we obtain:

$$\vartheta(\underline{T}(S)) \leq \vartheta(\underline{T}(T)),$$

$$\vartheta\left(\overline{T}(S)\right) \leq \vartheta\left(\overline{T}(T)\right).$$

Then,

$$|w'|(S) \leq |w'|(T),$$

$$\vartheta\left(\overline{T}(S)\right) - \vartheta\left(\underline{T}(S)\right) \leq \vartheta\left(\overline{T}(T)\right) - \vartheta\left(\underline{T}(T)\right),$$

$$\vartheta\left(\underline{T}(T)\right)+\vartheta\left(\bar{T}(S)\right)\leq\vartheta\left(\underline{T}(S)\right)+\vartheta\left(\bar{T}(T)\right),$$

for all $S,T\in 2^N$ with $S\subset T$. Since w' satisfies the condition

$$\vartheta\left(\underline{T}(T)\right)+\vartheta\left(\bar{T}(S)\right)\leq\vartheta\left(\underline{T}(S)\right)+\vartheta\left(\bar{T}(T)\right),$$

$\langle N,w'\rangle$ belongs to the class of $SM\mathcal{G}G^N$.

We know that if an grey game is belonging to $SM\mathcal{G}G^N$, then the grey Shapley value is always given (Alparslan Gök SZ et al., 2009a).

Remark 4.1.2. The grey Shapley value of the transportation grey games always exists.

The grey marginal operators and the grey Shapley value were defined on $SM\mathcal{G}G^N$ in Alparslan Gök SZ et al. (2009a) as follows.

The grey Shapley value assigns to each cooperative grey game a payoff vector whose components are compact greys of real numbers. Cooperative games in the additive cone on which we use the grey Shapley value arise from several OR and economic situations with grey data.

The grey Shapley value $\Phi: SM\mathcal{G}G^N \to \mathcal{G}(\mathbb{R})^N$ is defined by

$$\Phi'(w'):=\frac{1}{n!}\sum_{\sigma\in\Pi(N)}m^{\sigma}(w')\in\frac{1}{n!}\sum_{\sigma\in\Pi(N)}m^{\sigma}(\underline{A}),\frac{1}{n!}\sum_{\sigma\in\Pi(N)}m^{\sigma}(\bar{A})].$$

The following example shows the calculation of the grey Shapley value in the transportation grey game.

Example 4.1.3. Consider $\langle N,w'\rangle$ as the transportation grey game in Example 3.1.

Here, $N=\{1,2,3\}$ and the characteristic function w' is given as

$$w'(1)=w'(2)=w'(3)=w'(23)\in[0,0],$$

$$w'(12)=[6,20],w'(13)=[5,18],w'(123)\in[11,28].$$

Then, the grey marginal vectors are given in the Table 1. The set of permutations of N is

$$\pi(N)=\begin{Bmatrix}\sigma_1=(1,2,3),\sigma_2=(1,3,2),\sigma_3=(2,1,3),\\ \sigma_4=(2,3,1),\sigma_5=(3,1,2),\sigma_6=(3,2,1)\end{Bmatrix}.$$

Firstly, for σ_2=(1,3,2) we calculate the grey marginal vectors. Then,

$$m_1^{\sigma_2}(v')=w'(1)\in[0,0],$$

$$m_2^{\sigma_2}(v')=w'(123)-w'(13)\in[11,28]-[5,18]=[6,10],$$

$$m_3^{\sigma_2}(v')=w'(13)-w'(1)\in[5,18]-[0,0]=[5,18].$$

The others can be calculated similarly, which is shown in Table 5.

Table 5 illustrates the grey marginal vectors of the cooperative transportation grey game in

The average of the six grey marginal vectors is the grey Shapley value of this game, which can be written as:

$$\Phi(w')\in\left(\left[5\frac{1}{2},15\frac{2}{3}\right],\left[3,6\frac{2}{3}\right],\left[2\frac{1}{2},5\frac{2}{3}\right]\right).$$

Table 5. Grey marginal vectors

σ	$m_1^{\sigma}(w')$	$m_2^{\sigma}(w')$	$m_3^{\sigma}(w')$
σ_1=(1,2,3)	$m_1^{\sigma_1}(w')\in[0,0]$	$m_2^{\sigma_1}(w')\in[6,20]$	$m_3^{\sigma_1}(w')\in[5,8]$
σ_2=(1,3,2)	$m_1^{\sigma_2}(w')\in[0,0]$	$m_2^{\sigma_2}(w')\in[6,20]$	$m_3^{\sigma_2}(w')\in[5,18]$
σ_3=(2,1,3)	$m_1^{\sigma_3}(w')\in[6,20]$	$m_2^{\sigma_3}(w')\in[0,0]$	$m_3^{\sigma_3}(w')\in[5,8]$
σ_4=(2,3,1)	$m_1^{\sigma_4}(w')\in[11,28]$	$m_2^{\sigma_4}(w')\in[0,0]$	$m_3^{\sigma_4}(w')\in[0,0]$
σ_5=(3,1,2)	$m_1^{\sigma_5}(w')\in[5,18]$	$m_2^{\sigma_5}(w')\in[6,10]$	$m_3^{\sigma_5}(w')\in[0,0]$
σ_6=(3,2,1)	$m_1^{\sigma_6}(w')\in[11,28]$	$m_2^{\sigma_6}(w')\in[0,0]$	$m_3^{\sigma_6}(w')\in[0,0]$

The Grey Core of the Transportation Grey Game

In this section, we present some interesting results concerning the grey core of a transportation grey game. It is well known that the transportation games have non-empty cores (Branzei R et al., 2010b). Now, we check if this property can be extended to transportation grey situations.

The dual problem of the maximization problem for the pessimistic scenario (12) is given by the minimization problem:

$$\begin{aligned} T^D(S): \quad & \text{minimize} \quad \sum_{i \in S_P} \underline{p_i} u_i + \sum_{j \in S_Q} \underline{q_j} v_j \\ & \text{such that} \quad u_i + v_j \geq \underline{b_{ij}}, \ (i,j) \in S_P \times S_Q, \\ & \qquad\qquad\quad u_i, v_j \geq 0, \ i \in S_P, \ j \in S_Q, \end{aligned}$$

and the dual problem of the maximization problem for the optimistic scenario (13) is given by the minimization problem:

$$\begin{aligned} T^D(S): \quad & \text{minimize} \quad \sum_{i \in S_P} \overline{p_i} u_i + \sum_{j \in S_Q} \overline{q_j} v_j \\ & \text{such that} \quad u_i + v_j \geq \overline{b_{ij}}, \ (i,j) \in S_P \times S_Q, \\ & \qquad\qquad\quad u_i, v_j \geq 0, \ i \in S_P, \ j \in S_Q. \end{aligned}$$

Remark 4.2.1. An optimal solution to a linear program is a feasible solution with the largest objective function value. The value of the objective function for the optimal solution is said to be the value of the linear program. A linear program may have multiple optimal solutions but only one optimal solution value. In our case, there are optimal grey solutions. But some of them is not well-defined. We want to choose well-defined optimal grey solutions. After that, we only use the optimal grey solutions with well-defined.

Now, we give an example for better understand this situation.

Example 4.2.2. Consider the 3-person transportation grey situation (P, Q, B', p', q') in

$$P = \{1\}, Q = \{2,3\}, B' = ([1,3] \quad [2,4]), p' \in [3,5], q' \in ([2,4],[1,3]).$$

The dual problem of the maximization problem for the pessimistic scenario $T^D(\{1,2,3\})$ is:

$$\begin{aligned} \text{minimize} \quad & 3u_1 + 2v_2 + 1v_3 \\ \text{such that} \quad & u_1 + v_2 \geq 1, \\ & u_1 + v_3 \geq 2, \\ & u_1, v_2, v_3 \geq 0. \end{aligned} \tag{14}$$

When we solve the linear programming problem in Eq. (14), we see that there are many optimal solutions. One of them is (0,1,2) and the other solution is (1,0,1).

Example 4.2.3. The dual problem of the maximization problem for the optimistic scenario $T^D(\{1,2,3\})$ is:

$$\begin{aligned} \text{minimize} \quad & 5u_1 + 4v_2 + 3v_3 \\ \text{such that} \quad & u_1 + v_2 \geq 3, \\ & u_1 + v_3 \geq 4, \\ & u_1, v_2, v_3 \geq 0. \end{aligned} \tag{15}$$

Similarly, when we solve the linear programming problem in Eq. (15), (3,0,1) is the optimal solution of this linear programming problem. If we take (0;1,2) as an optimal solution of the linear programming problem in Eq. (14) and (3;0,1) as an optimal solution of the linear programming problem in Eq. (15), then the grey core imputation ([0,15], [1,0], [1,3]) is not well-defined. Then, we could not choose from the optimal solutions which are not well-defined. Moreover, we can select (1;0,1) as an optimal solution of the linear programming in Eq. (14) and (3;0,1) as an optimal solution of the linear programming problem in Eq. (15). Finally, if we take (1;0,1) as an optimal solution in Eq. (14) and (3;0,1) as an optimal solution in Eq. (15), then the grey core imputation ([0,15], [0,0], [1,3]) is well defined. As we can see that, there are optimal inverval solutions but some of them is well-defined. Then, we choose optimal grey solutions as well-defined optimal grey solutions.

The optimal grey value (well-defined the optimal grey solution) of $T^D(S)$ will be denoted by $\vartheta(T^D(S))$. Here,

$$\vartheta\left(T^D(S)\right) \in \left[\vartheta\left(\underline{T}^D(S)\right), \vartheta\left(\underline{T}^D(S)\right)\right] \in I(\mathbb{R})$$

such that $\vartheta(\underline{T}D(S))$ is the optimal value of the minimization problem of the pessimistic scenario. Here, $\vartheta(\overline{T}D(S))$ is the optimal value of the minimization problem of the optimistic scenario. In the next theorem, we make use of the well-known Duality Theorem from linear programming to show that every optimal solution of $T^D(S)$ determines an imputation in the core of the corresponding transportation game. It is clear that this theorem can be extended for transportation grey games. As it is well known, the Duality Theorem asserts that the objective functions of a linear programming problem and its dual attain the same optimal value, provided that both problems have finite optimal values, as it happens in our case (Schrijver A, 1986). Thus, we have $\vartheta(\underline{T}(S)) = \vartheta(\underline{T}D(S))$ and $\vartheta\left(\overline{T}(S)\right) = \vartheta\left(\overline{T^D}(S)\right)$, for every $S \subset N$ such that $S \cap P$ and $S \cap Q$ are non-empty. The following theorem is inspired by the paper Sánchez-Soriano J et al. (2001).

Theorem 4.2.4. Let (P, Q, B', p', q') be a transportation grey situation and consider its corresponding transportation grey game $\langle N, w' \rangle$. Let $(u^*; v^*) \in \mathcal{G}(\mathbb{R})^{n+m}$ be a well-defined grey optimal solution of $T^D(N)$ such that $(u^*; v^*) \in \left(\left[\underline{u}^*, \overline{u}^*\right]; \left[\underline{v}^*, \overline{v}^*\right]\right)$. Then,

$$\left(u_1^* p_1, \cdots, u_n^* p_n; v_1^* q_1, \cdots, v_m^* q_m\right)$$

$$\in \left(\left[\underline{u_1^* p_1}, \overline{u_1^* p_1}\right], \cdots, \left[\underline{u_n^* p_n}, \overline{u_n^* p_n}\right]; \left[\underline{v_1^* q_1}, \overline{v_1^* q_1}\right], \cdots, \left[\underline{v_m^* q_m}, \overline{v_m^* q_m}\right]\right)$$

belongs to the core $C(w')$.

Proof. From the Duality Theorem (Schrijver A, 1986) we conclude that for all $S \subset N$ such that $S \cap P$ and $S \cap Q$ are non-empty, it holds

$$w'(S) = \vartheta(T(S)) = \vartheta\left(T^D(S)\right).$$

In addition, it is clear that $\left(u_S^*; v_S^*\right)$, the restriction of $(u^*; v^*)$ to *S*, is a feasible grey solution of Eqns. (4) and (5). Thus,

$$\vartheta\left(T^{D}(S)\right)\in\left[\vartheta\left(\underline{T}^{D}(S)\right),\vartheta\left(\overline{T^{D}}(S)\right)\right]'$$

$$\left[\sum_{i\in S_P}\underline{u_i^* p_i}+\sum_{j\in S_Q}\underline{v_j^* q_j},\sum_{i\in S_P}\overline{u_i^* p_i}+\sum_{j\in S_Q}\overline{v_j^* q_j}\right].$$

Therefore,

$$\left(u_1^* p_1,\cdots,u_n^* p_n;v_1^* q_1,\cdots,v_m^* q_m\right)$$
$$\in\left(\left[\underline{u_1^* p_1},\overline{u_1^* p_1}\right],\cdots,\left[\underline{u_n^* p_n},\overline{u_n^* p_n}\right];\left[\underline{v_1^* q_1},\overline{v_1^* q_1}\right],\cdots,\left[\underline{v_m^* q_m},\overline{v_m^* q_m}\right]\right)$$

belongs to the grey core of the w', since for the other possible coalitions we know that $v'(S)=[0,0]$.

This theorem shows that transportation grey games have non-empty cores. For transportation grey games we show that the non-negative optimal dual solutions of the asscoiated transportation grey problems yield some imputations of their grey cores.

The set of all imputations obtained in this way through the different optimal solutions of $T^D(N)$ is denoted by $OS_d(w')$; i.e.,

$$OS_d(w')\in\left\{\left(u_1^* p_1,\cdots,u_n^* p_n;v_1^* q_1,\cdots,v_m^* q_m\right)\mid\left(u^*;v^*\right)\right.$$
$$\left.\text{is optimal for } T^D(N)\right\}.$$

The last theorem shows that $OS_d(w')$ is contained in the grey core of the transportation grey game w'; i.e., $OS_d(w')\subset C(w')$.

We would like to apply the idea of Owen set on transportation games (Llorca et al. 2004). Thus, for an arbitrary transportation grey game we can always select a grey core element, but with great difficulty. Owen (1975) introduced the class of linear production games and presented a method to find a non-empty subset of the core of these games. Since a transportation game may seen as a special case of linear production games, we can use this method to derive grey core elements. This set is the so-called *grey Owen set* of the transportation grey situation, which is defined by

$$OS(w') \in \left\{ (x,y) \in I(\mathbb{R})^P \times I(\mathbb{R})^Q \mid \begin{array}{c} x_i = \left[\underline{x_i}, \overline{x_i}\right] = \left[\underline{p_i u_i}, \overline{p_i u_i}\right], \forall i \in P, \\ y_j = \left[\underline{y_j}, \overline{y_j}\right] = \left[\underline{q_j v_j}, \overline{q_j v_j}\right], \forall j \in Q, \\ \text{with} \left(u^*; v^*\right) \in OS_d\left(v'\right) \end{array} \right\},$$

where $OS_d\left(v'\right)$ is the set of optimal solutions of the dual program of the transportation game for the grand coalition.

Corollary 4.2.5. Let $\left(P, Q, B', p', q'\right)$ be a transportation grey situation and consider its corresponding transportation grey game $\langle N, w' \rangle$. Then,

a) the grey Owen set of transportation grey game is non-empty,

b) $OS_d\left(w'\right) \subset C\left(w'\right)$,

c) transportation grey games are $\mathcal{I}$ -balanced.

Proof. From Theorem 5.1, the conditions a) and b) are satisfied. Transportation grey games have non-empty grey cores. We know that an grey game is $\mathcal{I}$ -balanced if and only if its grey core is non-empty of Theorem 3.1 in Alparslan Gök et al. (2008). Then, we can say that transportation grey games are $\mathcal{I}$ -balanced.

CONCLUSION

This paper examines two-sided transportation situations where the operators' unitary issue parameters (b_{ij}, p_i, q_j) in the transportation demonstrate are compact grey of real numbers. To begin with, we present the transportation grey situations. Second, we compute the grey Shapley value of a transportation grey game and show interesting results concerning the grey core of a transportation grey game. In addition, we recommend a methodology that changes a grey allocation into a payoff vector, under the presumption that lone the uncertainty with respect to the value of the stupendous coalition has been settled.

Let us say that a semi-limitless programming issue is an enhancement problem in which limitedly numerous factors show up in interminably numerous limitations (Hettich and Kortanek 1993). This model normally emerges in a bounteous number of applications in diverse fields of science, financial aspects, and building.

REFERENCES

Alparslan Gök, S. (2011). Set-valued solution concepts using interval-type payoffs for interval games. *Journal of Mathematical Economics, 47*(4-5), 621-626.

Alparslan Gök, S. Z., Branzei, R., & Tijs, S. (2008). *Cores and Stable Sets for Interval-Valued Games*. CentER Discussion Paper; Vol. 2008-17. Tilburg: Operations Research.

Alparslan Gök, S. Z., Branzei, R., & Tijs, S. (2009a). Convex interval games. *Journal of Applied Mathematics and Decision Sciences, 2009*, 342089. doi:10.1155/2009/342089

Alparslan Gök, S. Z., Miquel, S., & Tijs, S. (2009b). Cooperative under interval uncertainty. *Mathematical Methods of Operations Research, 69*(1), 99–109. doi:10.100700186-008-0211-3

Aparicio, J. (2010). *Cooperative logistics games. In SCIYO* (p. 129). COM.

Aziz, H., Cahan, C., Gretton, C., Kilby, P., Mattei, N., & Walsh, T. (2016). A study of proxies for shapley allocations of transport costs. *Journal of Artificial Intelligence Research, 56*, 573–611. doi:10.1613/jair.5021

Bai, G., Mao, J., & Lu, G. (2004). Grey transportation problem. *Kybernetes, 33*(2), 219–224. doi:10.1108/03684920410514148

Borm, Hamers, & Hendrickx. (2001). Operations research games: A survey. *Top, 9*(2), 139.

Branzei, R., Branzei, O., Gök, S. Z. A., & Tijs, S. (2010a). Cooperative interval games: A survey. *Central European Journal of Operations Research, 18*(3), 397–411. doi:10.100710100-009-0116-0

Branzei, R., Gök, S. Z. A., & Branzei, O. (2011). SZ Alparslan Gök, and O. Branzei. *Cooperative games under interval uncertainty: On the convexity of the interval undominated cores. Central European Journal of Operations Research, 19*(4), 523–532. doi:10.100710100-010-0141-z

Branzei, R., Tijs, S., & Alparslan Gök, S. Z. (2010b). How to handle interval solutions for cooperative interval games. *International Journal of Uncertainty, Fuzziness and Knowledge-based Systems, 18*(2), 123–132. doi:10.1142/S0218488510006441

Deng, J. (1988). *Grey Forecasting and Decision-Making*. Wuhan, China: Huazhong University of Science and Technology Press.

Deng, J.-L. (1982). Control problems of grey systems. *Systems & Control Letters, 1*(5), 288–294. doi:10.1016/S0167-6911(82)80025-X

Deng, J.-L. (1990). *Grey system theory tutorial.* Wuhan: Huazhong University of Science and Technology Press.

Frisk, M., Göthe-Lundgren, M., Jörnsten, K., & Rönnqvist, M. (2010). Cost allocation in collaborative forest transportation. *European Journal of Operational Research, 205*(2), 448–458. doi:10.1016/j.ejor.2010.01.015

Hettich, R., & Kortanek, K. O. (1993). Semi-infinite programming: Theory, methods, and applications. *SIAM Review, 35*(3), 380–429. doi:10.1137/1035089

James, K. S. (1989). Linear programming and its applications. Springer-Verlag World Publishing Corp.

Kyoto. (1997a). *Kyoto contract.* Retrieved from http://www.unfccc.org/resource/convkp.html

Kyoto. (1997c). *Kyoto protocol.* Retrieved from http://vitalgraphics.grida.no/kyoto

Llorca, N (2004). *On the Owen set of transportation situations. In Essays in Cooperative Games* (pp. 215–228). Boston, MA: Springer. doi:10.1007/978-1-4020-2936-3_17

Owen, G. (1975). On the core of linear production games. *Mathematical Programming, 9*(1), 358–370. doi:10.1007/BF01681356

Özener, O. Ö., & Ergun, Ö. (2008). Allocating costs in a collaborative transportation procurement network. *Transportation Science, 42*(2), 146–165. doi:10.1287/trsc.1070.0219

Palancı, O., Alparslan Gök, S. Z., Ergün, S., & Weber, G. W. (2015). Cooperative grey games and the grey Shapley value. *Optimization, 64*(8), 1657–1668. doi:10.1080/02331934.2014.956743

Palancı, O., Alparslan Gök, S. Z., Olgun, M. O., & Weber, G.-W. (2016). Transportation interval situations and related games. *OR-Spektrum, 38*(1), 119–136. doi:10.100700291-015-0422-y

Pickl, St., & Weber, G. W. (2000). An algorithmic approach by linear programming problems in generalized semi-infinite optimization, J. *Computing Techniques, 5*(3), 62–82.

Pickl, S. W., & Gerhard-W., W. (2002). *On Optimal Control of Heating Processes. In Operations Research Proceedings 2001.* Berlin: Springer.

Puerto, J., Fernández, F. R., & Hinojosa, Y. (2008). Partially ordered cooperative games: Extended core and Shapley value. *Annals of Operations Research, 158*(1), 143–159. doi:10.100710479-007-0242-9

Pulido, Sánchez-Soriano, & Llorca. (2002). Game theory techniques for university management: an extended bankruptcy model. *Annals of Operations Research, 109*(1-4), 129-142.

Pulido, M., Borm, P., Hendrickx, R., Llorca, N., & Sánchez-Soriano, J. (2008). Compromise solutions for bankruptcy situations with references. *Annals of Operations Research, 158*(1), 133–141. doi:10.100710479-007-0241-x

Sánchez-Soriano, J. (2003). The pairwise egalitarian solution. *European Journal of Operational Research, 150*(1), 220–231. doi:10.1016/S0377-2217(02)00503-9

Sánchez-Soriano, J. (2006). Pairwise solutions and the core of transportation situations. *European Journal of Operational Research, 175*(1), 101–110. doi:10.1016/j.ejor.2005.04.033

Sánchez-Soriano, J., Llorca, N., Meca, A., Molina, E., & Pulido, M. (2002). An integrated transport system for Alacant's students. *Annals of Operations Research, 109*(1-4), 41–60. doi:10.1023/A:1016387715917

Sánchez-Soriano, J., López, M. A., & Garc. (2001). On the core of transportation games. *Mathematical Social Sciences, 41*(2), 215–225. doi:10.1016/S0165-4896(00)00057-3

Schrijver, A. (1986). *Theory of Linear and Integer Programming*. Chichester, UK: John Wiley and Sons.

Soons, D. E., Slikker, M., & Fransoo, J. C. (2011). *The determination and division of benefits among partners of a horizontal cooperation in transportation*. Technische Universiteit Eindhoven.

Theys, C., Dullaert, W., & Notteboom, T. (2008). *Analyzing cooperative networks in intermodal transportation: a game-theoretic approach*. In Nectar Logistics and Freight Cluster Meeting, Delft, The Netherlands.

Xu, G. (1999). *Handbook of Operations Research Fundamentals*. Beijing, China: Science Press.

Chapter 7
On the Orthogonality of the q-Derivatives of the Discrete q-Hermite I Polynomials

Sakina Alwhishi
El Mergib University, Libya

Rezan Sevinik Adıgüzel
Atılım University, Turkey

Mehmet Turan
Atilim University, Turkey

ABSTRACT

Discrete q-Hermite I polynomials are a member of the q-polynomials of the Hahn class. They are the polynomial solutions of a second order difference equation of hypergeometric type. These polynomials are one of the q-analogous of the Hermite polynomials. It is well known that the q-Hermite I polynomials approach the Hermite polynomials as q tends to 1. In this chapter, the orthogonality property of the discrete q-Hermite I polynomials is considered. Moreover, the orthogonality relation for the k-th order q-derivatives of the discrete q-Hermite I polynomials is obtained. Finally, it is shown that, under a suitable transformation, these relations give the corresponding relations for the Hermite polynomials in the limiting case as q goes to 1.

DOI: 10.4018/978-1-7998-0134-4.ch007

INTRODUCTION

The so-called q.polynomials have great applications in several problems on theoretical and mathematical physics, for example, in the continued fractions, Eulerian series (Fine, 1988), algebras and quantum groups (Koornwinder, 1995, 1994; Vilenkin & Klimyk, 1998), discrete mathematics, algebraic combinatorics (coding theory, design theory, various theories of group representation) (Bannai, 1990), q.Schrödinger equation and q.harmonic oscillators (Askey et al., 1993; Askey & Suslov, 1993; Atakishiyev & Suslov, 1991; Berg & Ismail, 1994; Borzov & Damaskinsky, 2003; Macfarlane, 1989).

Discrete q.Hermite I polynomials form an important class of q.polynomials in the Hahn sense. They satisfy the following q.difference equation of hypergeometric type:

$$-D_q D_{q^{-1}} y + \frac{x}{1-q} D_{q^{-1}} y + \lambda y = 0, \tag{1}$$

where λ.is a constant which has a particular form depending on the parameter q. and D_q. is the q.Jackson derivative (Koekoek et al., 2010; Nikiforov et al., 1991; Nikiforov & Uvarov, 1986, 1988).

In fact, discrete q.Hermite polynomials are the q.analogous of the Hermite polynomials which are one of the important orthogonal family of the classical orthogonal polynomials (Koekoek et al., 2010; Nikiforov et al., 1991; Nikiforov & Uvarov, 1986, 1988; Szegö, 1939) that have enormous applications in mathematics and physics. They satisfy the following differential equation of hypergeometric type

$$y'' - 2zy' + 2ny = 0, \tag{2}$$

where $n \in \mathbb{N}_0$.The Hermite polynomials and their q.analogues can be obtained by using an appropriate limit relation from the other orthogonal polynomials (Andrews et al., 1999; Koekoek et al., 2010).

The classical orthogonal polynomials and their q.analogues have several properties. For example the differential and the difference equations that they satisfy have special forms, which are studied by Bochner in 1929 (Bochner, 1929) and by Routh in 1885 (Routh, 1885). Rodrigues formula is another property that provides a representation for the classical polynomials derived by Tricomi (Tricomi, 1955) and (Cryer, 1970). The three terms recurrence relation (TRRR) is also a way to construct classical orthogonal polynomials considered by Chihara (Chihara, 1978) and Szegö (Szegö, 1939). Moreover, the generating function is another option to define these polynomials. Using generating function, such polynomials can be obtained in a series form which is first studied by Abraham and Moivre in 1730 (Koekoek et al., 2010).

Another most practical property of the so-called classical polynomials are developed by Sonine in 1887 and by Hahn in 1939, which is stated in the following theorem:

Theorem 1.1 (Sonine-Hahn, (Alvarez-Nodarse, 2006; Hahn, 1935; Marcellan et al., 1994)) *A given sequence of orthogonal polynomials* $(P_n)_n$*.is a classical sequence if and only if the sequence of its derivatives* $(P'_n)_n$ *.is an orthogonal polynomial sequence.*

As a result, the orthogonality property with respect to a suitable inner product (Nikiforov et al., 1991; Nikiforov & Uvarov, 1986, 1988) is also an important characteristic property.

The main idea of this work is to study the orthogonality properties of discrete q.Hermite I polynomials and their q.derivatives which will also coincide with the classical Hermite polynomials with a suitable transformation in the limiting case as $q \to 1$.This chapter is organized as follows: In the next section, some basic definitions related to q.calculus are given. Later on, the authors introduce some properties of the discrete q.Hermite I polynomials and then the orthogonality properties of the discrete q.Hermite I polynomials and their k.th order q.derivatives are constructed. In the last part, the authors consider a suitable transformation after the application of which the presented orthogonality relations lead to the classical orthogonality relations in the limiting case as $q \to 1$.

BACKGROUND

In this section, some notations that are used in q.calculus will be presented. The definitions and notations that are given here can be found in (Andrews et al., 1999; Koekoek et al., 2010; Nikiforov et al., 1991; Nikiforov & Uvarov, 1986, 1988).

Let $q>0$.For any $n \in \mathbb{N}_0$, .the q.integer $[n]_q$.is defined by

$$[n]_q := 1 + q + \cdots + q^{n-1} = \frac{1-q^n}{1-q}, n = 1,2,\ldots,[0]_q := 0, \tag{3}$$

the q.factorial $[n]_q!$.by

$$[n]_q! = [1]_q [2]_q \ldots [n]_q,\ n = 1,2,\ldots,[0]_q! := 1.$$

and the q-Pochhammer's symbol (or q-shifted factorial) is defined by

$$(a;q)_0 := 1,\ (a;q)_k := \prod_{s=0}^{k-1}(1-aq^s),\ (a;q)_\infty := \prod_{s=0}^{\infty}(1-aq^s).$$

Note that for $q \neq 1$.one can write

$$[n]_q := \frac{1-q^n}{1-q},\ n=1,2,\ldots,[0]_q := 1.$$

Lemma 2.1 *For* $a \in \mathbb{C}$ *.and* $k \in \mathbb{N}_0$*,.the authors have* $(a;q^2)_k(aq;q^2)_k =: (a;q)_{2k}$.
Proof:

$$(a;q^2)_k(aq;q^2)_k = \prod_{s=0}^{k-1}\left(1-aq^{2s}\right)\prod_{s=0}^{k-1}\left(1-aq^{2s+1}\right) = \prod_{s=0}^{2k-1}\left(1-aq^s\right) = (a;q)_{2k}.$$

A simple corollary of this result is stated below which can be obtained by taking the limit as $k \to \infty$.

Corollary 2.2 *For* $a \in \mathbb{C}$*,.the authors have* $(a;q^2)_\infty(aq;q^2)_\infty =: (a;q)_\infty$.
Lemma 2.3 *For* $a \in \mathbb{C}$ *.and* $k \in \mathbb{N}_0$*,.the authors have* $(a;q)_k(-aq;q)_k =: (a^2;q^2)_k$.
Proof: It is easy to see that

$$(a;q)_k(-a;q)_k = \prod_{s=0}^{k-1}\left(1-aq^s\right)\prod_{s=0}^{k-1}\left(1+aq^s\right) = \prod_{s=0}^{k-1}\left(1-a^2q^{2s}\right) = (a^2;q^2)_k$$

which completes the proof.

Taking the limit as $k \to \infty$, one arrives at the following result:

Corollary 2.4 *For any* $a \in \mathbb{C}$*,.the authors have* $(a;q)_\infty(-aq;q)_\infty =: (a^2;q^2)_\infty$.

The Pochammer's symbol is given in general for complex numbers. Accordingly, the definition of -integer given by (3) is extended to the complex numbers and the authors have, for

The limit relation between the Pochammer's symbol and the q.Pochammer's symbol is given by the following lemma:

Lemma 2.5 *For* $\alpha \in \mathbb{C}$ *.and positive integer k,.the authors have*

$$\lim_{q\to 1}\frac{(q^{\alpha};q)_k}{(1-q)^k}=(\alpha)_k\,. \tag{4}$$

where $(\alpha)_k$.is the Pochammer symbol defined by $(\alpha)_k=\alpha(\alpha+1)\ldots(\alpha+k\text{-}1)$ for $k\geqslant 1$ .and $(\alpha)_0:=1$.

Proof: First of all, the authors note that

Now, since $\lim_{q\to 1}[m]_q = m$, .taking the limit of both sides as $q\to 1$ in the above expression, the authors get the stated result.

Next, the definition of the q.gamma function (Andrews et al., 1999) is given:

Definition 2.6 The q.gamma function is defined by

$$\Gamma_q(x)=\frac{(q;q)_\infty}{(q^x;q)_\infty}(1-q)^{1-x}. \tag{5}$$

This is a q.analogue of the gamma function and the authors have

$$\lim_{q\to 1}\Gamma_q(x)=\Gamma(x). \tag{6}$$

Note that, $\Gamma_q(1)=1$, $\Gamma_q(x+1)=[x]_q\Gamma_q(x)$.

Next, the q.derivative of a function when $q\neq 1$ is provided. The q.derivative which was first defined by Jackson is a q.analogue of the classical ordinary derivative. It is also called the q.Jackson derivative.

Definition 2.7 (q-Derivative) *The q.derivative of a function f, for q≠1, is defined as*

$$D_q f(x)=\begin{cases}\dfrac{f(x)-f(qx)}{(1-q)x}, & x\neq 0\\ f'(0), & x=0\end{cases}. \tag{7}$$

Observe that if f is differentiable, then

$$\lim_{q\to 1} D_q f(x) = \frac{df(x)}{dx}$$

It is not difficult to see that the operator D_q.is linear. That is, for constants a and b, and functions f and g, one has

$$D_q(af(x)+bg(x)) = aD_q f(x)+bD_q g(x).$$

If α is a nonzero scalar, then

$$D_q f(\alpha x) = \frac{f(\alpha x)-f(q\alpha x)}{(1-q)x} = \alpha \frac{f(\alpha x)-f(q\alpha x)}{(1-q)\alpha x} = \alpha D_q f(t)\big|_{t=\alpha x},$$

or, equivalently,

$$D_q f(t)\big|_{t=\alpha x} = \alpha^{-1} D_q f(\alpha x). \tag{8}$$

Also,

$$D_q f(q^{-1}x) = \frac{f(q^{-1}x)-f(x)}{(1-q)x} = \frac{f(q^{-1}x)-f(x)}{q(q^{-1}-1)x} = q^{-1} D_{q^{-1}} f(x). \tag{9}$$

However, the q.derivative of f at $q^{-1}x$ is the q^{-1}.derivative of f at x. More precisely, for $\alpha=q^{-1}$.in (8), the authors get

$$D_q f(t)\big|_{t=q^{-1}x} = qD_q f(q^{-1}x) = D_{q^{-1}} f(x). \tag{10}$$

Similarly,

$$D_{q^{-1}} f(t)\big|_{t=qx} = D_q f(x). \tag{11}$$

The higher order q derivative of f is defined in a similar way that the higher order derivative in classical sense is defined. The n.th order q derivative of f is

There is a nice relation between $D_q D_{q^{-1}}$ and $D_{q^{-1}} D_q$ as stated in the following lemma.

Lemma 2.8 *For any* $q\neq 1$, *the following relation holds:*

$$D_q D_{q^{-1}} = q^{-1} D_{q^{-1}} D_q. \tag{12}$$

Proof: For any f the authors have

$$D_q D_{q^{-1}} f(x) = \frac{D_{q^{-1}} f(x) - D_{q^{-1}} f(t)|_{t=qx}}{(1-q)x}.$$

Using (10) and (11), the authors get

$$D_q D_{q^{-1}} f(x) = \frac{D_q f(t)|_{t=q^{-1}x} - D_q f(x)}{(1-q)x} = q^{-1} D_{q^{-1}} D_q f(x)$$

which shows the validity of the claimed identity.

The product rule for q derivative slightly differs from that of ordinary derivative. More precisely, for functions f and g, the authors have

$$D_q\left(f(x)g(x)\right) = f(x)D_q g(x) + g(qx)D_q f(x) \tag{13}$$

and

$$D_q^n\left(f(x)g(x)\right) = \sum_{k=0}^{n} \begin{bmatrix} n \\ k \end{bmatrix}_q D_q^k f(x) D_q^{n-k} g(t)|_{t=q^k x}. \tag{14}$$

where $\begin{bmatrix} n \\ k \end{bmatrix}_q$ is the q binomial coefficient defined by

$$\begin{bmatrix} n \\ k \end{bmatrix}_q = \frac{[n]_q!}{[k]_q![n-k]_q!}$$

The authors now give the q binomial theorem (Andrews et al., 1999, Theorem 10.2.1).

Theorem 2.9 *For* $|x|<1$, $|q|<1$,

$$\sum_{k=0}^{\infty}\frac{(a;q)_k}{(q;q)_k}x^k=\frac{(ax;q)_\infty}{(x;q)_\infty}. \tag{15}$$

Some immediate consequences of this theorem are stated below:

Corollary 2.10 (Euler) *For* $|x|<1$, $|q|<1$

$$\sum_{k=0}^{\infty}\frac{x^k}{(q;q)_k}=\frac{1}{(x;q)_\infty} \tag{16}$$

Proof: Put a=0 in (15).

Corollary 2.11: $\sum_{k=0}^{n}\begin{bmatrix} n+k-1 \\ k \end{bmatrix}_q x^k=\frac{1}{(x;q)_n}$

Proof: First, let $a=q^n$ in (15):

$$\sum_{k=0}^{\infty}\frac{(q^n;q)_k}{(q;q)_k}x_k=\frac{(q^n x;q)_\infty}{(x;q)_\infty}$$

Note that $(q;q)_{n+k-1}=(q;q)_{n-1}(q^n;q)_k$. Therefore,

$$\frac{1}{(x;q)_n}=\sum_{k=0}^{\infty}\frac{(q^n;q)_k}{(q;q)_k}x^k=\sum_{k=0}^{\infty}\frac{(q;q)_{n+k-1}}{(q;q)_k(q;q)_{n-1}}x^k=\sum_{k=0}^{n}\begin{bmatrix} n+k-1 \\ k \end{bmatrix}_q x^k$$

as stated.

Corollary 2.12:

$$\lim_{q\to 1}\frac{1}{((1-q)x;q)_\infty}=e^x. \tag{17}$$

Proof: Replace x with (1-q)x in (16) to get:

$$\frac{1}{((1-q)x;q)_\infty}=\sum_{k=0}^{\infty}\frac{(1-q)^k x^k}{(q;q)_k}$$

Taking the limit as and using (4) with the authors get

$$\lim_{q\to 1}\frac{1}{((1-q)x;q)_\infty}=\sum_{k=0}^{\infty}\frac{x^k}{(1)_k}=\sum_{k=0}^{\infty}\frac{x^k}{k!}=e^x$$

which completes the proof.

Lemma 2.13: $\sum_{k=0}^{\infty}\frac{q^k}{(q^2;q^2)_k}=(-q;q)_\infty$.

Proof: In (16), replace q with q^2 and take $x=q$:

$$\sum_{k=0}^{\infty}\frac{q^k}{(q^2;q^2)_k}=\frac{1}{(q;q^2)_\infty}=\frac{(q^2;q^2)_\infty}{(q;q^2)_\infty(q^2;q^2)_\infty}$$

Using Lemma 2.2 with this becomes

$$\sum_{k=0}^{\infty}\frac{q^k}{(q^2;q^2)_k}=\frac{(q^2;q^2)_\infty}{(q;q)_\infty}$$

Now, using Corollary 2.4 with $a=q$.yields the result.

The q integral of a function f is defined to be the q antiderivative of a function F whose q derivative is f.

Definition 2.14 (q-Integral) *The q integral of f is defined as*

$$\int_0^a f(x)d_qx=(1-q)a\sum_{k=0}^{\infty}q^k f(q^k a) \tag{18}$$

for $a\geqslant 0$..

In fact, when $f(x)0$ for $x\in[0a]$, the right side of (18) is nothing but the Riemann sum as the sum of the areas under the graph of f on the interval $[0,a]$ with the base on the interval $\left[q^{k+1}a,q^k a\right]$.and the height $f\left(q^k a\right)$.

When $a<0$ the q-integral, on the interval $[a,0]$ is defined as

$$\int_a^0 f(x)d_qx = \int_0^{-a} f(-x)d_qx$$

For $0<a<b$, the authors have

$$\int_a^b f(x)d_qx = \int_0^b f(x)d_qx - \int_0^a f(x)d_qx,$$

and for $a<0<b$, the authors have

$$\int_a^b f(x)d_qx = \int_a^0 f(x)d_qx + \int_0^b f(x)d_qx$$

For functions f and g on an interval $[a,b]$ the authors have the q integration by parts

$$\int_a^b f(x)D_q g(x)d_qx = f(x)g(x)\,|_a^b - \int_a^b g(qx)D_q f(x)d_qx$$

As the authors are going to deal with the orthogonality of polynomials, an inner product is going to be needed. The inner product of f and g on the interval $[a,b]$ with respect to a q weight function $w_q(x)$ is given as

$$\langle f,g\rangle_q = \int_a^b w_q(x)f(x)g(x)d_qx.. \qquad (19)$$

MAIN FOCUS OF THE CHAPTER

The q Hermite difference equation has the form:

$$-D_qD_{q^{-1}}y + \frac{x}{1-q}D_{q^{-1}}y + \lambda y = 0. \qquad (20)$$

It should be noted here that the general property of the q difference equation of hypergeometric type is given in the book (Nikiforov et al., 1991, page 61) and also discussed in the book (Alvarez-Nodarse, 2003). Theorems 3.1 and 3.2 below are special cases of the results given in these books.

Theorem 3.1 *All q derivatives of a solution of (20) are also solutions of an equation of the same kind. More precisely, if* $v_k = D_q^k y(x)$ *.with* $v_0=y(x)$*,.then for any* $k=0,1,\ldots$*.the function* v_k*.is a solution of*

$$-D_q D_{q^{-1}} v_k + \frac{x}{1-q} D_{q^{-1}} v_k + \mu_k v_k = 0, \tag{21}$$

where

$$\mu_k = q^k \lambda + \frac{q[k]_q}{1-q}. \tag{22}$$

Proof: Taking the q-derivative of (20), and using the product rule for the q-derivatives given in (13) with $f(x)=z/(1-q)$ and $g(x) = D_{q^{-1}} y$, the authors get

$$D_q[-D_q D_{q^{-1}} y] + \frac{x}{1-q} D_q D_{q^{-1}} y + D_q\left(\frac{x}{1-q}\right) D_{q^{-1}} y(t)\,|_{t=qx} + D_q \lambda y = 0$$

Using (11) and (12), the last equation becomes

$$-q^{-1} D_q D_{q^{-1}} D_q y(x) + \frac{q^{-1}x}{1-q} D_{q^{-1}} D_q y(x) + \frac{1}{1-q} D_q y(x) + \lambda D_q y(x) = 0$$

Multiplying this equation by q and letting $v_1=D_q y$, the authors obtain

$$-D_q D_{q^{-1}} v_1 + \frac{x}{1-q} D_{q^{-1}} v_1 + \mu_1 v_1 = 0$$

where

$$\mu_1 = q\lambda + \frac{q}{1-q}$$

Suppose now that $v_k = D_q^k y$ .is a solution of

$$-D_q D_{q^{-1}} v_k + \frac{x}{1-q} D_{q^{-1}} v_k + \mu_k v_k = 0$$

for some $k \in \mathbb{N}_0$..Taking the q derivative of both sides, and using (11) and (12) yields

$$-q^{-1} D_q D_{q^{-1}} D_q v_k + \frac{q^{-1} x}{1-q} D_{q^{-1}} D_q v_k + \frac{1}{1-q} D_q v_k(x) + \mu_k D_q v_k = 0$$

Thus, $v_{k+1} = D_q^{k+1} y = D_q v_k$ .satisfies the equation

$$-D_q D_{q^{-1}} v_{k+1} + \frac{x}{1-q} D_{q^{-1}} v_{k+1} + \mu_{k+1} v_{k+1} = 0$$

where $\mu_{k+1} = q\mu_k + q/(1-q)$. .So, by mathematical induction, $v_k(z) := D_q^k y(z)$.satisfies the equation

$$-D_q D_{q^{-1}} v_k + \frac{x}{1-q} D_{q^{-1}} v_k + \mu_k v_k = 0$$

for all k=0,1,... where $v_0(z) := y(z)$, $\mu_0 := \lambda$, and

$$\mu_k = q\mu_{k-1} + \frac{q}{1-q} = q^k \mu_0 + \frac{q^k + q^{k-1} + \cdots + q}{1-q} = q^k \lambda + \frac{q[k]_q}{1-q}$$

which completes the proof.

Theorem 3.2 *If* $\lambda = \lambda_n := -\frac{q^{1-n}[n]_q}{1-q}$, *.then the equation (20) has a polynomial solution of degree n.*

Proof: Substituting $\lambda=\lambda_n$ into (22) leads to $\mu_n=0$. Then, for k=n (21) becomes

$$-D_q D_{q^{-1}} v_n + \frac{x}{1-q} D_{q^{-1}} v_n = 0$$

Clearly, this equation has a constant solution, say $v_n=c$. Since $v_n = D_q^n y(x)$, .the authors have $D_q^n y(x) = c$. .After q integrating n times, the authors see that y is a polynomial of degree n.

Definition 3.3 *For each* $n \in \mathbb{N}_0$, *.the monic polynomial solution of q Hermite difference equation is called q Hermite I polynomial and is denoted by* $h_{n,q}(x)$ *.and it satisfies*

$$-D_q D_{q^{-1}} y + \frac{x}{1-q} D_{q^{-1}} y - \frac{q^{1-n}[n]_q}{1-q} y = 0. \tag{23}$$

Remark 3.4 *Taking (21) into account, one can see that* $v_{kn} := D_q^k h_{n,q}(x)$ *.is a polynomial solution of the equation*

$$-D_q D_{q^{-1}} v_{kn} + \frac{x}{1-q} D_{q^{-1}} v_{kn} + \mu_{kn} v_{kn} = 0 \tag{24}$$

where

$$\mu_{kn} = q^k \lambda_n + \frac{q[k]_q}{1-q} = -\frac{q^{1-n+k}[n-k]_q}{1-q} = \lambda_{n-k} \tag{25}$$

for all $n \in \mathbb{N}_0$ .and k=0,1,…,n.

Let us consider the Sturm-Liouville or formal self-adjoint form of (23)

$$D_{q^{-1}}\left[-\rho_q(x) D_q y\right] + q\rho_q(x)\lambda_n y = 0, \tag{26}$$

where $\rho_q(x)$ satisfies the so-called Pearson equation

$$D_{q^{-1}}\left[-\rho_q(x)\right] = \frac{qx}{1-q}\rho_q(x). \tag{27}$$

From (27), one obtains $\rho_q(q^{-1}x) = (1-x^2)\rho_q(x)$ or $\rho_q(x) = (1-q^2x^2)\rho_q(qx)$. Using this relation repeatedly, the authors obtain…

Taking the limit as $n\to\infty$, the authors find $\rho_q(x)=(q^2x^2;q^2)_\infty\,\rho_q(0)$. Setting $\rho_q(0):=1$ and using Corollary 2.4 the authors obtain

$$\rho_q(x) = (qx, -qx; q)_\infty, \tag{28}$$

which is called the q weight function.

ORTHOGONALITY

Orthogonality of $h_{n,q}(x)$

Lemma 4.1 *The set of q Hermite I polynomials* $\{h_{n,q}(x)\}_{n=0}^{\infty}$ *.is orthogonal on the interval (a,b) with respect to the weight function* $\rho_q(x)$*.given in (28) if*

$$\rho_q(q^{-1}x)x^s\,|_{x=a,b}=0, s=0,1,2,\ldots. \tag{29}$$

Proof: From (26) one has

$$-D_{q^{-1}}\left[\rho_q(x)D_q h_{n,q}(x)\right]+q\rho_q(x)\lambda_n h_{n,q}(x)=0 \tag{30}$$

and

$$-D_{q^{-1}}\left[\rho_q(x)D_q h_{m,q}(x)\right]+q\rho_q(x)\lambda_m h_{m,q}(x)=0.. \tag{31}$$

Multiplying (30) by $h_{m,q}(x)$ and (31) by $h_{n,q}(x)$ and taking the difference of the equations, the authors get

$$-h_{m,q}(x)D_{q^{-1}}\left(\rho_q(x)D_q h_{n,q}(x)\right)$$
$$+h_{n,q}(x)D_{q^{-1}}\left(\rho_q(x)D_{q^{-1}}h_{m,q}(x)\right)+q\left(\lambda_n-\lambda_m\right)\rho_q(x)h_{n,q}(x)h_{m,q}(x)=0.$$

Take the q integral of both sides over the interval (a,b), and use (9):

$$(\lambda_n - \lambda_m)\int_a^b \rho_q(x)h_{n,q}(x)h_{m,q}(x)d_qx = \int_a^b h_{m,q}(x)D_q\left(\rho_q(q^{-1}x)D_qh_{n,q}(t)\,|_{t=q^{-1}x})d_qx\right)$$
$$-\int_a^b h_{n,q}(x)D_q\left(\rho_q(q^{-1}x)D_qh_{m,q}(t)\,|_{t=q^{-1}x}\right)d_qx$$

Using integration by parts and taking (10) into account, the first integral on the right side becomes

$$\int_a^b h_{m,q}(x)D_q\left(\rho_q(q^{-1}x)D_qh_{n,q}(t)\,|_{t=q^{-1}x}\right)d_qx$$
$$=\left(\rho_q(q^{-1}x)h_{m,q}(x)D_{q^{-1}}h_{n,q}(x)\right)|_a^b - \int_a^b \rho_q(x)D_qh_{n,q}(x)D_qh_{m,q}(x)d_qx$$

Similarly,

$$\int_a^b h_{n,q}(x)D_q\left(\rho_q(q^{-1}x)D_qh_{m,q}(t)\,|_{t=q^{-1}x}\right)d_qx$$
$$=\left(\rho_q(q^{-1}x)h_{n,q}(x)D_{q^{-1}}h_{m,q}(x)\right)|_a^b - \int_a^b \rho_q(x)D_qh_{m,q}(x)D_qh_{n,q}(x)d_qx$$

Therefore,

$$(\lambda_n - \lambda_m)\int_a^b \rho_q(x)h_{n,q}(x)h_{m,q}(x)d_qx = \rho_q(q^{-1}x)W_{q^{-1}}\left[h_{m,q}(x),h_{n,q}(x)\right]|_a^b, \tag{32}$$

where $W_{q^{-1}}\left[h_{m,q}(x),h_{n,q}(x)\right] = h_{m,q}(x)D_{q^{-1}}h_{n,q}(x) - h_{n,q}(x)D_{q^{-1}}h_{m,q}(x)$ is a polynomial of degree $n+m$-1. Because of (29), the right side of (32) vanishes and the authors obtain

$$(\lambda_n - \lambda_m)\int_a^b \rho_q(x)h_{n,q}(x)h_{m,q}(x)d_qx = 0. \tag{33}$$

Now, $n \neq m$ if and only if $\lambda_n - \lambda_m = (q^{1-m} - q^{1-n})/(1-q)^2 \neq 0$. Therefore, (33) gives us

$$\int_a^b \rho_q(x)h_{n,q}(x)h_{m,q}(x)d_qx = 0 \text{ for all } n \neq m$$

When $n=m$, the integral in (33) becomes a positive number whose value will be specified later.

Remark 4.2 *Since* $\rho_q(q^{-1}x) = (x,-x;q)_\infty$, *it is easy to see that (29) is true if one chooses* $a=-1$, $b=1$. *Moreover, it should be pointed out that (29) provides only a sufficient condition.*

In fact, the authors have proved the following theorem:

Theorem 4.3 *The set* $\{h_{n,q}(x)\}_{n=0}^{\infty}$ *.of q Hermite I polynomials is an orthogonal set on the interval* (-1,1) *with respect to the q weight function* $\rho(x) = (qx,-qx;q)_\infty$. *More precisely, the q Hermite I polynomials* $h_{n,q}(x)$ *satisfy*

$$\int_{-1}^{1} \rho_q(x) h_{n,q}(x) h_{m,q}(x) d_q x = \mathcal{M}_n^2 \delta_{nm}, \tag{34}$$

where $\mathcal{M}_n$ .is the norm of $h_{n,q}(x)$.

Orthogonality of $D_q^k h_{n,q}(x)$

Consider the self-adjoint form of (24)

$$-D_{q^{-1}}\left[\rho_q(x) D_q v_{kn}(x)\right] + q\mu_{kn}\rho_q(x) v_{kn}(x) = 0, \tag{35}$$

in which $v_{kn}(x) = D_q^k h_{n,q}(x)$, $k=0,1,\ldots,n$ and μ_{kn} is given by (25). The same form for $v_{km}(x)$ is

$$-D_{q^{-1}}\left[\rho_q(x) D_q v_{km}(x)\right] + q\mu_{km}\rho_q(x) v_{km}(x) = 0. \tag{36}$$

As it was done before, multiply (35) by $v_{km}(x)$, (36) by $v_{kn}(x)$ and subtract the resulting equations from each other to obtain

$$\begin{aligned} &q(\mu_{kn} - \mu_{km})\rho_q(x) v_{kn}(x) v_{km}(x) \\ &= v_{km}(x) D_{q^{-1}}\left(\rho_q(x) D_q v_{kn}(x)\right) - v_{kn}(x) D_{q^{-1}}\left(\rho_q(x) D_q v_{km}(x)\right), \end{aligned}$$

or, using (9),

$$\left(\mu_{kn}-\mu_{km}\right)\rho_q\left(x\right)v_{kn}\left(x\right)v_{km}\left(x\right)$$
$$=v_{km}\left(x\right)D_q\left(\rho_q\left(q^{-1}x\right)D_qv_{kn}\left(t\right)\big|_{t=q^{-1}x}\right)-v_{kn}\left(x\right)D_q\left(\rho_q\left(q^{-1}x\right)D_qv_{km}\left(t\right)\big|_{t=q^{-1}x}\right).$$

Take the q integral of both sides over the interval (-1,1):

$$\left(\mu_{kn}-\mu_{km}\right)\int_{-1}^{1}\rho_q\left(x\right)v_{kn}\left(x\right)v_{km}\left(x\right)d_qx$$
$$=\int_{-1}^{1}v_{km}\left(x\right)D_q\left(\rho_q\left(q^{-1}x\right)D_qv_{kn}\left(t\right)\big|_{t=q^{-1}x}\right)d_qx-\int_{-1}^{1}v_{kn}\left(x\right)D_q\left(\rho_q\left(q^{-1}x\right)D_qv_{km}\left(t\right)\big|_{t=q^{-1}x}\right)d_qx.$$

Using integration by parts and taking (29) into account, the first integral on the right side becomes

$$\int_{-1}^{1}v_{km}\left(x\right)D_q\left(\rho_q\left(q^{-1}x\right)D_qv_{kn}\left(t\right)\big|_{t=q^{-1}x}\right)d_qx=-\int_{-1}^{1}\rho_q\left(x\right)D_qv_{km}\left(x\right)D_qv_{kn}\left(x\right)d_qx.$$

Similarly,

$$\int_{-1}^{1}v_{kn}\left(x\right)D_q\left(\rho_q\left(q^{-1}x\right)D_qv_{km}\left(t\right)\big|_{t=q^{-1}x}\right)d_qx=-\int_{-1}^{1}\rho_q\left(x\right)D_qv_{kn}\left(x\right)D_qv_{km}\left(x\right)d_qx.$$

Therefore,

$$\left(\mu_{kn}-\mu_{km}\right)\int_{-1}^{1}\rho_q\left(x\right)v_{kn}\left(x\right)v_{km}\left(x\right)d_qx=0. \tag{37}$$

Now,if$n\neq m$thenfrom(25)oneseesthat $\mu_{kn}-\mu_{km}=q^{1-k}\left(q^{-m}-q^{-n}\right)/(1-q)^2\neq 0$. Therefore, (37) yields

$$\int_{-1}^{1}\rho_q(x)v_{kn}(x)v_{km}(x)d_qx=0 \text{ for all } n\neq m.$$

When $n=m\geqslant k$, the left side of (37) will be a positive number, and hence, the authors have shown that the following result holds:

Theorem 4.4 *For all k=0,1,…, the set* $\left\{D_q^k h_{n,q}(x)\right\}_{n=0}^{\infty}$ *is orthogonal with respect to the inner product defined by (19) over the interval (-1,1) with the q-weight function* $\rho_q(x)$*. More precisely,*

$$\int_{-1}^{1}\rho_q(x)D_q^k h_{n,q}(x)D_q^k h_{m,q}(x)d_qx=\mathcal{M}_{kn}^2\delta_{nm} \tag{38}$$

where δ_{nm} is the Kronecker's delta and $\mathcal{M}_{kn}$ is the norm of $D_q^k h_{n,q}(x)$.

Remark 4.5 *When k>n the authors have* $D_q^k h_{n,q}(x)=0$*. Thus, it is clear from (38) that* $\mathcal{M}_{kn}=0$ *for k>n.*

Evaluation of The Norms

First, the authors note that $\mathcal{M}_n=\mathcal{M}_{0n}$ for all n=0,1,… Also, as observed before, the authors have $\mathcal{M}_{kn}=0$ for $k>n$.

Lemma 4.6 *Let* $\mathcal{M}_{kn}$ *denote the norm of* $D_q^k h_{n,q}(x)$*. That is,*

$$\mathcal{M}_{kn}^2=\int_{-1}^{1}\rho_q(x)v_{kn}^2(x)d_qx. \tag{39}$$

Then, for all k=0,1,…,n-1, the authors have

$$\mathcal{M}_{kn}^2=\frac{-1}{\mu_{kn}}\mathcal{M}_{k+1,n}^2. \tag{40}$$

Proof: Multiply the self-adjoint form (35) by $v_{kn}(x)$ and use (9):

$$\mu_{kn}\rho_q(x)v_{kn}^2(x)=v_{kn}(x)D_q\left(\rho_q\left(q^{-1}x\right)v_{k+1,n}\left(q^{-1}x\right)\right).$$

Then, q-integrate both sides over the interval (-1,1) and use the integration by parts on the right side:

$$\mu_{kn}\int_{-1}^{1}\rho_q(x)v_{kn}^2(x)d_qx=\rho_q\left(q^{-1}x\right)v_{kn}(x)v_{k+1,n}\left(q^{-1}x\right)|_{-1}^{1}-\int_{-1}^{1}\rho_q(x)v_{k+1,n}(x)D_qv_{k,n}(x)d_qx.$$

Using (29) and the fact that $D_qv_{kn}(x)=v_{k+1,n}(x)$, the authors arrive at

$$\mu_{kn}\mathcal{M}_{kn}^2=-\int_{-1}^{1}\rho_q(x)v_{k+1,n}^2(x)d_qx=-\mathcal{M}_{k+1,n}^2,$$

which completes the proof.

To continue evaluating the norms $\mathcal{M}_{kn}$ the authors shall need the following auxiliary results:

Lemma 4.7

$$\int_{-1}^{1}\rho_q(x)d_qx=(1-q)(q,-1,-q;q)_\infty.$$

Proof: First of all, note that

$$\rho_q(1)=(q;q)_\infty(-q;q)_\infty=(q;q)_k(q^{k+1};q)_\infty(-q;q)_k(-q^{k+1};q)_\infty=(q^2;q^2)_k\rho_q\left(q^k\right) \tag{41}$$

for all k=0,1,… Thus, since $\rho_q(-x)=\rho_q(x)$,

$$\int_{-1}^{1}\rho_q(x)d_qx=2\int_{0}^{1}\rho_q(x)d_qx=2(1-q)\sum_{k=0}^{\infty}q^k\rho_q\left(q^k\right).$$

Using (41) and Lemma 2.13, the above equality becomes

$$\int_{-1}^{1}\rho_q(x)d_qx = 2(1-q)\rho_q(1)\sum_{k=0}^{\infty}\frac{q^k}{(q^2;q^2)_k} = 2(1-q)\rho_q(1)(-q;q)_\infty.$$

As $2(-q;q)_\infty=(-1;q)_\infty$, the authors get

$$\int_{-1}^{1}\rho_q(x)d_qx = (1-q)\rho_q(1)(-1;q)_\infty = (1-q)(q,-1,-q;q)_\infty$$

which completes the proof.

Lemma 4.8 $\mathcal{M}_{nn} = (1-q)(q,-1,-q;q)_\infty\left([n]_q!\right)^2$ *for all n=0,1,…*

Proof: Since $v_{nn}(x) = D_q^n h_{n,q}(x)$ and $h_{n,q}(x)$ is a monic polynomial of degree n, the authors have $v_{nn}=[n]_q!$. Thus, for $k=n$, in (39), the authors get

$$\mathcal{M}_{nn}^2 = \int_{-1}^{1}\rho_q(x)v_{nn}^2d_qx = \left([n]_q!\right)^2\int_{-1}^{1}\rho_q(x)d_qx = (1-q)(q,-1,-q;q)_\infty\left([n]_q!\right)^2.$$

Lemma 4.9 *For* $n\in\mathbb{N}_0$ *and k=0,1,…,n-1, the authors have*

$$\mathcal{M}_{kn}^2 = (1-q)^{n-k+1}q^{\binom{n-k}{2}}(q,-1,-q;q)_\infty\frac{\left([n]_q!\right)^2}{[n-k]_q!}.$$

Proof: Repeated application of (40) gives us

$$\mathcal{M}_{kn}^2 = \frac{-1}{\mu_{kn}}\mathcal{M}_{k+1,n}^2 = \frac{(-1)^2}{\mu_{kn}\mu_{k+1,n}}\mathcal{M}_{k+2,n}^2 = \frac{(-1)^{n-k}}{\mu_{kn}\mu_{k+1,n}\cdots\mu_{n-1,n}}\mathcal{M}_{nn}^2.$$

Since

$$\prod_{s=k}^{n-1}\mu_{sn} = \prod_{s=k}^{n-1}\frac{-q^{1-n+s}[n-s]_q}{1-q} = \prod_{s=0}^{n-k-1}\frac{-q^{-s}[s+1]_q}{1-q} = \frac{(-1)^{n-k}[n-k]_q!}{(1-q)^{n-k}q^{\binom{n-k}{2}}},$$

using Lemma 4.8, the authors arrive at the stated result.

Thus, the authors proved the following orthogonality relation for $D_q^k h_{n,q}(x)$.

Theorem 4.10 *Let $v_{kn}(x)$ be the kth order q-derivative of $h_{n,q}(x)$ and $p=\min\{k,n\}$ Then, for all $k,m,n \in \mathbb{N}_0$, the authors have*

$$\int_{-1}^{1} \rho_q(x) v_{kn}(x) v_{km}(x) d_q x = (1-q)^{n-k+1} q^{\binom{n-k}{2}} (q,-1,-q;q)_\infty \frac{([n]_q!)^2}{[n-k]_q!} \delta_{kp}\delta_{mn}. \tag{42}$$

Corollary 4.11 *For all $m,n \in \mathbb{N}_0$, the authors have*

$$\int_{-1}^{1} \rho_q(x) h_{n,q}(x) h_{m,q}(x) d_q x = \mathcal{M}_n^2 \delta_{nm}, \tag{43}$$

where

$$\mathcal{M}_n^2 = (1-q)(q,q)_n q^{\binom{n}{2}} (q,-1,-q;q)_\infty. \tag{44}$$

Proof: Let $k=0$ in (42) and use the relation $[n]_q! = (1-q)^{-n}(q;q)_n$.

LIMIT RELATIONS

Limit Relation Between the Differential Equations

Consider the equation (23). Let $x = \sqrt{1-q^2}\,z$ and set $u(z) = y\left(\sqrt{1-q^2}\,z\right)$. Using (8) with $\alpha = \sqrt{1-q^2}$, the authors see that

$$D_{q^{-1}} y(x)\big|_{x=\sqrt{1-q^2}z} = \frac{1}{\sqrt{1-q^2}} D_{q^{-1}} y\left(\sqrt{1-q^2}\,z\right) = \frac{1}{\sqrt{1-q^2}} D_{q^{-1}} u(z),$$

and,

$$D_q D_{q^{-1}} y(x)|_{x=\sqrt{1-q^2}z} = \frac{1}{1-q^2} D_q D_{q^{-1}} y\left(\sqrt{1-q^2}z\right) = \frac{1}{1-q^2} D_q D_{q^{-1}} u(z).$$

Thus, (23) becomes

$$-\frac{1}{1-q^2} D_q D_{q^{-1}} u(z) + \frac{z}{1-q} D_{q^{-1}} u(z) - \frac{q^{1-n}}{1-q} [n]_q u(z) = 0. \tag{45}$$

Multiplying this equation by $-(1-q^2)$, gives us

$$D_q D_{q^{-1}} u(z) - (1+q) z D_{q^{-1}} u(z) + q^{1-n} (1+q)[n]_q u(z) = 0.$$

Finally, taking the limit as $q \to 1$ yields

$$u''(z) - 2zu'(z) + 2nu(z) = 0, \tag{46}$$

which is the Hermite differential equation. Since monic polynomial solutions of (45) and (46) are

$$\frac{h_{n,q}\left(\sqrt{1-q^2}z\right)}{(1-q^2)^{n/2}} \text{ and } \frac{H_n(z)}{2^n},$$

respectively, in fact, the authors have shown that the following result is true:

Theorem 5.1 *For all* $n \in \mathbb{N}_0$,

$$\lim_{q \to 1} \frac{h_{n,q}\left(\sqrt{1-q^2}z\right)}{(1-q^2)^{n/2}} = \frac{H_n(z)}{2^n}. \tag{47}$$

Limit Relation Between the Weight Functions

In the q-weight function $\rho_q(x) = (q^2x^2;q^2)_\infty$, let $x = \sqrt{1-q^2}z$:

$$\rho_q\left(\sqrt{1-q^2}\,z\right)=\left(q^2\left(1-q^2\right)z^2;q^2\right)_\infty=\frac{\left(\left(1-q^2\right)z^2;q^2\right)_\infty}{1-\left(1-q^2\right)z^2}.$$

Now, take the limit as $q\to1$ and use the relation (17) with q replaced by q^2:

$$\lim_{q\to1}\rho_q\left(\sqrt{1-q^2}\,z\right)=\lim_{q\to1}\frac{\left(\left(1-q^2\right)z^2;q^2\right)_\infty}{1-\left(1-q^2\right)z^2}=e^{-z^2}.$$

This tells us that

$$\lim_{q\to1}\rho_q\left(\sqrt{1-q^2}\,z\right)=\rho(z) \tag{48}$$

where $\rho_q(x)$ is the q-weight function for the q-Hermite I polynomials and $\rho(z)$ is the weight function for the Hermite polynomials.

Limit Relation Between the Orthogonalities

Using the substitution $x=z\sqrt{1-q^2}$ in the orthogonality relation (43), the authors get

$$\int_{-\frac{1}{\sqrt{1-q^2}}}^{\frac{1}{\sqrt{1-q^2}}}\rho_q\left(\sqrt{1-q^2}\,z\right)h_{n,q}\left(\sqrt{1-q^2}\,z\right)h_{m,q}\left(\sqrt{1-q^2}\,z\right)\sqrt{1-q^2}\,d_qx=\mathcal{M}_n^2\delta_{nm},$$

where $\mathcal{M}_n$ is the norm of $h_{n,q}(x)$ given in (44). Dividing both sides by $(1-q^2)^{\frac{n+m+1}{2}}$ gives us

$$\int_{-\frac{1}{\sqrt{1-q^2}}}^{\frac{1}{\sqrt{1-q^2}}}\rho_q\left(z\sqrt{1-q^2}\right)\frac{h_{n,q}\left(z\sqrt{1-q^2}\right)}{(1-q^2)^{\frac{n}{2}}}\frac{h_{m,q}\left(z\sqrt{1-q^2}\right)}{(1-q^2)^{\frac{m}{2}}}d_qz=\frac{\mathcal{M}_n^2}{(1-q^2)^{\frac{n+m+1}{2}}}\delta_{nm}.$$

Taking the limit as $q\to 1$, the authors obtain

$$\int_{-\infty}^{\infty} e^{-z^2} \frac{H_n(z)}{2^n}\frac{H_m(z)}{2^m}dz = \lim_{q\to 1}\frac{\mathcal{M}_n^2}{(1-q^2)^{\frac{n+m+1}{2}}}\delta_{nm}.$$

Since $\delta_{mn}=0$ for $m\neq n$, the above relation can be written as

$$\int_{-\infty}^{\infty} e^{-z^2} H_n(z)H_m(z)dz = \lim_{q\to 1}\frac{2^{2n}\mathcal{M}_n^2}{(1-q^2)^{\frac{2n+1}{2}}}\delta_{nm}. \tag{49}$$

Now,

$$\lim_{q\to 1}\frac{\mathcal{M}_n^2}{(1-q^2)^{\frac{2n+1}{2}}} = \lim_{q\to 1}\frac{(1-q)(q;q)_n(q,-1,-q;q)_\infty}{(1-q^2)^{\frac{2n+1}{2}}} = \lim_{q\to 1}\frac{(1-q)(q;q)_n(q^2;q^2)_\infty(-1,q)_\infty}{(1-q^2)^{\frac{2n+1}{2}}}$$

$$= \lim_{q\to 1}(1-q)\frac{(q;q)_n}{(1-q)^n}\frac{(q^2;q^2)_\infty(-1;q)_\infty}{(1+q)^n\sqrt{1-q^2}} = \lim_{q\to 1}(1-q)[n]_q!\frac{(q^2;q^2)_\infty(-1;q)_\infty}{(1+q)^n\sqrt{1-q^2}}.$$

From (5), one can see that

$$\Gamma_{q^2}\left(\frac{1}{2}\right) = \frac{(q^2;q^2)_\infty}{(q;q^2)_\infty}\sqrt{1-q^2}.$$

Therefore,

$$\lim_{q\to 1}\frac{\mathcal{M}_n^2}{(1-q^2)^{\frac{2n+1}{2}}} = \lim_{q\to 1}[n]_q!\frac{\Gamma_{q^2}\left(\frac{1}{2}\right)(q;q^2)_\infty(-1;q)_\infty}{(1+q)^{n+1}} = \frac{n!}{2^n}\lim_{q\to 1}\Gamma_{q^2}\left(\frac{1}{2}\right)(q;q^2)_\infty(-q;q)_\infty$$

where the authors have used the fact that $(-1;q)_\infty=2(-q;q)_\infty$. Using Lemma 2.2 and Corollary 2.4 with $a=q$, one obtains $(q;q^2)_\infty(-q;q)_\infty=1$. Also, from (6) one has

$$\lim_{q\to 1}\Gamma_{q^2}\left(\frac{1}{2}\right)=\Gamma\left(\frac{1}{2}\right)=\sqrt{\pi}.$$

Hence,

$$\lim_{q\to 1}\frac{\mathcal{M}_n^2}{(1-q^2)^{\frac{2n+1}{2}}}=\frac{n!\sqrt{\pi}}{2^n}$$

As a result, (49) gives us

$$\int_{-\infty}^{\infty} e^{-z^2} H_n(z) H_m(z)\,dz = 2^n n!\sqrt{\pi}\,\delta_{nm}$$

which is the orthogonality relation for the Hermite polynomials.

FUTURE RESEARCH DIRECTIONS

The discrete q-Hermite I polynomials constitute only one subclass of the q-polynomials. The other class of q-polynomials can also be considered in the same direction in the limiting case as $q\to 1$.

CONCLUSION

A complete study for the orthogonality of the discrete q-Hermite I polynomials has been introduced using the q-Sturm-Liuville approach starting from the second order difference equation of hypergeometric type that they satisfy. Moreover, the limit relation as $q\to 1$ has been considered.

The authors notice that the family of discrete q-Hermite I polynomials is a special case of the Al-Salam and Carlitz I polynomials with a=-1 (Koekoek et al., 2010, page 547) and the main algebraic properties, that characterize them, were considered in Table 3 of (Medem et al., 2001) as well as in the aforesaid page of the book (Koekoek et al., 2010).

ACKNOWLEDGMENT

The authors express their sincere gratitude to the referees for their valuable comments and suggestions in preparing this chapter.

REFERENCES

Alvarez-Nodarse, R. (2003). *Polinomios hipergemétricos y q-polinomios. Monografías del Seminario Matemático "García de Galdeano" Número 26*. Spain: Prensas Universitarias de Zaragoza.

Alvarez-Nodarse, R. (2006). On characterizations of classical polynomials. *Journal of Computational and Applied Mathematics, 196*(1), 320–337. doi:10.1016/j.cam.2005.06.046

Andrews, G. E., Askey, R., & Roy, R. (1999). *Special functions, Encyclopedia of Mathematics and Its Applications*. Cambridge, UK: The University Press.

Askey, R. A., Atakishiyev, N. M., & Suslov, S. K. (1993). An analog of the Fourier transformation for a q-harmonic oscillator. *Symmetries in Science, VI*, 57–63. doi:10.1007/978-1-4899-1219-0_5

Askey, R. A., & Suslov, S. K. (1993). The q-harmonic oscillator and an analogue of the Charlier polynomials. *Journal of Physics. A, Mathematical and General, 26*(15), L693–L698. doi:10.1088/0305-4470/26/15/014

Askey, R. A., & Suslov, S. K. (1993). The q-harmonic oscillator and the Al-Salam and Carlitz polynomials. *Letters in Mathematical Physics, 29*(2), 123–132. doi:10.1007/BF00749728

Atakishiyev, N. M., & Suslov, S. K. (1991). Difference analogs of the harmonic oscillator. *Theoretical and Mathematical Physics, 85*, 442–444.

Atakishiyev, N. M., & Suslov, S. K. (1991). Realization of the q-harmonic oscillator. *Theoretical and Mathematical Physics, 87*, 1055–1062.

Bannai, E. (1990). Orthogonal polynomials in coding theory and algebraic combinatorics. In Theory and Practice (pp. 25–54). NATO, ASI Series.

Berg, C., & Ismail, M. E. H. (1996). Q-Hermite polynomials and classical orthogonal Polynomials. *Can. J. Math., 48*(1), 43–63. doi:10.4153/CJM-1996-002-4

Bochner, S. (1929). Über sturm-liouvilleesche polynomsysteme. *Mathematische Zeitschrift, 29*(1), 730–736. doi:10.1007/BF01180560

Borzov, V. V., & Damaskinsky, E. V. (2006). Generalized coherent states for q-oscillator connected with q-Hermite polynomials. *Journal of Mathematical Sciences*, *132*(1), 26–36. doi:10.100710958-005-0472-9

Chihara, T. S. (1978). *An Introduction to orthogonal polynomials.* New York: Gordon and Breach.

Cryer, C. W. (1970). Rodrigues formula and the classical orthogonal polynomials. *Bol. Un. Mat. Ital.*, *25*(3), 1–11.

Fine, N. J. (1988). Basic hypergeometric series and applications. American Mathematical Society.

Hahn, W. (1935). Über die jacobischen polynome und zwei verwandte polynomklassen. *Mathematische Zeitschrift*, *39*(1), 634–638. doi:10.1007/BF01201380

Koekoek, R., Lesky, P. A., & Swarttouw, R. F. (2010). *Hypergeometric orthogonal polynomials and their q-analogues. Springer Monographs in Mathematics.* Berlin: Springer-Verlag. doi:10.1007/978-3-642-05014-5

Koornwinder, T. H. (1994). Compact quantum groups and q-special functions. Representations of lie groups and quantum groups, 46–128.

Koornwinder, T. H. (1995). Orthogonal polynomials in connection with quantum groups. Orthogonal polynomials, theory and practice, 257–292.

Macfarlane, A. J. (1989). On q-analogues of the quantum harmonic oscillator and the quantum group SU_q. *Journal of Physics. A, Mathematical and General*, *22*(21), 4581–4588. doi:10.1088/0305-4470/22/21/020

Marcellan, F., Branquinho, A., & Petronilho, J. (1994). Classical orthogonal polynomials: A functional approach. *Acta Applicandae Mathematicae*, *34*, 283–303. doi:10.1007/BF00998681

Medem, J. C., Álvarez-Nodarse, R., & Marcellán, F. (2001). On the q-polynomials: A distributional study. *Journal of Computational and Applied Mathematics*, *135*(2), 157–196. doi:10.1016/S0377-0427(00)00584-7

Nikiforov, A., Suslov, S. K., & Uvarov, V. (1991). *Classical orthogonal polynomials of a discrete variable.* Berlin: Springer-Verlag. doi:10.1007/978-3-642-74748-9

Nikiforov, A., & Uvarov, V. (1986). Classical orthogonal polynomials of a discrete variable on nonuniform lattices. *Letters in Mathematical Physics*, *11*(1), 27–34. doi:10.1007/BF00417461

Nikiforov, A., & Uvarov, V. (1988). *Special functions of mathematical physics.* Basel: Birkhäuser. doi:10.1007/978-1-4757-1595-8

Routh, R. J. (1885). On some properties of certain solutions of a differential equation of the second order. *Proceedings of the London Mathematical Society, 16*, 245–261.

Szegö, G. (1939). *Orthogonal polynomials.* American Mathematical Society Colloquium Publication Volume XXIII.

Tricomi F. (1955). Vorlesungenüber orthogonalreihen. *Grundlehren der mathematischen wissenschafen,* 76.

Vilenkin, N. J., & Klimyk, A. U. (1998). Representations of lie groups and special functions. *Bulletin of the American Mathematical Society, 35*(3), 265–270.

Chapter 8
Spectral Problem for a Polynomial Pencil of the Sturm-Liouville Equations:
On the Completeness of the System of Eigenfunctions and Associated Eigenfunction, Asymptotic Formula

Anar Adiloğlu-Nabiev
https://orcid.org/0000-0001-5602-5272
Süleyman Demirel University, Turkey

ABSTRACT

A boundary value problem for the second order differential equation -y''+∑_{m=0}$^{N-1}$λ^{m}q_{m}(x)y=λ^{2N}y with two boundary conditions a_{i1}y(0)+a_{i2}y'(0)+a_{i3}y(π)+a_{i4}y'(π)=0, i=1,2 is considered. Here n>1, λ is a complex parameter, $q_0(x),q_1(x)$,...,q_{n-1}(x) are summable complex-valued functions, a_{ik} (i=1,2; k=1,2,3,4) are arbitrary complex numbers. It is proved that the system of eigenfunctions and associated eigenfunctions is complete in the space and using elementary asymptotical metods asymptotic formulas for the eigenvalues are obtained.

DOI: 10.4018/978-1-7998-0134-4.ch008

INTRODUCTION

Consider the boundary value problem generating in the interval by the Sturm-Liouville pencil

$$-y''+\sum_{m=0}^{n-1}\lambda^m q_m(x)y=\lambda^{2n}y \tag{1}$$

and two boundary conditions

$$a_{i1}y(0)+a_{i2}y'(0)+a_{i3}y(\pi)+a_{i4}y'(\pi)=0,\ i=1,2, \tag{2}$$

where $n>1$, λ is a complex parameter, $q_0(x), q_1(x),\ldots,q_{n-1}(x)$ are summable complex-valued functions, a_{ik} (i=1,2: k=1,2,3,4) are arbitrary complex numbers. In the case $n>1$ the equation (1) is a classical Sturm-Liouville equation and the spectral problem under boundary conditions (2) was completely studied in (Marchenko, 1997) where special integral representations called transformation operators for linearly independent solutions of the Sturm-Liouville equation are applied to investigate the spectral problem. In this monograph, using some important properties of the transformation operators, especially the relations between the potential of the Sturm-Liouville equation and the kernel of the transformation operators, the boundary value problem generated in a finite interval by the Sturm-Liouville equation and by regular boundary conditions of type (2) was investigated, the completeness of the system of eigenfunctions and associated eigenfunctions was proved and asymptotical formulas as $\lambda\to\infty$ was obtained for the solutions of the Sturm-Liouville equation. These asymptotic formulas of solutions are important to obtain the asymptotical formulas for the eigenvalues. The transformation operators allow to estimate the remainder part of the asymptotic formulas for the eigenvalues in relation with smoothness of the potential.

Second order differential equation (1) or its general form

$$-y''+\sum_{m=0}^{2n-1}\lambda^m q_m(x)y=\lambda^{2n}y,\ n\in \mathrm{N} \tag{1'}$$

arises in relation with some integrable nonlinear systems of differential equations (Jaulent and Jean, 1982; Alonso, 1980). Such equations are called as a polynomial pencil of the Sturm-Liouville equation. In the paper (Alonso, 1980) authors considered full line inverse scattering problem for the equation (1') with $q_{n+k}(x)=0$, k=1,…,

n-1 and examined some spectral properties of this equation under some regularity conditions of the potential functions $q_0(x),q_1(x),\ldots,q_n(x)$. They reduced this equation to the second order pencil of the matrix Sturm-Liouville equation and gave the method of solution of the inverse scattering problem in a special case. But a vector form of solutions and strong regularity conditions on the potential functions are not suitable for applications and further investigations. The first attempt to construct special solutions of equation (1') on the reel line was considered in (Jordanov, 1984). But form of these solutions also is not of type integral transformations as Fourier or Laplace, so they are not also suitable for applications. Later in (Guseinov, 1997) the author has found a significant relation between special solutions of the equation (1) on the semiaxis and Riemann-Liouville fractional integro- differential operators. It was opened the fact that Jost type solutions of equation (1) which satisfy a condition at infinity may be represented by transformation operators similarly to the case n=1 (Marchenko, 1997), but with essentially different properties of the kernel function of the transformation operator. Namely, the kernel function of the transformation operator has not ordinary derivatives with respect to the integration variable, but it has some integrable fractional derivatives of order $\frac{1}{n}$. After work (Guseinov, 1997), the same author with his group have constructed transformation operators for the eqaution (1) in a finite interval and they investigated important properties of the solution of equation (1) widely using Riemann-Liouville fractional integrals and derivatives (Guseinov et al., 2000). Later the integral representation for the Jost solutions of equation (1') with $q_{n+k}(x)=0$, k=1,…,n-1 on the real line were constructed in (Nabiev and Guseinov, 2005) where relations between potential functions and fractional integrals of the kernel of the transformation operator have been obtained. And these results motivated the author to investigate the full line inverse scattering problem for the equation (1) without eigenvalues (Nabiev, 2006). Note that in the case n=1 equation (1') is the second order pencil called a quadratical pencil of the Sturm-Liouville equation and it is known integral representations of the special solutions of the equation (1') for this case in both infinite and finite intervals (see (Jaulent, 1972; Jaulent and Jean, 1976; Guseinov, 1985)). In this case a direct and inverse spectral problems have been widely investigated by many authors (see (Gasymov and Guseinov, 1981; Gasymow and Gasymow, 1982; Tsutsmi, 1981; Aktosun et al., 1998; Sattinger and Szmigielski, 1995; Van der Mee and Pivovarchik, 2001; Yurko, 2000; Nabiev, 2004; Maksudov and Guseinov, 1986; Kamimura, 2007) for details). Spectral analysis and inverse spectral problems of the quadratical pencil of the Sturm-Liouville equation in a finite interval with various boundary conditions can be found in (Guseinov, 1985; Gasymov and Guseinov, 1981; Gasymow and Gasymow, 1982; Nabiev, 2004). Note also direct and inverse scattering problems for the quadratic pencil of the Sturm-Liouville equation with various settings have been considered in studies (Aktosun et al., 1998; Sattinger and Szmigielski, 1995; Van der Mee and Pivovarchik, 2001; Yurko, 2000; Maksudov and Guseinov, 1986; Kamimura, 2007). It is important to investigate direct and inverse spectral problems for equation (1') without any limits on potentials. Namely, it is important to construct the Jost solutions of equation (1') similarly for the case (1) and then to investigate the relationships between potential functions and the kernel functions of the integral representations of Jost solutions. These problems are open and difficult problems as in the entire real line and halfline either in a finite interval.

In the present work some important spectral properties of the problem (1)-(2) are studied. As eigenvalues of the problem (1)-(2) we will understand those complex values of the parameter λ for which the equation (1) has a nontrivial solutions. This nontrivial solution which correspond to the eigenvalue λ of the problem (1)-(2) will be called an eigenfunction related with this eigenvalue. Investigating the boundary value problem (1)-(2) the question of multiple completeness of the system eigenfuntions and associated eigenfunctions are studied and asymptotic formulas for the series of eigenvalues from the sectors $S_m = \left\{\lambda : \frac{m\pi}{n} \leq \arg\lambda \leq \frac{(m+1)\pi}{n}\right\}$, $m = \overline{0, 2n-1}$ are obtained. The spectral problem for the equation (1) with simple conditions $y(0)=y(\pi)=0$ was studied in (Guseinov et al., 2000) where Fourier type integral representations for the linearly independent solutions of the equation (1) were constructed and applied to investigate this spectral problem. As it is known (Guseinov et al., 2000) for each $\lambda \in S_m$ equation (1) has two linearly independent solutions

$$y_j(x,\lambda) = e^{(-1)^{j+1} i\lambda^n x}\left(1 + \int_{\left[(-1)^{j+m}-1\right]\frac{x}{2}}^{+\infty} K_{v,m}(x,t) e^{(-1)^m 2i\lambda^n t} dt\right) \tag{3}$$

where $K_{v,m}(x,t)$ $\left(v = j + \frac{1}{2}\left[(-1)^{j+m} - (-1)^j\right]\right)$ is a summable function with respect to t in the interval $\left[(-1)^{j+m} - (-1)^j\right]\frac{x}{2}$ for each $x \in [0,\pi]$. Moreover,

$$\int_{\left[(-1)^{j+m}-1\right]\frac{x}{2}}^{+\infty} \left|K_{v,m}(x,t)\right| dt \leq e^{\sigma(x)} - 1 \tag{4}$$

where $\sigma(x) = \sum_{k=0}^{n-1} \frac{2^{1-\frac{k}{n}}}{\Gamma(1-\frac{k}{n})} \int_0^x (x-s)^{1-\frac{k}{n}} \left|q_k(s)\right| ds$ and $\Gamma(.)$is the Euler gamma function. Morover, the kernel functions $K_{j,m}(x,.)$ $(j=1,2)$ have summable derivatives with respect to x. Namely, if $q_0(x) \in C[0,a]$, $q_k(x) \in C^1[0,a]$ $k = \overline{1, n-1}$ then $\frac{\partial}{\partial x} K_{1,m}(x,.) \in L_1(-x,+\infty)$ and $\frac{\partial}{\partial x} K_{2,m}(x,.) \in L_1(0,+\infty)$ for each $x \in [0,a]$. The next important property of the kernel functions is the fact that they have not derivatives with recpect to the variable t. But they have some Riemann-Liouville fractional derivatives (Samko et al., 1997).

Definition 1.1. The Riemann-Liouville fractional integral of order a $(0<a<1)$ with respect to the variable t of a function $f(x,.)\in L_1(a,+\infty)$ is the integral

$$I^a_{a,t}f(x,t)=\frac{1}{\Gamma(a)}\int_a^t (t-s)^{a-1}f(x,s)ds.$$

The Riemann-Liouville fractional derivative of order a $(0<a<1)$ with respect to the variable t of a function $f(x,.)\in L_1(\alpha,+\infty)$ is defined as

$$D^a_{a,t}f(x,t)=\frac{\partial}{\partial t}\frac{1}{\Gamma(1-a)}\int_a^t (t-s)^{-a}f(x,s)ds.$$

It was proved in (Guseinov et al., 2000) that if $q_0(x)\in C[0,a]$, $q_k(x)\in C^1[0,a]$ $k=\overline{1,n-1}$ then the fractional derivatives $(D^{\frac{1}{n}}_{-x,t})^p$ $K_{1,m}(x,t)$ and $(D^{\frac{1}{n}}_{0,t})^p$ $K_{2,m}(x,t)$, where $p=1,2,\ldots,n$, are summable with respect to on the intervals $(-x,+\infty)$ and $(0,+\infty)$ respectively. For more details related with other properties of the kernel functions we refer to (Guseinov et al., 2000) Further using the integral representations of the system of fundamental solutions of the equation (1) in (Guseinov et al., 2000) it was shown that the spectral problem generated by the equation (1) and simple Drichlet boundary conditions $y(0)=y(\pi)=0$ has an infinite number of eigenvalues and all these eigenvalues lie in the sectors $T^m_\varepsilon=\left\{\lambda:\left|\arg\lambda-\frac{m\pi}{n}\right|<\frac{\varepsilon}{n}\right\}$, where $\varepsilon>0$ is sufficiently small and $m=\overline{0,2n-1}$. Moreover, the eigenvalues consist of $2n$ series and m^{th} series of them can be arranged as $\lambda_{1,m}$, $\lambda_{2,m},\ldots,\lambda_{s,m},\ldots;\left|\lambda_{s,m}\right|<\left|\lambda_{s+1,m}\right|$ and they can be defined asymptotically by the formula

$$\lambda_{k,m}=e^{\frac{im\pi}{n}}\sqrt[n]{k}+\sum_{j=1}^{n}\frac{d_j^{(m)}}{k^{1+\frac{j-1}{n}}}+o\left(\frac{1}{k^{1+\frac{n-1}{n}}}\right),\ k\to+\infty,\ m=0,1,\ldots,2n-1,$$

where $d_j^{(m)}$ are constants not depending on k.

Considering the integral representations (3) of the solutions

$$y_j(x,\lambda)=e^{(-1)^{j+1}i\lambda^n x}+o(1),\ |\lambda|\to+\infty$$

one can define solutions $s(x,\lambda)$ and $c(x,\lambda)$ of the equation (1) with initial conditions $s(0,\lambda)=c'(0,\lambda)=0$ and $s'(0,\lambda)=c(0,\lambda)=1$. Namely, we have

$$s(x,\lambda)=\frac{y_1(x,\lambda)-y_2(x,\lambda)}{2i\lambda^n},\ c(x,\lambda)=\frac{y_1(x,\lambda)+(x,\lambda)}{2}$$

which are also two linearly independent solutions of the equation (1). In (Agamaliyev and Nabiyev, 2005) other type special boundary conditions were considered for the equation (1) and asymptotic formulas were obtained for the eigenvalues of the considered problems. Note that in (Nabiev and Guseinov, 2005) it is also given more detailed asymptotic formulas for the solutions $s(x,\lambda)$ $c(x,\lambda)$, and their derivatives. We will use above results and other results of (Guseinov et al., 2000) and (Agamaliyev and Nabiyev, 2005) in the investigation of the spectral problem (1)-(2).

ON THE COMPLETENESS OF THE SYSTEM OF EIGENFUNCTIONS AND ASSOCIATED EIGENFUNCTIONS

Eigenvalues of the sspectral problem (1)-(2) coincide with the roots of the characteristic function

$$\Delta(\lambda)=d_{12}+d_{34}+d_{13}s(\pi,\lambda)+d_{14}s'(\pi,\lambda)+d_{32}c(\pi,\lambda)+d_{42}c'(\pi,\lambda), \tag{5}$$

where $d_{ij}=a_{1i}a_{2j}-a_{2i}a_{1j}$ $(i,j=1,2)$. Further, let us consider the solutions $\omega_i(x,\lambda)$ $(i=1,2)$ of the equation (1) defined as

$$\omega_i(x,\lambda)=\left(a_{i2}+a_{i3}s(\pi,\lambda)+a_{14}s'(\pi,\lambda)\right)c(x,\lambda)-$$
$$-\left(a_{i1}+a_{i3}c(\pi,\lambda)+a_{14}c'(\pi,\lambda)\right)s(x,\lambda).$$

It is easy to check out that if the left hand-side of the boundary conditions (2) denote by $U_i(y)$ then $U_1(\omega_1)=U_2(\omega_2)=0$, $U_1(\omega_2)=-U_2(\omega_1)=\Delta(\lambda)$.

Definition 2.1. An eigenvalue of the boundary value problem (1)-(2) is called an eigenvalue with multiplicity p if it is a root of multiplicity p of the characteristic equation $\Delta(\lambda)=0$.

It is easy to show that if $\lambda=\mu$ is an eigenvalue with multiplicity p then the function $\omega_{i,k}(x)=\frac{(-1)^k}{k!}\frac{\partial^k}{\partial\lambda^k}\omega_i(x,\mu)$ $(0\le k\le p-1)$ satisfies the boundary conditions (2). The first nonzero function of the chain $\omega_{i,k}(x)$ (i=1,2; k=0,1,…,p-1) is the eigenfunction and the others are associated eigenfunction. Let know denote $\lambda^K y(x,\lambda)=z_k(x,\lambda)$, $k=\overline{0,2n-1}$, where $y(x,\lambda)$ is a solution of the problem (1)-(2). It is easy to obtain that the boundary value problem (1)-(2) is equivalent to the following spectral problem:

$$L(\frac{d}{dx})z=\lambda z,\ U_1(z)=0,\ U_2(z)=0, \tag{6}$$

where $z=(z_1,z_2,...,z_{2n-1})^T$, $L(\frac{d}{dx})$ is the square matrix with elements $L_{ij}(\frac{d}{dx})$, $i,j=\overline{0,2n-1}$ such that

$$L_{ij}(\frac{d}{dx})=0(\text{if } i=\overline{0,2n-2},\ j\neq i+1 \text{ and if } i=2n-1,\ j=\overline{n,2n-1}),$$

$$L_{i,i+1}(\frac{d}{dx})=1, i=\overline{0,2n-2},\ L_{2n-1,0}(\frac{d}{dx})=-\frac{d^2}{dx^2}+q_0(x),$$

$$L_{2n-1,j}(\frac{d}{dx})=q_j(x),\ j=\overline{1,n-1}.$$

Lemma 2.2. The set of eigenvalues of the problems (1)-(2) and (5)-(6) are coincide. If $y_0(x)$ is the eigenfunction of the problem (1)-(2) corresponding to an eigenvalue λ_0 then $z^0(x)=(y_0(x),\lambda y_0(x),...,\lambda^{2n-1}y_0(x))^T$ is the eigenvector of the problem (5)-(6) corresponding to the same eigenvalue. Conversely, if $z^0(x)=(z_0^0(x),z_1^0(x),...,z_{2n-1}^0(x))^T$ is an eigenvector of the problem (5)-(6) related to some eigenvalue, then $y_0(x)=z_0^0(x)$ is an eigenfunction of the problem (1)-(2) with the same eigenvalue.

From the above lemma it is obvious that if is an eigenvalue, the vector functions

$$W_1(x,\lambda)=\left(\omega_1(x,\lambda),\lambda\omega_1(x,\lambda),...,\lambda^{2n-1}\omega_1(x,\lambda)\right)^T,$$

$$W_2(x,\lambda)=\left(\omega_2(x,\lambda),\lambda\omega_2(x,\lambda),...,\lambda^{2n-1}\omega_2(x,\lambda)\right)^T$$

satisfy the equation

$$L(\frac{d}{dx})=W_i=\lambda W_i(i=1,2) \tag{7}$$

and the boundary conditions

$$U_1(W_i) = U_2(W_i) = 0,\ i=1,2 \tag{8}$$

for $\lambda=\lambda_m$. Hence we have that if $\lambda=\lambda_m$ is an eigenvalue of multiplicity p for the boundary value problem (1)-(2) then, this is also an eigenvalue with the same multiplicity for the problem (5)-(6). In this case the vector functions

$$W_{i,k}(x) = \frac{(-1)^k}{k!}\frac{\partial^k}{\partial\lambda^k}W_i(x,\lambda)\ (0 \le k \le p-1, \lambda = \lambda_m)$$

are the chain the first nonzero vector of which is an eigenvector and the others are associated eigenvectors of the problem (5)-(6).

Definition 2.3. If the system of eigenvectors and associated eigenvectors $\left\{Z^{(k,m)}(x)\right\}$ corresponding to all eigenvalues $\left\{\lambda_{k,m}\right\}_{m=\overline{0,2n-1}}$ of the problem (5)-(6) is complete in the space $L_2^{2n}(0,\pi)$ then the system of eigenfunctions and associated eigenfunctions $\left\{y_k^m(x)\right\} = \left\{z_{k,m}^0(x)\right\}$ of the problem (1)-(2) will be called $2n$ -fold complete in the space $L_2(0,\pi)$.

Here $L_2^{2n}(0,\pi)$ is the $2n$-dimensional vector space with vectors whose coordinates belong to $L_2(0,\pi)$.

Let S is the spectrum, i.e. the set of all eigenvalues $\left\{\lambda_{k,m}\right\}_{m=\overline{0,2n-1}}$ of the boundary value problem (1)-(2) and $p_k^{(m)}$ is the multiplicity of the eigenvalue $\lambda_{k,m}$. According to the previous discussions the vector functions $\frac{(-1)^s}{s!}\frac{\partial^s}{\partial\lambda^s}W_i(x,\lambda)\,|_{\lambda=\lambda_{k,m}}$ $(0 \le s \le p_k^{(m)} - 1$, $i=1,2$, $\lambda_{k,m} \in S)$ are equal to zero or are eigenvectors or associated eigenvectors of the problem (5)-(6). Therefore, for the completeness of the system of eigenvectors and associated eigenvectors it is enough to prove the fact that if $F(x) \in L_2^{2n}(0,\pi)$ and

$$\left\langle \frac{\partial^s}{\partial\lambda^s}W_i(x,\lambda), F(x)\right\rangle|_{\lambda=\lambda_{k,m}} = 0 \tag{9}$$

for all $\lambda_{k,m} \in S$, $0 \le s \le p_k^{(m)} - 1$, $i=1,2$ then $F(x)=0$ almost everywhere in $L_2^{2n}(0,\pi)$. Here, $\langle h(x), g(x)\rangle$ is a scaler production in the space $L_2^{2n}(0,\pi)$ and it means

$$\langle h(x), g(x)\rangle = \int_0^\pi \sum_{j=0}^{2n-1} h_j(x)\overline{g_j(x)}dx,\ h(x), g(x) \in L_2^{2n}(0,\pi). \tag{10}$$

In other hand the function $W_i(F,\lambda)=\langle W_i(x,\lambda),F(x)\rangle$ is an entire function of the parameter λ. Therefore if (9) is valid then every $p_k^{(m)}$ – fold root $\lambda_{k,m}(m=\overline{0,2n-1})$ of the characteristic function Δ(λ) is also a root (at least with the same multiplicity) of the functions $W_i(F,\lambda)(i=1,2)$. Consequently we have that (9) is satisfied if and only if both the functions $W_i(F,\lambda)[\Delta(\lambda)]^{-1}(i=1,2)$ are entire. Hence, to prove the completeness of the system of eigenvectors and associated eigenvectors it is enough to show that the functions $W_i(F,\lambda)[\Delta(\lambda)]^{-1}(i=1,2)$ are entire if and only if $F(x)=0$ almost everywhere in $L_2^{2n}(0,\pi)$.

Using integral representations, (3) as it is shown in (Guseinov et al., 2000; Agamaliyev and Nabiyev, 2005), we obtain the following asimptotic formulas for the solutions and:

$$s(x,\lambda)=\frac{\sin\lambda^n x}{\lambda^n}+O(\frac{e^{|\mathrm{Im}\,\lambda^n x|}}{|\lambda|^{n+1}}),|\lambda|\to+\infty,$$

$$s'(x,\lambda)=\cos\lambda^n x+O(\frac{e^{|\mathrm{Im}\,\lambda^n x|}}{|\lambda|}),|\lambda|\to+\infty,$$

$$c(x,\lambda)=\cos\lambda^n x+O(\frac{e^{|\mathrm{Im}\,\lambda^n x|}}{|\lambda|}),|\lambda|\to+\infty,$$

$$c'(x,\lambda)=-\lambda^n\sin\lambda^n x+O(|\lambda|^{n-1}e^{|\mathrm{Im}\,\lambda^n x|}),|\lambda|\to+\infty.$$

Using the above asymptotic formulas it is easy to prove the following statement.

Lemma 2.4. For all functions $f(x)\in L_1(0,\pi)$ we have

$$\lim_{|\lambda|\to+\infty} e^{-|\mathrm{Im}\,\lambda^n\pi|}\int_0^\pi f(x)\omega_i(x,\lambda)dx)=0.$$

The last lemma yields that if $F(x)\in L_1^{2n}(0,\pi)$, where $L_1^{2n}(0,\pi)$ is the $2n$-dimensional vector space with vectors whose coordinates belong to $L_1(0,\pi)$ then

$$\lim_{|\lambda|\to+\infty} e^{-|\mathrm{Im}\,\lambda^n\pi|}W_i(F,\lambda). \tag{11}$$

Examining the characteristic function Δ(λ) we have

$$\Delta(\lambda)=\begin{cases} -d_{42}\lambda^n \sin\lambda^n\pi + o(\lambda^n e^{\left|\operatorname{Im}\lambda^n\pi\right|}), \text{ if } d_{42}\neq 0 \\ h_1\cos\lambda^n\pi + h_2 + o(e^{\left|\operatorname{Im}\lambda^n\pi\right|}), \text{ if } d_{42}=0, h_1\neq 0 \\ d_{13}\lambda^{-n}\sin\lambda^n\pi + h_2 + o(\lambda^{-n}e^{\left|\operatorname{Im}\lambda^n\pi\right|}), \text{ if } h_1=d_{42}=0, d_{13}\neq 0 \end{cases} \tag{12}$$

where $h_1=d_{14}+d_{32}$, $h_2=d_{12}+d_{34}$.

From the formula (12) and Lemma 1.3.2 in (Marchenko, 1997) we have that there exist a sequence unboundely extending countour K_m on which the inequality

$$\left|\Delta(\lambda)\right| > C\left|\lambda\right|^{-n} e^{\left|\operatorname{Im}\lambda^n\pi\right|} \tag{13}$$

is hold for some constant $C>0$. Now (11) and (13) imply

$$\lim_{m\to\infty}\max_{\lambda\in K_m}\left|\frac{W_i(F,\lambda)}{\lambda^n\Delta(\lambda)}\right| = 0, i=1,2$$

which allows us to confirm that if $W_i(F,\lambda)[\Delta(\lambda)]^{-1}(i=1,2)$ is an entire function then it is growing slower than $|\lambda|^n$ as $|\lambda|\to\infty$. Then

$$\int_0^\pi \omega_i(x,\lambda)\sum_{j=0}^{2n-1}\lambda^j\overline{f_j(x)}dx = \Delta(\lambda)\sum_{j=0}^{n-1}a_j^{(i)}\lambda^j$$

$$W_i(F,\lambda)=\Delta(\lambda)\sum_{j=0}^{n-1}a_j^{(i)}\lambda^j$$

with some constants $a_j^{(i)}$. If we remember here the definition of the function $W_i(F,\lambda)$ we can write from which it is easy to obtain

$$\sum_{j=0}^{2n-1}\lambda^j s(f_j,\lambda)dx = Q_{n-1}^{(1)}(\lambda)\Delta_{22}(\lambda) - Q_{n-1}^{(2)}(\lambda)\Delta_{12}(\lambda)$$

where

$$\Delta_{i2}(\lambda)=a_{i2}+a_{i3}s(\pi,\lambda)+a_{i4}s'(\pi,\lambda),\ Q_{n-1}^{(i)}(\lambda)=\sum_{j=0}^{n-1}a_j^{(i)}\lambda^j\,(i=1,2).$$

Therefore,

$$\sum_{j=0}^{2n-1} \lambda^j s(f_j,\lambda)dx = Q_{n-1}^{(1)}(\lambda)a_{22} - Q_{n-1}^{(2)}(\lambda)a_{12}$$
$$+\left(Q_{n-1}^{(1)}(\lambda)a_{23} - Q_{n-1}^{(2)}(\lambda)a_{13}\right)s(\pi,\lambda) + \left(Q_{n-1}^{(1)}(\lambda)a_{24} - Q_{n-1}^{(2)}(\lambda)a_{14}\right)s'(\pi,\lambda). \quad (14)$$

Now using expressions for the functions $s(x,\lambda)$ and $s'(x,\lambda)$ from (Agamaliyev and Nabiyev, 2005) it is easy to derive that

$$s(x,\lambda) = \lambda^{-n}\sin\lambda^n x + \frac{\delta_1(x,\lambda)}{\lambda}e^{i\lambda^n x} + \frac{\delta_2(x,\lambda)}{\lambda}e^{-i\lambda^n x} + \frac{\delta_3(x,\lambda)}{\lambda^n}, \quad (15)$$

$$s'(x,\lambda) = \cos\lambda^n x + i\frac{\delta_1(x,\lambda)}{\lambda}e^{i\lambda^n x} - i\frac{\delta_2(x,\lambda)}{\lambda}e^{-i\lambda^n x} + \frac{\delta_4(x,\lambda)}{\lambda^n}, \quad (16)$$

where $\lim_{|\lambda|\to+\infty}\max_{0\le x\le\pi}\delta_i(x,\lambda)e^{-|\operatorname{Im}\lambda^n\pi|} = 0$ and $\delta_i(x,\lambda)$, $i=1,2,3,4$ is integrable over $[0,\pi]$ for each value of the parameter λ. From the expression of $s(x,\lambda)$ we immediately have that if $\operatorname{Im}\lambda^n=0$ then $s(f_j,\lambda) = \lambda^{-n}\delta_j(\lambda)$ with $\delta_j(\lambda\to 0)$ as $|\lambda|\to+\infty$. Consequently, from the equation (14) we have

$$\sum_{j=0}^{2n-1}\lambda^j\delta_j(\lambda) = \sum_{j=0}^{n-1}\left(a_j^{(1)}a_{22} - a_j^{(2)}a_{12}\right)\lambda^{n+j}$$
$$+\sum_{j=0}^{n-1}\left(a_j^{(1)}a_{23} - a_j^{(2)}a_{13}\right)\lambda^{n+j}s(\pi,\lambda) + \sum_{j=0}^{n-1}\left(a_j^{(1)}a_{24} - a_j^{(2)}a_{14}\right)\lambda^{n+j}s'(\pi,\lambda).$$

Now taking into our account equations (15) and (16) again for the entire functions $s(\pi,\lambda)$ and $s'(\pi,\lambda)$ we obtain

$$a_j^{(1)}a_{22} - a_j^{(2)}a_{12} = a_{23} - a_j^{(2)}a_{13} = a_{24} - a_j^{(2)}a_{14} = 0,\ j = 0,1,...,n-1.$$

Hence it is obtained that

$$s(f_j,\lambda) = 0,\ j=0,1,\ldots,2n\text{-}1. \quad (17)$$

Using the integral expressions of the solution $s(x,\lambda)$ and the definition of $s(f_j,\lambda)$ we find that equalities (17) are equivalent to

$$\int_0^\pi f_j(t)e^{i\lambda^n t}dt-\int_{-\pi}^0 f_j(-t)e^{i\lambda^n t}dt+\int_{-\pi}^{+\infty}e^{i\lambda^n t}dt\int_{-t}^{+\infty}f_j(x)K(x,t)dx=0,$$

where $K(x,.)$ is summable function on the interval $(-\pi,+\infty)$ for each $x\in[0,\pi]$. From the last equality we find the equations

$$\begin{cases} -f_j(-t)+\int_{-t}^{+\infty}f_j(x)K(x,t)dx=0, \text{ if } -\pi<t<0, \\ f_j(t)+\int_{-t}^{+\infty}f_j(x)K(x,t)dx=0, \text{ if } 0<t<\pi, \\ \int_{-t}^{+\infty}f_j(x)K(x,t)dx, \text{ if } t>\pi \end{cases}$$

which imply that $F(x)=0$ almost everywhere in $L_2^{2n}(0,\pi)$. Hence we have proved the following theorem.

Theorem 2.5. The system of eigenfunctions and associated eigenfunctions of the boundary value problem (1)-(2) is $2n$-fold complete in the space $L_2(0,\pi)$.

ASYMPTOTIC FORMULAS FOR THE EIGENVALUES OF THE BOUNDARY VALUE PROBLEM

Consider the characteristic equation

$\Delta(\lambda)=0,\ \lambda\in S_m,$ (18)

where

$$\Delta(\lambda)=d_{12}+d_{34}s(\pi,\lambda)+d_{14}s'(\pi,\lambda)+d_{32}c(\pi,\lambda)+d_{42}c'(\pi,\lambda),$$

which is examining for the cases shown in the formula (12). We will consider these cases separately.

Case 1. Let $d_{42}\neq 0$. In this case

$$\Delta(\lambda)=-d_{42}\lambda^n\sin\lambda^n\pi+\lambda^n e^{|\mathrm{Im}\,\lambda^n\pi|}\varepsilon_1(\lambda),\ \lim_{|\lambda\to+\infty|}\varepsilon_1(\lambda)=0$$

and as an unboundly extending contours we can choose the circles

$$D_k = \left\{ \lambda \ : \ |\lambda| = \sqrt[n]{k + \frac{1}{2}} \right\}, \ k = 0,1,2,...$$

on which the inequality (13) is satisfied. From here we conclude that there exists the constant $C>0$ such that

$$\left|\Delta(\lambda) + d_{42}\lambda^n \sin \lambda^n \pi\right| \le C|\lambda|^{n-1} e^{\left|\operatorname{Im} \lambda^n \pi\right|} \tag{19}$$

is satisfied for sufficiently large $|\lambda|$. In particular, the inequality (19) is satisfied on D_k for sufficiently large k.

Fix a real number $\varepsilon>0$ $(0 < \varepsilon < \frac{\pi}{2})$ and consider the angular domains

$$S_{\varepsilon}^{(m)} = \left\{ \lambda : \frac{m\pi}{n} + \frac{\varepsilon}{n} \le \arg \lambda \le \frac{(m+1)\pi}{n} - \frac{\varepsilon}{n} \right\}, \ m = \overline{0, 2n-1}\,.$$

When $\lambda \in S_{\varepsilon}^{(m)}$ and λ has sufficiently large module the inequality

$$\left|\lambda^n \sin \lambda^n \pi\right| > \frac{|\lambda|^n}{4} e^{\left|\operatorname{Im} \lambda^n \pi\right|} \tag{20}$$

easily can be verified. Therefore from the estimates (19) and (20) we conclude that

$$\left|\Delta(\lambda) + d_{42}\lambda^n \sin \lambda^n \pi\right| < \left|d_{42}\lambda^n \sin \lambda^n \pi\right| \tag{21}$$

in the domain $S_{\varepsilon}^{(m)} \cap \{\lambda \ : \ |\lambda| > \mathbb{R}\}$, where $\mathbb{R} > 0$ is sufficiently large number. Now the inequality (21) implies that there is not any zero of the function $\Delta(\lambda)$ in the domain $S_{\varepsilon}^{(m)} \cap \{\lambda \ : \ |\lambda| > \mathbb{R}\}$. Hence, by the uniqueness theorem of analytical functions there may be at most a finite number of zeros of the function $\Delta(\lambda)$ in the domain $S_{\varepsilon}^{(m)}$. Consequently, we can claim that for sufficiently large $|\lambda|$, where $\lambda \in S_{\varepsilon}^{(m)}$, the zeros of the function $\Delta(\lambda)$ may be concentrated near by the rays $\arg \lambda = \frac{m\pi}{n}$ and $\arg \lambda = \frac{(m+1)\pi}{n}$, $m = \overline{0, 2n-1}$. Moreover, since for sufficiently large k we have

$$\left|\lambda^n \sin \lambda^n \pi\right| > \frac{|\lambda|^n}{4} e^{-\pi} e^{\left|\operatorname{Im} \lambda^n \pi\right|} \tag{22}$$

on the arcs $D_k \cap (S_m \setminus S_\varepsilon^{(m)})$ then the inequality (21) also takes place on the circles D_k for large values of the integer k. Applying the Rouche's theorem we obtain that the number of zeros in D_k of the function $\Delta(\lambda)$ is equal to the number of zeros of the function $\lambda^n \sin\lambda^n\pi$. Since the function $\lambda^n \sin\lambda^n\pi$ has $2n$ series of zeros inside the circle D_k then the zeros of $\Delta(\lambda)$ consists of $2n$ series with $k+1$ zeros in each series and $m^{th} (m = \overline{0, 2n-1})$ seriés is concentrated near by the ray $\arg \lambda = \frac{m\pi}{n}$. By the argument similar previous we can show that the function $\Delta(\lambda)$ has exactly $2n+1$ zeros inside the ring $\sqrt[n]{k-\frac{1}{2}} \le |\lambda| \le \sqrt[n]{k+\frac{1}{2}}$ and each zero belongs to the different series. Consequently, we can arrange the eigenvalues of the problem (1)-(2) in sequel by the series:

$$\lambda_{1,m}, \lambda_{2,m}, \ldots, \lambda_{s,m}, \ldots; \left|\lambda_{s,m}\right| < \left|\lambda_{s+1,m}\right|.$$

Now let us solve asymptotically the eigenvalue equation $\Delta(\lambda)=0$ in the sector $T_\varepsilon^{(m)} = \left\{\lambda : \left|\arg \lambda = \frac{m\pi}{n}\right| < \frac{\varepsilon}{n}\right\}$. Using asymptotic expressions of the functions $s(x,\lambda)$, $s'(x,\lambda)$, $c(x,\lambda)$, $c'(x,\lambda)$ (Agamaliyev and Nabiyev, 2005) we can reduce the equation $\Delta(\lambda)=0$ in the following asymptotical form:

$$e^{2i\lambda^n \pi} = \frac{1 + \sum_{k=0}^{n-1} \frac{\gamma_{k+1}}{\lambda^{k+1}} a_k^{(2)} + \lambda^{-n} d_{42}^{-1} i (d_{31} + d_{14}) + o(\lambda^{-n})}{1 - \sum_{k=0}^{n-1} \frac{\gamma_{k+1}}{\lambda^{k+1}} a_k^{(1)}}, \ \lambda \in S_{2\nu} \tag{23}$$

and

$$e^{-2i\lambda^n \pi} = \frac{1 - \sum_{k=0}^{n-1} \frac{\gamma_{k+1}}{\lambda^{k+1}} a_k^{(1)} - \lambda^{-n} d_{42}^{-1} i (d_{31} + d_{14}) + o(\lambda^{-n})}{1 + \sum_{k=0}^{n-1} \frac{\gamma_{k+1}}{\lambda^{k+1}} a_k^{(1)}}, \ \lambda \in S_{2\nu+1}, \tag{24}$$

where $\gamma_k = 2^{-\frac{k}{n}} e^{\frac{i\pi k}{2n}}$, $a_k^{(i)} = a_k^{(i)}(\pi) (i = 1, 2)$ are constant related with integrals of the functions $q_0(x)$, $q_1(x)$, …,$q_{n-1}(x)$:

$$a_0^{(i)}(x)=\gamma_{n-1}\int_0^x q_{n-1}(s)ds,\ a_k^{(i)}(x)$$
$$=\gamma_{n-k-1}\int_0^x q_{n-k-1}(s)ds+(-1)^i\sum_{p=1}^{k}\gamma_{n-p}\int_0^x q_{n-p}(s)a_{n-p}^{(i)}(s)ds,$$
$$k=\overline{1,n-1}.$$

Finally, solving asimptoticaly the equations (23) and (24) we obtain the following asymptotic formulas for the series of eigenvalues of the boundary value problem(1)-(2):

$$\lambda_{k,m}=e^{\frac{im\pi}{n}}\sqrt[n]{k}+\sum_{j=1}^{n}\frac{a_j^{(m)}}{k^{1+\frac{j-1}{n}}}+o(\frac{1}{k^{1+\frac{n-1}{n}}}),\ k\to+\infty,\ m=0,1,...,2n-1. \tag{25}$$

Here $a_j^{(m)}$ are some constants not depending on k.

Case 2. Let $d_{42}=0$, $d_{32}+d_{14}\neq 0$. In this case the equation $\Delta(\lambda)=0$ can be written in the asymptotical form

$$\Delta(\lambda)=h_1\cos\lambda^n\pi+h_2+e^{\left|\operatorname{Im}\lambda^n\pi\right|}\varepsilon_2(\lambda),\ \lim_{|\lambda\to+\infty|}\varepsilon_2(\lambda)=0,$$

where $h_1=d_{14}+d_{32}$, $h_2=d_{12}+d_{34}$.

As an unboundly extending contours we can choose the circles

$$D_{k1}=\left\{\lambda\ :\ |\lambda|=\sqrt[n]{k}\right\},\ k=0,1,2,...\text{ if }\theta=\pi^{-1}\arccos(-\tfrac{h_2}{h_1})\neq 0,$$
$$D_{k2}=\left\{\lambda\ :\ |\lambda|=\sqrt[n]{2k+1}\right\},\ k=0,1,2,...\text{ if }\theta=0,$$
$$D_{k2}=\left\{\lambda\ :\ |\lambda|=\sqrt[n]{2k}\right\},\ k=0,1,2,...\text{ if }\theta=1$$

on which the inequality (13) is satisfied. By the similar arguments as in the Case 1 using The Rouche's theorem we can show that there are infinitely many eigenvalues of the problem (1)-(2) and sufficiently large in module eigenvalues are replaced near by the rays $\arg\lambda=\frac{m\pi}{n}\ (m=\overline{0,2n-1})$. Reducing the eigenvalue equation $\Delta(\lambda)=0$ to the form

$$\cos\lambda^n\pi-\frac{1}{2h_1}e^{i\lambda^n\pi}\sum_{k=0}^{n-1}\frac{\gamma_{k+1}}{\lambda^{k+1}}a_k^{(1)}+\frac{1}{2h_1}e^{-i\lambda^n\pi}\sum_{k=0}^{n-1}\frac{\gamma_{k+1}}{\lambda^{k+1}}a_k^{(2)}+\frac{h_2}{h_1}\frac{\sin\lambda^n\pi}{\lambda^n}o(\lambda^{-n})e^{\left|\operatorname{Im}\lambda^n\pi\right|}=0 \tag{26}$$

as in the Case1 we solve the equation (26) asymptotically and get the following asymptotical formulas for the eigenvalue series $\lambda_{k,m}$:

$$\lambda_{k,m} = e^{\frac{im\pi}{n}} \sqrt[n]{2k} + \theta_{\pm} + \sum_{j=1}^{n} \frac{b_j^{(m)}}{k^{1+\frac{j-1}{n}}} + o(\frac{1}{k^{1+\frac{n-1}{n}}}),\ k \to +\infty,\ m = 0,1,...,2n-1 \tag{27}$$

where $\theta = \pm\pi^{-1} \arccos\left(-\frac{h_2}{h_1}\right) = \frac{1}{i\pi} Ln(-\frac{h_2}{h_1} \pm \sqrt{\left(\frac{h_2}{h_1}\right)^2 - 1})$ and $b_j^{(m)}$ are some constants not depending on k.

Case 3. $d_{42}=d_{32}+d_{14}=0$, $d_{13}\neq 0$. In this case

$$\Delta(\lambda) = h_2 + d_{13}\lambda^{-n} \sin \lambda^n \pi + \lambda^{-n} e^{|\mathrm{Im}\,\lambda^n \pi|} \varepsilon_3(\lambda),\quad \lim_{|\lambda \to +\infty|} \varepsilon_3(\lambda) = 0.$$

As an unboundedly extending contour we choose the circles D_k if $h_2=0$ and the contours

$$D_{k3} = \cup_{v=0}^{n-1} \left[C^{(2v)} \cup C^{(2v+1)} \cup P^{(v)} \right] \text{ if } h_2 \neq 0.$$

Here

$$C^{(2v)} = \left\{ \lambda \in S_{2v}\ :\ |\lambda| = \sqrt[n]{2k + \tfrac{1}{2} - \gamma} \right\},$$

$$C^{(2v+1)} = \left\{ \lambda \in S_{2v+1}\ :\ |\lambda| = \sqrt[n]{2k + \tfrac{1}{2} - \gamma} \right\},$$

$$P^{(v)} = \left\{ \lambda\ :\ \arg \lambda = \tfrac{(v+1)\pi}{n},\ 2k + \tfrac{1}{2} - \gamma \le |\lambda| \le 2k + \tfrac{1}{2} + \gamma \right\}.$$

By the same way and arguments as in the previous cases using the Rouche's theorem we can show that there are infinitely many eigenvalues of the problem (1)-(2) for this case and sufficiently large in module eigenvalues are replaced near by the rays $\arg \lambda = \frac{m\pi}{n}\ (m = \overline{0, 2n-1})$. The eigenvalue equation $\Delta(\lambda)=0$ can be written as

$$\frac{\sin \lambda^n \pi}{\lambda^n} - \frac{1}{2i\lambda^n} e^{i\lambda^n \pi} \sum_{k=0}^{n-1} \frac{\gamma_{k+1}}{\lambda^{k+1}} a_k^{(1)} - \frac{1}{2i\lambda^n} e^{-i\lambda^n \pi} \sum_{k=0}^{n-1} \frac{\gamma_{k+1}}{\lambda^{k+1}} a_k^{(2)} + \frac{h_2}{d_{13}} + o(\lambda^{-2n}) e^{|\mathrm{Im}\,\lambda^n \pi|} = 0 \tag{28}$$

and as in the previous cases we solve the equation (28) asymptotically and obtain the following asymptotical formulas for the eigenvalue series $\lambda_{k,m}$:

$$\lambda_{k,m} = e^{\frac{2im\pi}{n}} \sqrt[n]{k} + \sum_{j=1}^{n} \frac{c_j^{(m)}}{k^{1+\frac{j-1}{n}}} + o(\frac{1}{k^{1+\frac{n-1}{n}}}),\ k \to +\infty,\ m = 0,1,...,2n-1 \quad (29)$$

if $h_2=0$ and

$$\lambda_{k,m} = e^{\frac{im\pi}{n}} \left(\sqrt[n]{2k} - \frac{1}{2n} \frac{1}{(2k)^{1-\frac{1}{n}}} - \frac{i}{n\pi} \frac{1}{(2k)^{1-\frac{1}{n}}} \ln(2k\frac{h_2}{d_{13}} + o(\frac{\ln k}{k})) \right),\ k \to +\infty,\ m = \overline{0,2n-1} \quad (30)$$

if $h_2 \neq 0$. In the formula (29) $c_j^{(m)}$ are some constants not depending on k.

Thus we have proved the following theorem.

Theorem 3.1. Boundary value problem (1)-(2) has infinite number of eigenvalues. The eigenvalues with sufficiently large modules are placed near by rays $\arg\lambda = \frac{m\pi}{n} (m = \overline{0,2n-1})$ and for the mth series of eigenvalues the asymptotic formulas (25), (27), (29)-(30) are satisfied.

REFERENCES

Agamaliyev, A., & Nabiyev, A. (2005). On eigenvalues of some boundary value problems for a polynomial pencil of Sturm-Liouville equation. *J. Applied Math. and Computations*, *165*(3), 503–515. doi:10.1016/j.amc.2004.04.116

Aktosun, T., Klaus, M., & van der Mee, C. (1998). Wave scattering in one dimension with absorption. *Journal of Mathematical Physics*, *39*(4), 1957–1992. doi:10.1063/1.532271

Alonso, M. (1980). Schrödinger spectral problems with energy-dependent potentials as sources of nonlinear Hamiltonian evolution equations. *Journal of Mathematical Physics*, *21*(9), 2342–2349. doi:10.1063/1.524690

Gasymov, M. G., & Guseinov, G. Sh. (1981). Determining the diffusion operator from spectral data [Russian]. *Akad. Nauk Azerbaijan SSR Dokl.*, *37*(2), 19–23.

Gasymow, M.G., Gasymow. (1982). On the spectral theory of differential operators polnomially depending on a parameter [Russian]. *Uspekhi Matematicheskhi Nauk, 37*(4), 99.

Guseinov, G. S. (1985). On spectral analysis of a quadratic pencil of Stırm-Liouville operators. *Soviet Math. Dokl., 32*(3), 859–862.

Guseinov, I. M. (1997). On a transformation operatör. *J. Mat. Notes, 62*(2), 172–180. doi:10.1007/BF02355905

Guseinov, I. M., Nabiev, A. A., & Pashayev, R. T. (2000). Transformation operators and asymptotic formulas for the eigenvalues of a polynomial pencil of Sturm-Liouville operators. *Sib. Math. J., 41*(3), 453–464. doi:10.1007/BF02674102

Jaulent, M. (1972). On an inverse scattering problem with an energy dependent potential. *Ann. Inst. Henri Poincare, 27*(4), 363–378.

Jaulent, M., Jean, C. (1976). The inverse scattering problem for the one dimensional Schrödinger equation with an energy-dependent potential I, II. *Ann. Inst. Henri Poincare, 25*, 105-137.

Jaulent, M., & Jean, C. (1982). Solution of a Schrödinger inverse scattering problem with a polynomial spectral dependence in the potential. *Journal of Mathematical Physics, 23*(2), 258–266. doi:10.1063/1.525347

Jordanov, R. (1984). About some spectral properties of the Schrödinger equation with an energy dependent potential generating fully integrable Hamiltonian systems. *Ann. de L'universite de Sofia 'Kliment Ohridski', 78*, 1–29.

Kamimura, Y. (2007). An inversion formula in energy dependent scattering. *Journal of Integral Equations and Applications, 19*(4), 473–512. doi:10.1216/jiea/1192628620

Maksudov, F. G., & Guseinov, G. Sh. (1986). On the solution of the inverse scattering problem for the quadratic pencil of the Schrödinger equation on the full-line. *Dokl. Akad. Nauk USSR, 289*(1), 42–46.

Marchenko, V. A. (1997). *Sturm-Liouville's operators and their applications*. Kiev: Naukova Dumko.

Nabiev, A. A. (2006). Inverse scattering problem for the Schrödinger-type equation with a polynomial energy-dependent potential. *Inverse Problems, 22*(6), 2055–2068. doi:10.1088/0266-5611/22/6/009

Nabiev, A. A., & Guseinov, I. M. (2005). On the Jost solutions of the Schrödinger-type equations with a polynomial energy-dependent potential. *Inverse Problems*, *22*(1), 55–67. doi:10.1088/0266-5611/22/1/004

Nabiev, I. M. (2004). The inverse spectral problem for the diffusion operator on an interval [Russian]. *Mat. Fiz.Anal. Geom.*, *11*(3), 302–313.

Samko, S. G., Kilbas, A. A., & Marichev, O. M. (1987). *Integral and derifatives of fractional order and its applications*. Minsk: Nauka and Tekhnika.

Sattinger, D. H., & Szmigielski, J. (1995). Energy dependent scattering theory. *Differential and Integral Equations*, *8*, 945–959.

Tsutsmi, M. (1981). On the inverse scattering problem for the one -dimensional Schrödinger equation with an energy dependent potential. *Journal of Mathematical Analysis and Applications*, *83*(1), 316–350. doi:10.1016/0022-247X(81)90266-3

Van der Mee, C., & Pivovarchik, V. (2001). Inverse scattering for a Schrödinger equation with energy dependent potential. *Journal of Mathematical Physics*, *42*(1), 158–181. doi:10.1063/1.1326921

Yurko, V. A. (2000). An inverse problem for differential operator pencils, *Mat. Sb.*, *191*(10), 137-160; English transl. *Sbornik Mathematics*, *191*(10), 1561–1586. doi:10.1070/SM2000v191n10ABEH000520

Chapter 9
Stability Analysis of a Nonlinear Epidemic Model With Generalized Piecewise Constant Argument

Duygu Aruğaslan-Çinçin
Süleyman Demirel University, Turkey

Nur Cengiz
Süleyman Demirel University, Turkey

ABSTRACT

The authors consider a nonlinear epidemic equation by modeling it with generalized piecewise constant argument (GPCA). The authors investigate invariance region for the considered model. Sufficient conditions guaranteeing the existence and uniqueness of the solutions of the model are given by creating integral equations. An important auxiliary result giving a relation between the values of the unknown function solutions at the deviation argument and at any time t is indicated. By using Lyapunov-Razumikhin method developed by Akhmet and Aruğaslan for the differential equations with generalized piecewise constant argument (EPCAG), the stability of the trivial equilibrium is investigated in addition to the stability examination of the positive equilibrium transformed into the trivial equilibrium. Then sufficient conditions for the uniform stability and the uniform asymptotic stability of trivial equilibrium and the positive equilibrium are given.

DOI: 10.4018/978-1-7998-0134-4.ch009

INTRODUCTION

In this chapter, the authors take a differential equation in population dynamics mentioned by Huang, Liu, and Fory´s (2016) into account. The considered equation represents a generalization of the SIS model established by Cooke (1979). The authors develop this model with a GPCA since change between the GPCA can be arbitrarily chosen. Since the taken model has a nonlinear function with GPCA, the model and the examinations performed for it are remarkable. As the beginning of these examinations, invariance region for the considered equation with GPCA is investigated by the authors. Besides, the authors give sufficient conditions for the existence and uniqueness of the trivial equilibrium and the positive equilibrium of the proposed model. Next, the authors give sufficient conditions guaranteeing the existence and uniqueness of the solutions of the nonlinear epidemic model by creating integral equations. The authors indicate an important auxiliary result giving a relation between the values of the unkown function at the deviation argument and at any time t of the proposed model. By using Lyapunov-Razumikhin method developed by Akhmet and Aruğaslan (2009) for EPCAG, the authors investigate the stability of the trivial equilibrium for the considered nonlinear epidemic model with GPCA. Moreover, based on the theoretical results in the paper (Akhmet & Aruğaslan, 2009), the authors investigate the stability of the positive equilibrium point by transforming it into the trivial equilibrium. Then, sufficient conditions for uniform stability and uniform asymptotic stability of the trivial equilibrium and the transformed positive equilibrium are given. Thus, the authors have reached the results depending on the parameters of the considered equation. During all these investigations, the nature of the solutions is evaluated within the biologically meaningful range $[0,1]$ as required by the examination performed for the positive invariance region.

BACKGROUND

Differential equations are very valuable in understanding the real life problems since they allow the mathematical expression of the real phenomena. However, modeling of problems with ordinary differential equations is often not enough. Because, while problems are set up mathematically by neglecting the discontinuous effects, the models and thus the results of their qualitative analysis are far distance from the reality. This necessitated the introduction and development of the theory of differential equations with discontinuities. One type of equation developed as a result of this requirement is the differential equations with deviating arguments. Differential equations with deviating arguments host many classes of equations, such as functional differential equations, differential equations with delay, piecewise constant argument and generalized piecewise constant argument. Differential equations in which the highest order derivative of the unknown function depends on the previous values of the unknown function are called delay differential equations. A delay differential equation is most fundamentally given by

$$x'(t) = f\left(t, x(t), x(t-\tau)\right).$$

Here, τ is the positive constant specified in the amount of delay, and it can be seen that the development of the state vector depends on the current time, the value of the state vector at this moment and also on the previous values of the process.

In the early 1980s, differential equations with piecewise constant argument which are in the class of delay differential equations are defined by Cooke and Wiener (1984) in the form of

$$x'(t) = f\left(t, x(t), x([t])\right).$$

The literature knowledge is intense for studies based on the transformation of such equations into discrete equations. While models are established by differential equations with piecewise constant argument, change between the arguments is constant and always one unit since the greatest integer function is taken as the deviation argument. However, this approach can contradict with the reality phenomenon. Because it is important to approach the fact of reality when establishing the real life problems. In the light of this idea, the theory of differential equations with deviating arguments have been developed over time. Differential equations with piecewise constant argument are generalized by Akhmet (2007a, 2007b) by taking an arbitrary piecewise constant function instead of the greatest integer function as deviation argument. In this way, differential equations with GPCA are given by the following equations

$$x'(t) = f\left(t, x(t), x(\beta(t))\right) \text{ or } x'(t) = f\left(t, x(t), x(\gamma(t))\right).$$

Piecewise constant functions $\beta(t)$ or $\gamma(t)$, $t \in R$, taken as a deviation argument in such equations allow arbitrary choice of time intervals. Thus, models can be addressed closer to reality phenomenon and be considered with a better approach. The piecewise function $\beta(t)$ is defined by a real valued sequence $\{\theta_i\}$, $i \in Z$, such as $|\theta_i| \to \infty$ while $i \to \infty$, and it is assumed that $\beta(t) = \theta_i$ in the interval $t \in [\theta_i, \theta_{i+1})$. If $\theta_i = i$, $i \in Z$, is chosen then the piecewise function $\beta(t)$ corresponds to the greatest integer function

$[t]$. Moreover, the piecewise function $\gamma(t)$, $t \in R$, is defined by real valued sequences $\{\theta_i\}$ and $\{\zeta_i\}$, $i \in Z$, such as $\theta_i \leq \zeta_i \leq \theta_{i+1}$, $i \in Z$. It is assumed that $\gamma(t)=\zeta_i$ in the interval $t \in [\theta_i, \theta_{i+1})$. Generally, Akhmet has examined these equations by writing equivalent integral equations, and the studies on the qualitative analysis of the problems modeled with these equations is intense in the literature (Akhmet, 2008a, 2008b, 2010, 2011, 2014; Akhmet & Aruğaslan, 2009; Akhmet, Aruğaslan, & Yılmaz, 2010a, 2010b; Aruğaslan, 2009; Aruğaslan & Cengiz, 2017, 2018). These studies provide important contributions to the investigations of stability and periodicity properties of the solutions of EPCAG. The literature on the investigation of differential equations with GPCA by transforming them into discrete equations is very rare. Due to the arbitrary choice of the piecewise constant functions, the discretization process converts a differential equation with GPCA into a nonautonomous discrete equation. In this respect, the results achieved by Aruğaslan and Güzel (2015) in the stability analysis for a logistic population model are noteworthy.

It is very important that the models are developed and handled with the best and most natural approach. Differential equations with deviation arguments allow us to realize this approach. In addition, the analysis of the qualitative characteristics of their solutions is very important for understanding the real life problems. However, it is difficult to analyze the dynamic structure of the models established with such discontinuous effects. Knowledge of the behavior of these structures is possible with the help of qualitative theory. Within this broad theory, methods that provide information about their solutions without having to solve such equations are very useful. Lyapunov-Razumikhin and Lyapunov-Krasovskii methods, which are developed in this direction and shed light on the stability analysis of the solutions, attract the attention of the scientific world. The basics of these methods are based on the second Lyapunov method (1949). These methods have been remarkable in the stability analysis of ecological, biological, epidemiological, mechanical, economic, financial, etc. models (Arnold & Schmalfuss, 2001; Challamel & Gilles, 2007; Gopalsamy & Liu, 1998; Hahn, 1967; Korobeinikov, 2006; Michel & Hu, 2000; Miller & Michel, 1982). In Lyapunov method, analysis is carried out by establishing a function which provides certain features. Sometimes it is difficult to construct these functions, but the fact that we don't need exact solutions to analyze the behaviour of solutions attracts notice to the usefulness of the method. Moreover, since models are considered as more complex structures when established with deviation arguments, trying to reach the solutions of these structures brings many difficulties and problems with them. As a remedy, Lyapunov-Razumikhin (1956) and Lyapunov-Krasovskii (1963) methods developed respectively by Razumikhin and Krasovskii for functional differential equations made a great contribution to the world of science. In Lyapunov-Razumikhin method, the analysis is carried out by constructing a function which provides certain features. Furthermore, the development of this method for differential equations with generalized piecewise constant argument by Akhmet and Aruğaslan (2009) has been a gain in scientific knowledge. Because stability analysis with other methods brings about more intensive operations and calculations for such equations. In Lyapunov-Krasovskii method, analysis is carried out by constructing a functional which provides certain features. The analyzes conducted with the help of these methods are extensively available in the literature.

The real processes established by differential equations allow the understanding of the nature of the problems encountered in many areas. In this respect, population dynamics is a branch of science which examines the biological problems in the mathematical framework. Moreover, differential equations with deviating arguments make the time delays caused by various biological factors mathematically visible on the models. It is very important to construct the structures that give proportional definitions of epidemic outbreaks, to perform their dynamical analysis and to control them. Because, considering the danger of outbreaks that could seriously affect the population and looking at the history, it is essential to ensure the control of public health and to analyze the behavioral characteristics of the emerging diseases in terms of mathematics. In this respect, considerable studies are required. Although it is important to determine the variables that define the epidemic perspective, it is essential to establish time-dependent changes of these variables well. Because, it is obvious that the spreading process of the disease will vary depending on the persons and their environment. This process can be severely affected by a situation in the past. Evaluating these effects is possible by differential equations with deviating argument, and with the help of the qualitative theory of the differential equations with deviating argument, information about the behavior of the models can be obtained. Therefore, efforts to build and develop models representing real processes, has been and continues to be an ongoing phenomenon for many years. From this perspective, the delayed SIS model

$$y'(t) = by(t-T)\left[1-y(t)\right]-cy(t) \tag{1}$$

proposed by Cooke (1979) is notable. Here, b,c are positive and constant parameters corresponding to infection and recovery rates, respectively. The proportion of humans in the community who are infectious at time t and the proportion who are susceptible are denoted by $y(t)$ and $S(t)$, respectively. Then, Cooke (1979) takes $y(t)+S(t)=1$ by assuming that the infection in humans confers negligible immunity and does not result in death or isolation. At the end of this time expressed in this assumption, it is assumed that the vector can infect a susceptible human (Cooke, 1979; Huang et al., 2016). Additionally, denote by $z(t)$ the number of infectious vectors in the community at time t according to Cooke (1979). The population is evaluated in two classes: susceptible and infectious. This evaluation supports homogeneous mixing of the vector and human populations (Busenberg & Cooke, 1980).

According to Cooke (1979), the Equation 1 with delay has representation of the proportion of infective individuals by considering the following assumptions arising from the dynamic of the spread of a communicable disease:

1. The infection is transmitted by a vector like mosquito to individuals.
2. The infection in the individual gives an immunity, but does not cause death or isolation.
3. The population is fixed without considering the change in births, deaths and immigration to the population.
4. When a susceptible vector is infected by a person, there is a fixed time T during which the infectious agent develops in the vector. At the end of this time, the vector can infect a susceptible human.
5. Human and vector populations have a homogeneous mixing.
6. The rate of recovery of infected people is positive constant c.
7. The vector population is very large and $z(t)$ is simply proportional to $y(t\text{-}T)$.
8. The infection is transmitted by a vector like mosquito to individuals.

In the light of above descriptions and assumptions, the multiple of $S(t)z(t)$ means the number of new infections per unit time, and so the time-dependent change of the proportion of infectious humans in the community is expressed by the differential equation (1) (Cooke, 1979).

Additionally, Huang et al. (2016) proposed the following equation

$$x'(t) = f\left(1 - x(t), x(t-\tau)\right) - cx(t), \ c > 0$$

which is a generalization of the nonlinear epidemic model (1). They investigated the global stability for equilibrium points of this equation by constructing appopriate Lyapunov functionals. In the related paper, Huang et al. (2016) give the assumptions on f for the existence and uniqueness of the trivial and positive equilibriums.

As the subject of the present chapter, the authors propose a differential equation in population dynamics by modeling it with GPCA. In other words, the following equation

$$x'(t) = f\left(1 - x(t), x(\beta(t))\right) - cx(t)$$

is considered by developing the model with EPCAG which is introduced by Akhmet (2007a, 2007b). Due to the considered model, the analyzes performed in this chapter will shed light on the behavior of many epidemic models. Moreover, it is considerable that the present chapter addresses a stability examination with the help of Lyapunov-Razumikhin method for EPCAG (Akhmet & Aruğaslan, 2009; Aruğaslan, 2009). Because, this method requires less operations and calculations than other methods developed for stability analysis in the literature, and makes it easier to reach the desired results, as can be understood by the examination of the chapter. As a result, the analysis of models with complex structure and strong discontinuity effects can be easier by Lyapunov-Razumikhin method introduced for EPCAG (Akhmet & Aruğaslan, 2009).

MAIN FOCUS OF THE CHAPTER

Let R, N_0 and R^+ be the sets of all real numbers, non-negative integers and non-negative real numbers, respectively, i.e., $R=(-\infty,\infty)$, $N_0=\{1,2,\dots\}$ and $R^+=[0,\infty)$. Denote the n-dimensional real space by R^n, $n\in N$. Fix a real-valued sequence $\{\theta_i\}$, $i\in N_0$ such that $0=\theta_0<\theta_1<\dots<\theta_i<\dots$ with $\theta_i\to\infty$ as $i\to\infty$. In the present chapter, the authors propose a differential equation in population dynamics by modelling it with GPCA. In this context, the authors consider the following model

$$x'(t)=f\big(1-x(t),x(\beta(t))\big)-cx(t). \tag{2}$$

In model (2), c is a positive and constant parameter corresponding to recovery rates. This equation has a representation of the proportion of infective individuals by considering the expressions and the assumptions (A)-(H) in the Background above. Here, $x\in R$, $t\in R^+$, $\beta(t)=\theta_i$ if $t\in[\theta_i,\theta_{i+1})$, $i\in N_0$. For the sequence $\{\theta_i\}$, $i\in N_0$, let $0\le\theta_i<\theta_{i+1}$ for all $i\in N_0$ and $\theta_i\to\infty$ as $i\to\infty$. Let us assume without loss of generality that $\theta_i<t_0\le\theta_{i+1}$ for some $i\in N_0$. Let $\mathcal{D}$ be a subset of the product $R\times R$ and let us assume $f:\mathcal{D}\to R$ to be of class $C^1(\mathcal{D})$.

Basically, the authors investigate uniform stability and uniform asymptotic stability of the trivial equilibrium and the positive equilibrium by using Lyapunov-Razumikhin method developed by Akhmet and Aruğaslan (2009) for EPCAG. Before the stability examinations, the authors indicate an important auxiliary result giving a relation between the values of the unknown function solutions at the deviation argument $\beta(t)$ and any time t of the proposed model. This auxiliary result is useful in the proofs of the stability theory based on Lyapunov-Razumikhin method. Next, the authors give sufficient conditions guaranteeing the existence and uniqueness of the solutions of the considered model by creating integral equation. Based on the results in the paper (Akhmet & Aruğaslan, 2009; Aruğaslan, 2009), the authors investigate the stability of the trivial equilibrium and the positive equilibrium after transforming it into the trivial equilibrium. Then, sufficient conditions for uniform stability and uniform asymptotic stability of the trivial equilibrium and the transformed positive equilibrium are given. Thus, the authors have reached the results depending on the parameters of the considered equation. During all these investigations, the nature of the solutions is evaluated within the biologically meaningful range [0,1] as required by the examination performed for the positive invariance region. Therefore, before these investigations, invariance region for the equation (2) with GPCA is investigated. Besides, the authors express sufficient conditions for the existence and uniqueness of the trivial equilibrium and the positive equilibrium of (2) by considering the nonlinear function f with GPCA in (2).

For these purposes, the following assumptions will be needed throughout the chapter:

(A1) there exists a positive number $\bar{\theta}$ such that $\theta_{i+1} - \theta_i \le \bar{\theta}$, $i \in N_0$;
(A2) $f(u,v) \in C(\mathcal{D})$, $\mathcal{D} = R \times R$, is a real-valued function, and let $f : \mathcal{D} \to R$ be of class $C^1(\mathcal{D})$;
(A3) $f(u,v) \geq 0$ for any $u,v \in [0,1]$ and $f(0,v) < c$ for any $v \in [0,1]$;
(A4) $f(u,0) = 0$ for any $u \in [0,1]$;
(A5) $f_v(1,0) > c$ for the function $f(u,v)$, $u,v \in [0,1]$;
(A6) $f_u(1,0) + f_v(1,0) > c$ and $f_{uu}(u,v) + 2f_{uv}(u,v) + f_{vv}(u,v) < 0$ for any $u,v \in [0,1]$;
(A7) $f(u,v)$ satisfies the condition

$$\left|f(u_1,v_1) - f(u_2,v_2)\right| \le \ell\left(\left|u_1 - u_2\right| + \left|v_1 - v_2\right|\right)$$

for all $t \in R^+$ and $u_i, v_i \in [0,1]$, $i=1,2$, where $\ell > 0$ is a Lipschitz constant;

(A8) $\bar{\theta}\left(\ell + (\ell + c)\left(1 + \bar{\theta}\ell\right)e^{\bar{\theta}(\ell+c)}\right) < 1$;
(A9) $\bar{\theta}\left(3\ell + c\right)e^{(\ell+c)\bar{\theta}} < 1$;
(A10) $2\ell \le c$.

It is obvious that $x^* = 0$ is an equilibrium point of (2) while f satisfies (A4) (Huang et al., 2016). The conditions (A3)-(A5) and (A6) are required for the existence and uniqueness of the positive equilibrium, respectively, similar to the results given by

Huang et al. (2016). Detailed information related to these assumptions can be found in paper (Huang et al., 2016) and by Lemma 2 given below.

Now, let us describe the crucial sets of functions:

$$\mathcal{K} = \{a \in C(R^+, R^+) : a \text{ is strictly increasing and } a(0) = 0\},$$

$$\mathcal{M} = \{d \in \mathbf{C}(R^+, R^+) : d(0) = 0,\ d(s) > 0 \text{ for } s > 0\},$$

which will be used in the stability examinations.

In order to investigate stability of the positive equilibrium of the nonlinear epidemic equation (2) with the help of Lyapunov-Razumikhin method (Akhmet & Aruğaslan, 2009), the positive equilibrium point $x^*>0$ of (2) is transformed into the trivial equilibrium by $y=x-x^*$, then the following equation is reached:

$$y'(t) = f\left(1 - y(t) - x^*, y(\beta(t)) + x^*\right) - cy(t) - cx^*. \tag{3}$$

The definitions related to solutions of the nonlinear epidemic models (2) and (3) are given by Definition 1 and Definiton 2, respectively.

Definiton 1 A function $x(t)$ is a solution of (2) on R^+ if:

1. $x(t)$ is continuous on R^+;
2. the derivative $x'(t)$ exists for $t \in R^+$ with the possible exception of the points θ_i, $i \in N_0$, where one-sided derivatives exist;
3. equation (2) is satisfied by $x(t)$ on each interval (θ_i, θ_{i+1}), $i \in N_0$, and it holds for the right derivative of $x(t)$ at the points θ_i, $i \in N_0$.

Definiton 2 A function $y(t)$ is a solution of (3) on R^+ if:

1. $y(t)$ is continuous on R^+;
2. the derivative $y'(t)$ exists for $t \in R^+$ with the possible exception of the points θ_i, $i \in N_0$, where one-sided derivatives exist;
3. equation (3) is satisfied by $y(t)$ on each interval (θ_i, θ_{i+1}), $i \in N_0$, and it holds for the right derivative of $y(t)$ at the points θ_i, $i \in N_0$.

Basic aim of the chapter is to investigate stability of the trivial equilibrium and the positive equilibrium for the proposed model (2) with GPCA. Accordingly, chapter is organized as follows: First section gives the results concerning the positive invariance of the solutions of (2). So, the invariance region of the solutions of (3) becomes obvious. In second section, important auxiliary results for (2) are indicated by Lemma 3 and Lemma 4. These results give a relation between the values of the unknown function at the deviation argument $\beta(t)$ and any time t of the models (2) and (3). Then, Section 3 addresses sufficient conditions guaranteeing the existence and uniqueness of the solutions of the models (2) and (3), by creating integral equations. Next, by using Lyapunov-Razumikhin method developed by Akhmet and Aruğaslan (2009) for EPCAG, stability of the trivial equilibrium and the positive equilibrium for (2) with GPCA is investigated in Section 4. Then, sufficient conditions for the uniform stability are given by Theorem 4 and Theorem 5. Moreover, sufficient conditions for uniform asymptotic stability of the trivial equilibrium and the positive equilibrium of (2) are obtained as seen in Theorem 6 and Theorem 7. The results depending on the parameters of the taken equation have been reached.

The Positive İnvariance for the Solutions of (2)

Now, positive invariance for (2) shall be investigated. So, the results concerning positive invariance for (2) is given by the lemma and the theorems below. So, this results give an information about the invariance region of the solutions for (3). Take only solutions $x(t)$ with $0 \le x(\theta_0)=x_0 \le 1$.

Lemma 1 The equation (2) with $x(\theta_0)=x_0$ is equivalent to the following integral equation

$$x(t)=e^{-c(t-\theta_0)}x_0+\int_{\theta_0}^{t}e^{-c(t-s)}f\left(1-x(s),x\left(\beta(s)\right)\right)ds,\ t\in\left[\theta_0,\alpha\right). \tag{4}$$

Proof: *Necessity.* Let $x(t)$ be the solution of (2) with $x(\theta_0)=x_0$. Based on Definition 1, (4) satisfies the equation (2) on each interval $\left[\theta_i,\theta_{i+1}\right)$, $i\in N_0$. For $t\in[\theta_0,\theta_1)$, the solution is

$$x(t)=e^{-c(t-\theta_0)}x_0+\int_{\theta_0}^{t}e^{-c(t-s)}f\left(1-x(s),x\left(\beta(s)\right)\right)ds.$$

Letting $t\to\theta_1$, by the continuity of the solutions, it is obtained that

$$x_1=x(\theta_1)=e^{-c(\theta_1-\theta_0)}x_0+\int_{\theta_0}^{\theta_1}e^{-c(\theta_1-s)}f\left(1-x(s),x\left(\beta(s)\right)\right)ds.$$

Therefore, (4) holds on $[\theta_0,\theta_1]$. Assume that (4) is valid on the interval $[\theta_0,\theta_k]$ for some $k\geq 1$, with the initial condition

$$x_k = x(\theta_k) = e^{-c(\theta_k-\theta_0)}x_0 + \int_{\theta_0}^{\theta_k} e^{-c(\theta_k-s)} f\left(1-x(s), x(\beta(s))\right)ds.$$

Then, for $t\in[\theta_k,\theta_{k+1})$, it is true that

$$x(t) = e^{-c(t-\theta_k)}x_k + \int_{\theta_k}^{t} e^{-c(t-s)} f\left(1-x(s), x(\beta(s))\right)ds = e^{-c(t-\theta_k)}e^{-c(\theta_k-\theta_0)}x_0$$

$$+e^{-c(t-\theta_k)}\int_{\theta_0}^{\theta_k} e^{-c(\theta_k-s)} f\left(1-x(s), x(\beta(s))\right)ds + \int_{\theta_k}^{t} e^{-c(t-s)} f\left(1-x(s), x(\beta(s))\right)ds$$

$$= e^{-c(t-\theta_0)}x_0 + \int_{\theta_0}^{t} e^{-c(t-s)} f\left(1-x(s), x(\beta(s))\right)ds.$$

Letting $t\to\theta_{k+1}$, it can be seen that

$$x_{k+1} = x(\theta_{k+1}) = e^{-c(\theta_{k+1}-\theta_0)}x_0 + \int_{\theta_0}^{\theta_{k+1}} e^{-c(\theta_{k+1}-s)} f\left(1-x(s), x(\beta(s))\right)ds.$$

Thus, (4) holds on $[\theta_0,\theta_{k+1}]$. Based on the induction method, the result can be observed for all $t\geq\theta_0$.

Sufficiency. Let $x(t)$ be the solution of (2). Fix $i\in N_0$ and consider the interval $[\theta_i,\theta_{i+1})$. Differentiating (4), it can be seen that $x(t)$ satisfies (2). Letting $t\to\theta_i$ from right and considering that $x(\beta(t))$ is a right continuous function, it is seen that $x(t)$ satisfies (2) on $[\theta_i,\theta_{i+1})$.

Theorem 1 If $x:[\theta_0,\alpha)\to R$ is a solution of (2) for $\theta_0 < t < \theta_{i+1} \leq \alpha$ satisfying the initial condition $0\leq x(\theta_0)=x_0\leq 1$ and (A3) holds, then the set $\Omega=\{x\in R: 0\leq x\leq 1\}$ is positively invariant for (2).

Proof: Let us assume without loss of generality that $\theta_i < t_0 \leq \theta_{i+1}$ for some $i\in N_0$. Let $x(t):[\theta_0,\alpha)\to R$ be a solution of (2) through the initial condition $x(\theta_0)=x_0$ satisfying $0\leq x_0\leq 1$. Then, the solution of (2) is equivalent to the following integral equation

$$x(t) = e^{-c(t-\theta_0)}x_0 + \int_{\theta_0}^{t} e^{-c(t-s)} f\left(1 - x(s), x\left(\beta(s)\right)\right)ds, \ t \in \left[\theta_0, \alpha\right).$$

Now, assume that there exists a $\bar{t}$ such that $0 \le \theta_0 \le \bar{t} \le \alpha$ and $x(\bar{t}) = 0$. Hence, by (A3),

$$x'(\bar{t}) = f\left(1 - x(\bar{t}), x_0\right) - cx(\bar{t}) = f\left(1, x_0\right) \ge 0$$

is found, which means that $x(t)$ does not exceed the value $x=0$. Thus, $x(t)$ satisfies $0 \le x(t)$ for all t in $[\theta_0,\alpha)$ while $0 \le x(\theta_0) = x_0 \le 1$.

Next, assume that there exists a $\bar{t}$ such that $0 \le \theta_0 \le \bar{t} \le \alpha$ and $x(\bar{t}) = 1$. Hence, by (A3),

$$x'(\bar{t}) = f\left(1 - x(\bar{t}), x_0\right) - cx(\bar{t}) = f\left(0, x_0\right) - c < c - c = 0$$

is found, which means that $x(t)$ does not exceed the value $x=1$. Thus, $x(t)$ satisfies $x(t) \le 1$ for all t in $[\theta_0,\alpha)$ while $0 \le x(\theta_0) = x_0 \le 1$.

Thus, it is seen that $x(t)$ satisfies $0 \le x(t) \le 1$ for all t in $[\theta_0,\alpha)$ while $0 \le x(\theta_0) = x_0 \le 1$.

Since (3) is reached by a linear tranformation of (2), the set $\Omega^* = \{y \in R: -x^* \le y \le 1 - x^*, -x^* > 0\}$ obviously shows the invariance region for (3).

Lemma 2 gives sufficient conditions for the existence and uniqueness of the positive equilibrium of (2).

Lemma 2 Let the conditions (A2)-(A6) hold true. Then, the equation (2) has a positive equilibrium $x^*>0$ in Ω.

Proof: Define the following function $g(u,v) = f(u,v) - c(1-u)$ or $g(v) = f(1-v,v) - cv$ for any $v \in [0,1]$.

It can be seen that $g(0)=0$ and $g(1)<0$ by the assumption (A3). Based on these values, there exists a positive equilibrium point if $g(v)>0$ is valid for $v \in (0,\delta)$, $0<\delta<1$. Moreover, it is obtained that

$$\lim_{v\to 0^+} \frac{g(v)}{v} = \lim_{v\to 0^+} \frac{f(1-v,v)}{v} - c = \lim_{v\to 0^+} \frac{f(1-v,v) - f(1-v,0)}{v} - c = f_v(1,0) - c.$$

Thus, by (A5), it is seen that $f_v(1,0)>c$ and then $\frac{g(v)}{v}>0$. Hence, there exists a positive number v^* in $(\delta,1]$ satisfying $g(v^*)=0$.

Besides, geometrically, sufficient conditions for uniquness of the positive equilibrium point can be determined by assuming that the function g is concave for $\forall v\in(0,1)$ and strictly increasing at $v=0$. In detail, this assumption gives $f_{uu}(u,v)+2f_{uv}(u,v)+f_{vv}(u,v)<0$. Besides, it is known that g is a continuous function on $[0,1]$ and is differentiable on $(0,1)$. So, assume that the function g has not an inflection point for $v\in(0,1)$. For this aim, it is sufficient to take $g''(v)<0$ which does not contradict with the concavity, and to take $f_u(1,0)+f_v(1,0)>c$.

The Auxiliary Results for the Solutions of (2)

Now, the lemmas given for the nonlinear epidemic model (2) contain an auxiliary result which is crucial for the proof of stability analysis in the sense of Lyapunov-Razumikhin method.

Lemma 3 Let the assumptions (A1)-(A4), (A7) and (A8) be satisfied. Then, the following inequality

$$x(\beta(t))\leq \hbar_1 x(t) \tag{5}$$

holds for (2) and for all $t\geq 0$, $x\in\Omega$, where $\hbar_1=\left\{1-\bar{\theta}\left(\ell+c\left(1+\bar{\theta}\ell\right)e^{\bar{\theta}c}\right)\right\}^{-1}$.

Proof: Let us fix $t\in R^+$. Then there exists $k\in N_0$ such that $t\in[\theta_k,\theta_{k+1})$. The solution $x(t)$ of (2) is equivalent to the integral equation

$$x(t)=x(\theta_k)+\int_{\theta_k}^{t}\left(f\left(1-x(s),x(\theta_k)\right)-cx(s)\right)ds, t\in[\theta_k,\theta_{k+1}),$$

and thus, by adding the term $f(1\text{-}x(s),0)$ due to the condition (A4), it can be written as

$$\begin{aligned}|x(t)| = x(t) &\leq |x(\theta_k)| + \int_{\theta_k}^{t} \left| f\left(1-x(s), x(\theta_k)\right) - cx(s) - f\left(1-x(s), 0\right)\right| ds \\ &\leq x(\theta_k) + \int_{\theta_k}^{t} \left(\left| f\left(1-x(s), x(\theta_k)\right) - f\left(1-x(s), 0\right)\right| + cx(s)\right) ds \\ &\leq x(\theta_k) + \int_{\theta_k}^{t} \left(\ell x(\theta_k) + cx(s)\right) ds \leq x(\theta_k)\left(1+\bar{\theta}\ell\right) + \int_{\theta_k}^{t} cx(s)\, ds\end{aligned}.$$

Based on the Gronwall-Bellman Lemma and then the condition (A1), the following inequality

$$x(t) \leq x(\theta_k)\left(1+\bar{\theta}\ell\right)e^{(t-\theta_k)c} \leq x(\theta_k)\left(1+\bar{\theta}\ell\right)e^{\bar{\theta}c} \tag{6}$$

is obtained. Besides, for $t \in [\theta_k, \theta_{k+1})$, it can be written

$$x(\theta_k) = x(t) - \int_{\theta_k}^{t} \left(f\left(1-x(s), x(\theta_k)\right) - cx(s)\right) ds.$$

The last equality gives

$$|x(\theta_k)| = x(\theta_k) \leq |x(t)|$$
$$+ \int_{\theta_k}^{t} \left| f\left(1-x(s), x(\theta_k)\right) - cx(s) - f\left(1-x(s), 0\right)\right| ds \leq x(t) + \int_{\theta_k}^{t} \left(cx(s) + \ell x(\theta_k)\right) ds.$$

The inequality (6) leads to

$$x(\theta_k) \leq x(t) + \bar{\theta}\left(c\left(1+\bar{\theta}\ell\right)e^{\bar{\theta}c} + \ell\right)x(\theta_k).$$

Then,

$$x(\theta_k) \leq \left\{1 - \bar{\theta}\left(\ell + c\left(1+\bar{\theta}\ell\right)e^{\bar{\theta}c}\right)\right\}^{-1} x(t).$$

It follows from the condition (A8) that the inequality (5) is true for $t \in [\theta_k, \theta_{k+1})$. Hence, (5) holds for all $t \geq 0$.

Similarly, Lemma 4 gives an auxiliary result for (3).

Lemma 4 Let the assumptions (A1)-(A6), (A7) and (A8) be satisfied. Then, the following inequality

$$y(\beta(t)) \leq \hbar_2 y(t) \tag{7}$$

holds for (3) and for $y \in \Omega^*$, $t \geq 0$, where $\hbar_2 = \left\{1 - \bar{\theta}\left(\ell + (\ell + c)\left(1 + \bar{\theta}\ell\right)e^{\bar{\theta}(\ell+c)}\right)\right\}^{-1}$.

The Existence and Uniqueness of the Solutions

Now, sufficient conditions are given for the existence and uniqueness of the solutions of the nonlinear epidemic model (2) with GPCA.

Lemma 5 Let (A1)-(A3) and (A7)-(A9) be satisfied and $i \in N_0$ be fixed. Then for every $(\xi, x_0) \in [\theta_i, \theta_{i+1}] \times \Omega$, there exists a unique solution $x(t) = x(t, \xi, x_0)$ of (2) on $[\theta_i, \theta_{i+1}]$.

Proof: *Existence.* Fix $i \in N_0$ and assume without loss of generality that $\theta_i \leq \xi \leq \theta_{i+1}$. Define a norm $\left|x(t)\right|_0 = \max_{[\theta_i, \xi]} \left|x(t)\right|$. Take $x^0(t) = x^0$ and define a sequence

$$x^{p+1}(t) = x^0 + \int_{\xi}^{t} \left(f\left(1 - x^p(s), x^p(\theta_i)\right) - cx^p(s)\right) ds, p \geq 0, t \in [\theta_i, \theta_{i+1}).$$

Then, for $p=0$, it is obtained that

$$
\begin{aligned}
\left|x^1(t)-x^0(t)\right|_0 &= \max_{[\theta_i,\xi]}\left|\int_\xi^t \left(f\left(1-x^0(s),x^0(\theta_i)\right)-cx^0(s)\right)ds\right| \\
&= \max_{[\theta_i,\xi]}\left|\int_\xi^t \left(f\left(1-x^0(s),x^0(\theta_i)\right)-cx^0(s)-f\left(1-x^0(s),0\right)\right)ds\right| \\
&\le \max_{[\theta_i,\xi]}\left|\int_\xi^t \left(\left|f\left(1-x^0(s),x^0(\theta_i)\right)-f\left(1-x^0(s),0\right)\right|+c\left|x^0(s)\right|\right)ds\right| \\
&\le \max_{[\theta_i,\xi]}\left|\int_\xi^t \left(\ell\left|x^0(\theta_i)\right|+c\left|x^0(s)\right|\right)ds\right| \le \bar{\theta}(\ell+c)\left|x^0\right| \le \bar{\theta}(2\ell+c)x^0.
\end{aligned} \tag{8}
$$

Second, for p=0 and p=1, the following inequality is obtained:

$$
\begin{aligned}
\left|x^2(t)-x^1(t)\right|_0 &\le \max_{[\theta_i,\xi]}\left|\int_\xi^t \left(\left|f\left(1-x^1(s),x^1(\theta_i)\right)-f\left(1-x^0(s),x^0(\theta_i)\right)\right|+c\left|x^1(s)-x^0(s)\right|\right)ds\right| \\
&\le \ell\bar{\theta}\left|x^1(\theta_i)-x^0(\theta_i)\right|+\max_{[\theta_i,\xi]}\left|\int_\xi^t (\ell+c)\left|x^1(s)-x^0(s)\right|ds\right|.
\end{aligned}
$$

By (8), it is seen that

$$
\left|x^2(t)-x^1(t)\right|_0 \le (2\ell+c)\bar{\theta}\bar{\theta}(2\ell+c)x^0 = \left(\bar{\theta}(2\ell+c)\right)^2 x^0.
$$

Similarly, for p=1 and p=2 for p=2 and p=3,…, $\left|x^{p+1}(t)-x^p(t)\right|_0$ can be evaluated. Then, by the induction method, it can be seen that

$$
\left|x^{p+1}(t)-x^p(t)\right|_0 \le \left(\bar{\theta}(2\ell+c)\right)^{p+1} x^0.
$$

Then, the condition (A9) implies that the sequence $x^p(t)$ is convergent and its limit $x(t)$ satisfies the following integral equation

$$
x(t)=x^0+\int_\xi^t \left(f\left(1-x(s),x(\theta_i)\right)-cx(s)\right)ds
$$

on $t \in [\theta_i, \xi]$ based on the result in the paper (Akhmet & Aruğaslan, 2009). The existence is proved.

Uniqueness: Let $x_j(t) = x(t, \xi, x_j^0), x_j(\xi) = x_j^0, j = 1, 2,$ denote the solutions of (2) where $\theta_i \leq \xi \leq \theta_{i+1}$. Now, it is shown that $x_1(t) \neq x_2(t)$ while $x_1^0 \neq x_2^0$ for every $t \in [\theta_i, \theta_{i+1}]$. For all $t \in [\theta_i, \theta_{i+1}]$, the solutions $x_1(t)$ and $x_2(t)$ satisfy the following integral equations

$$x_1(t) = x_1^0 + \int_{\xi}^{t} \left(f\left(1 - x_1(s), x_1(\theta_i)\right) - cx_1(s) \right) ds$$

and

$$x_2(t) = x_2^0 + \int_{\xi}^{t} \left(f\left(1 - x_2(s), x_2(\theta_i)\right) - cx_2(s) \right) ds,$$

respectively. It is true that

$$x_1(t) - x_2(t) = x_1^0 - x_2^0 + \int_{\xi}^{t} \left(f\left(1 - x_1(s), x_1(\theta_i)\right) - cx_1(s) - f\left(1 - x_2(s), x_2(\theta_i)\right) + cx_2(s) \right) ds.$$

So,

$$\begin{aligned} \left| x_1(t) - x_2(t) \right| &\leq \left| x_1^0 - x_2^0 \right| + \int_{\xi}^{t} \left(\left| f\left(1 - x_1(s), x_1(\theta_i)\right) - f\left(1 - x_2(s), x_2(\theta_i)\right) \right| + c\left| x_1(s) - x_2(s) \right| \right) ds \\ &\leq \left| x_1^0 - x_2^0 \right| + \int_{\xi}^{t} \left(\ell\left(\left| x_1(\theta_i) - x_2(\theta_i) \right| + \left| x_1(s) - x_2(s) \right| \right) + c\left| x_1(s) - x_2(s) \right| \right) ds \\ &\leq \left| x_1^0 - x_2^0 \right| + \bar{\theta}\,\ell \left| x_1(\theta_i) - x_2(\theta_i) \right| + \int_{\xi}^{t} (\ell + c) \left| x_1(s) - x_2(s) \right| ds. \end{aligned}$$

Then, by the Gronwall-Bellman inequality, it can be written as

$$\left| x_1(t) - x_2(t) \right| \leq \left\{ \left| x_1^0 - x_2^0 \right| + \bar{\theta}\,\ell \left| x_1(\theta_i) - x_2(\theta_i) \right| \right\} e^{(\ell + c)\bar{\theta}}. \tag{9}$$

Moreover, for $t=\theta_i$, it is obvious that

$$\left|x_1(\theta_i)-x_2(\theta_i)\right| \le \left\{\left|x_1^0-x_2^0\right|+\bar{\theta}\ell\left|x_1(\theta_i)-x_2(\theta_i)\right|\right\}e^{(\ell+c)\bar{\theta}}$$

and so

$$\left|x_1(\theta_i)-x_2(\theta_i)\right| \le \left\{1-\bar{\theta}\ell e^{(\ell+c)\bar{\theta}}\right\}^{-1}\left|x_1^0-x_2^0\right|e^{(\ell+c)\bar{\theta}}. \tag{10}$$

Substituting (10) into (9), the following inequality

$$\left|x_1(t)-x_2(t)\right|$$
$$\le \left\{1+\bar{\theta}\ell\left\{1-\bar{\theta}\ell e^{(\ell+c)\bar{\theta}}\right\}^{-1}e^{(\ell+c)\bar{\theta}}\right\}\left|x_1^0-x_2^0\right|e^{(\ell+c)\bar{\theta}} \le \left\{1-\bar{\theta}\ell e^{(\ell+c)\bar{\theta}}\right\}^{-1}e^{(\ell+c)\bar{\theta}}\left|x_1^0-x_2^0\right| \tag{11}$$

is reached. Now, on the contrary, assume that there exists a $t^* \in [\theta_i,\theta_{i+1}]$ such that $x_1(t^*)=x_2(t^*)$. Then, it can be written that

$$x_1^0-x_2^0=\int_{\xi}^{t^*}-\left(f\left(1-x_1(s),x_1(\theta_i)\right)-cx_1(s)-f\left(1-x_2(s),x_2(\theta_i)\right)+cx_2(s)\right)ds,$$

so

$$\left|x_1^0-x_2^0\right| \le \int_{\xi}^{t^*}\left(\ell\left|x_1(\theta_i)-x_2(\theta_i)\right|+(\ell+c)\left|x_1(s)-x_2(s)\right|\right)ds.$$

By (11), the last inequality takes the following form

$$\begin{aligned}\left|x_1^0-x_2^0\right| &\le \int_{\xi}^{t^*}(2\ell+c)\left\{1-\ell\bar{\theta}e^{(\ell+c)\bar{\theta}}\right\}^{-1}e^{(\ell+c)\bar{\theta}}\left|x_1^0-x_2^0\right|ds\\ &\le \bar{\theta}(2\ell+c)\left\{1-\ell\bar{\theta}e^{(\ell+c)\bar{\theta}}\right\}^{-1}e^{(\ell+c)\bar{\theta}}\left|x_1^0-x_2^0\right| < \left|x_1^0-x_2^0\right|,\end{aligned}$$

which is a contradiction due to the condition (A9). The uniqueness is proved.

Then, based on Theorem 1.3 given by Akhmet and Aruğaslan (2009), Theorem 2 gives sufficient conditions guaranteeing the existence and uniqueness of the solutions of (2) in Ω.

Theorem 2 Assume that the conditions (A1)-(A3) and (A7)-(A9) hold true. Then for every $(t_0,x_0)\in R^+\times\Omega$, there exists a unique solution $x(t)=x(t,t_0,x_0)$ of (2) on R^+ in the sense of Definition 1 such that $x(t_0)=x_0$.

Additionally, the following lemma and theorem give sufficient conditions guaranteeing the existence and uniqueness of the solutions of (3) in Ω^* based on the paper (Akhmet & Aruğaslan, 2009).

Lemma 6 Let (A1)-(A6) and (A7)-(A9) be satisfied and $i\in N_0$ be fixed. Then for every $(\xi, y_0)\in[\theta_i,\theta_{i+1}]\times\Omega^*$, there exists a unique solution $y(t)=y(t,\xi,x_0)$ of (3) on $[\theta_i,\theta_{i+1}]$.

Theorem 3 Assume that the conditions (A1)-(A6) and (A7)-(A9) hold true. Then for every $(t_0,y_0)\in R^+\times\Omega$, there exists a unique solution $y(t)=y(t,t_0,y_0)$ of (3) on R^+ in the sense of Definition 2 such that $y(t_0)=y_0$.

The Stability Analysis of the Solutions for (2)

In this section, stability of the trivial equilibrium and the positive equilibrium of (2), i.e., the trivial equilibrium of (3), shall be investigated. While investigating the stability, Lyapunov-Razumikhin method developed by Akhmet and Aruğaslan (2009) for EPCAG will be taken into account. Since the epidemic equation (2) with GPCA corresponds to many logistic equations, the stability examinations performed in current section have an importance.

From now on, based on the Lyapunov-Razumikhin method developed by Akhmet and Aruğaslan (2009), the next theorems give sufficient conditions for the uniform stability of the trivial equilibrium and the positive equilibrium of (2), respectively.

Theorem 4 Assume that the conditions (A1)-(A4) and (A7)-(A10) are satisfied. Then, the trivial equilibrium of (2) is uniformly stable in Ω.

Proof: Based on Definition 1.4 in the paper which contains Lyapunov-Razumikhin method developed by Akhmet and Aruğaslan (2009), construct the following positive definite Lyapunov function

$$V(x(t))=x^2(t) \tag{12}$$

Functions $u, v \in \mathcal{K}$, can be found such that $u(|x|) \le V(x) \le v(|x|), x \in \Omega$. Now, let us evaluate the derivative of (12) for $t \neq \theta_i$ $i \in N_0$:

$$
\begin{aligned}
&V'(x(t), x(\beta(t))) = 2x(t) f(1 - x(t), x(\beta(t))) - 2cx^2(t) \\
&= 2x(t) f(1 - x(t), x(\beta(t))) - 2x(t) f(1 - x(t), 0) - 2cx^2(t) \\
&\le 2x(t) \left| f(1 - x(t), x(\beta(t))) - f(1 - x(t), 0) \right| - 2cx^2(t) \\
&\le 2\ell x(t) x(\beta(t)) - 2cx^2(t) \le 2\ell x(t) x(t) - 2cx^2(t) \le -2(c - \ell) x^2(t)
\end{aligned}
$$

whenever $x(\beta(t)) \le x(t)$. Thus, by Theorem 2.4 in (Akhmet & Aruğaslan, 2009), the trivial equilibrium of (2) is uniformly stable if $\frac{c}{\ell} \ge 1$. Thus, the condition (A10) signs that the trivial equilibrium of (2) is uniformly stable.

Theorem 5 Assume that the conditions (A1)-(A10) are satisfied. Then, the trivial equilibrium of (3) (i.e., positive equilibrium of (2)) is uniformly stable.

Proof: Based on Definition 1.4 in the paper which contains Lyapunov-Razumikhin method developed by Akhmet and Aruğaslan (2009), construct the following Lyapunov function

$$V(y(t)) = y^2(t). \tag{13}$$

It is obvious that the Lyapunov function (13) is positive definite. Functions $u, v \in \mathcal{K}$, can be found such that $u(|y|) \le V(y) \le v(|y|), y \in \Omega^*$. Now, let us evaluate the derivative of (13) for $t \neq \theta_i$ $i \in N_0$,

$$
\begin{aligned}
&V'(y(t), y(\beta(t))) = 2y(t) y'(t) = 2y(t) f(1 - y(t) - x^*, y(\beta(t)) + x^*) \\
&-2cy^2(t) - 2y(t) cx^* = 2y(t) f(1 - y(t) - x^*, y(\beta(t)) + x^*) - 2cy^2(t) \\
&-2y(t) f(1 - x^*, x^*) \le 2|y(t)| \left| f(1 - y(t) - x^*, y(\beta(t)) + x^*) - f(1 - x^*, x^*) \right| - 2cy^2(t) \\
&\le 2\ell y^2(t) + 2\ell |y(t)| |y(\beta(t))| - 2cy^2(t) \le 2\ell y^2(t) + 2\ell y^2(t) - 2cy^2(t) \le -2(c - 2\ell) y^2(t)
\end{aligned}
$$

whenever $|y(\beta(t))| \le |y(t)|$. Therefore, based on Theorem 2.4 in (Akhmet & Aruğaslan, 2009), the trivial equilibrium of (3) is uniformly stable, in other words, the positive equilibrium point x^* of (2) is uniformly stable if $\frac{c}{\ell} \ge 2$. Thus, by condition (A10), the trivial equilibrium of (3), in other words, the positive equilibrium of (2), is uniformly stable (Akhmet & Aruğaslan, 2009).

From now on, based on the Lyapunov-Razumikhin method developed by Akhmet and Aruğaslan (2009), the next theorems give sufficient conditions for the uniform asymptotic stability of the trivial equilibrium and the positive equilibrium of (2).

Theorem 6 Assume that the conditions (A1)-(A4) and (A7)-(A10) are satisfied. Then, the trivial equilibrium of (2) is uniformly asymptotically stable.

Proof: Based on Definition 1.4 in the paper which contains Lyapunov-Razumikhin method developed by Akhmet and Aruğaslan (2009), construct the following positive definite Lyapunov function

$$V(x(t))=x^2(t) \tag{14}$$

Functions $u, v \in \mathcal{K}$, can be found such that $u(|x|) \le V(x) \le v(|x|), x \in \Omega$. Based on Theorem 2.5 developed by Akhmet and Aruğaslan (2009), it can be said that the trivial equilibrium of (2) is uniformly asymptotically stable. In detail, let us take a constant $\hbar_1$ such that $1 < \hbar_1 \le \frac{c}{\ell}$. Then, for $\varphi(s) = \hbar_1^2 s$, $w(s) = 2(c - \ell\hbar_1)s^2$, let us evaluate the derivative of (14) for $t \neq \theta_i$ $i \in N_0$.

$$V'(x(t), x(\beta(t))) \le 2\ell x(t)x(\beta(t)) - 2cx^2(t)$$
$$\le 2\ell x(t)\hbar_1 x(t) - 2cx^2(t) = -2(c - \ell\hbar_1)x^2(t)$$

whenever $x(\beta(t)) < \hbar_1 x(t)$. Here, φ is a continuous nondecreasing function such that $\varphi(s)>s$ for $s>0$, and $w \in \mathcal{M}$. Thus, condition (A10) signs that the trivial equilibrium of (2) is uniformly asymptotically stable.

Theorem 7 Assume that the conditions (A1)-(A10) are satisfied. Then, the trivial equilibrium of (3) (i.e., positive equilibrium of (2)) is uniformly asymptotically stable.

Proof: Based on Definition 1.4 in the paper which contains Lyapunov-Razumikhin method developed by Akhmet and Aruğaslan (2009), construct the following Lyapunov function

$$V(y(t))=y^2(t). \tag{15}$$

It is obvious that the Lyapunov function is positive definite. It can be found functions $u, v \in \mathcal{K}$ such that $u(|y|) \le V(y) \le v(|y|), y \in \Omega^*$. Based on Theorem 2.5 developed by Akhmet and Aruğaslan (2009), it is said that the trivial equilibrium of (3) is uniformly asymptotically stable. In detail, let us take a constant $\hbar_2$ such that $1 < \hbar_2 \le \frac{c}{\ell} - 1$. Then, for $\varphi(s) = \hbar_2^2 s$, $w(s) = 2(c - \ell(1 + \hbar_2))s^2$, let us evaluate the derivative of (15) for $t \ne \theta_i$ $i \in N_0$.

$$V'(y(t), y(\beta(t))) \le 2\ell y^2(t) + 2\ell |y(t)||y(\beta(t))| - 2cy^2(t)$$
$$\le 2\ell y^2(t) + 2\ell\hbar_2 y^2(t) - 2cy^2(t) \le -2(c - \ell(1 + \hbar_2))y^2(t)$$

whenever $|y(\beta(t))| < \hbar_2 |y(t)|$. Here, φ is a continuous nondecreasing function such that $\varphi(s) > s$ for $s > 0$, and $w \in \mathcal{M}$. Thus, by the condition (A10), the trivial equilibrium of (3), in other words, the positive equilibrium of (2), is uniformly asymptotically stable (Akhmet and Aruğaslan, 2009).

FUTURE RESEARCH DIRECTIONS

In this chapter, a nonlinear epidemic model is developed by using the piecewise function $\beta(t)$ as deviating argument. As a subject of future research, this model can be developed by EPCAG taking the piecewise function $\gamma(t)$ as deviating argument, or by functional differential equations. Alternatively, different models may be subjected to similar analysis by developing them with the help of EPCAG or of functional differential equations.

CONCLUSION

The present chapter addresses a nonlinear epidemic equation modeled by differential equations with GPCA. The importance of developing and analyzing this model can be seriously understood by looking at the literature. Because, it is obvious that the model studied in this chapter is remarkable in order to control the health of the population and to examine the spreading behaviors of diseases. Furthermore, modeling this equation by a generalized piecewise constant argument which makes it possible to have knowledge about effects on the structures of past behaviors is a sign of the value of the results achieved in the chapter. Because, the effect of a past value of real life problems on current behavior can be very serious. Even, the past situation of systems can change their current situation seriously. Therefore, the arguments chosen during the construction of models are very important. In this respect, the deviation argument considered for the nonlinear epidemic model in the present chapter is remarkable and it is obvious that it contributes to the development of model. The generalized piecewise constant argument provides a more natural approach. In the chapter, the fact that model contains such an argument makes it difficult to attain an explicit equation for its behavior. Therefore, analyzing the nonlinear epidemic model without reaching an explicit form of its solution has a facilitating effect. In this direction, the analysis of the equation studied in the chapter is performed with the help of Lyapunov-Razumikhin method without the need to reach its exact solution. It is seen that this method developed by Akhmet and Aruğaslan (2009) for EPCAG is very useful. Because, for the analysis of the model with GPCA, computations and operations performed in the sense of this method can be preferred conveniently compared to other methods in the literature.

ACKNOWLEDGMENT

This research received no specific grant from any funding agency in the public, commercial, or not-for-profit sectors.

REFERENCES

Akhmet, M. U. (2007a). Integral manifolds of differential equations with piecewise constant argument of generalized type. *Nonlinear Analysis: Theory, Methods and Applications*, *66*, 367–383.

Akhmet, M. U. (2007b). On the reduction principle for differential equations with piecewise constant argument of generalized type. *Journal of Mathematical Analysis and Applications*, *336*(1), 646–663. doi:10.1016/j.jmaa.2007.03.010

Akhmet, M. U. (2008a). Asymptotic behavior of solutions of differential equations with piecewise constant arguments. *Applied Mathematics Letters*, *21*(9), 951–956. doi:10.1016/j.aml.2007.10.008

Akhmet, M. U. (2008b). Stability of differential equations with piecewise constant arguments of generalized type. *Nonlinear Analysis*, *68*(4), 794–803. doi:10.1016/j.na.2006.11.037

Akhmet, M. U. (2010). *Principles of Discontinuous Dynamical Systems*. New York: Springer. doi:10.1007/978-1-4419-6581-3

Akhmet, M. U. (2011). *Nonlinear Hybrid Continuous Discrete-Time Models*. Amsterdam: Atlantis Press. doi:10.2991/978-94-91216-03-9

Akhmet, M. U. (2014). Quasilinear retarded differential equations with functional dependence on piecewise constant argument. *Communications on Pure and Applied Analysis*, *13*(2), 929–947. doi:10.3934/cpaa.2014.13.929

Akhmet, M. U., & Aruğaslan, D. (2009). Lyapunov-Razumikhin method for differential equations with piecewise constant argument. *Discrete and Continuous Dynamical Systems. Series A*, *25*(2), 457–466. doi:10.3934/dcds.2009.25.457

Akhmet, M. U., Aruğaslan, D., & Yılmaz, E. (2010a). Stability analysis of recurrent neural networks with piecewise constant argument of generalized type. *Neural Networks*, *23*(7), 805–811. doi:10.1016/j.neunet.2010.05.006 PMID:20605400

Akhmet, M. U., Aruğaslan, D., & Yılmaz, E. (2010b). Stability in cellular neural networks with piecewise constant argument. *Journal of Computational and Applied Mathematics*, *233*(9), 2365–2373. doi:10.1016/j.cam.2009.10.021

Arnold, L., & Schmalfuss, B. (2001). Lyapunov's second method for random dynamical systems. *Journal of Differential Equations*, *177*(1), 235–265. doi:10.1006/jdeq.2000.3991

Aruğaslan, D. (2009). *Süreksizlikleri Olan Diferansiyel Denklemler ve Populasyon Dinamiği*. Ankara, Turkey: Orta Doğu Teknik Üniversitesi.

Aruğaslan, D., & Cengiz, N. (2017). Green's Function and Periodic Solutions of a Spring-Mass System in which the Forces are Functionally Dependent on Piecewise Constant Argument. *Süleyman Demirel University Journal of Natural and Applied Sciences*, *21*(1), 266–278.

Aruğaslan, D., & Cengiz, N. (2018). Existence of periodic solutions for a mechanical system with piecewise constant forces. *Hacettepe Journal of Mathematics and Statistics*, *47*(3), 521–538.

Aruğaslan, D., & Güzel, L. (2015). Stability of the logistic population model with generalized piecewise constant delays. Adv. Difference Equ., 2015(173).

Busenberg, S., & Cooke, K. L. (1980). The effect of integral conditions in certain equations modelling epidemics and population growth. *Journal of Mathematical Biology*, *10*(1), 13–32. doi:10.1007/BF00276393 PMID:7205075

Challamel, N., & Gilles, G. (2007). Stability and dynamics of a harmonically excited elastic-perfectly plastic oscillator. *Journal of Sound and Vibration*, *301*(3-5), 608–634. doi:10.1016/j.jsv.2006.10.014

Cooke, K. L. (1979). Stability analysis for a vector disease model. *The Rocky Mountain Journal of Mathematics*, *7*, 253–263.

Cooke, K. L., & Wiener, J. (1984). Retarded differential equations with piecewise constant delays. *Journal of Mathematical Analysis and Applications*, *99*(1), 265–297. doi:10.1016/0022-247X(84)90248-8

Gopalsamy, K., & Liu, P. (1998). Persistence and global stability in a population model. *Journal of Mathematical Analysis and Applications*, *224*(1), 59–80. doi:10.1006/jmaa.1998.5984

Hahn, W. (1967). *Stability of Motion*. Berlin: Springer-Verlag. doi:10.1007/978-3-642-50085-5

Huang, G., Liu, A., & Fory's, U. (2016). Global Stability Analysis of Some Nonlinear Delay Differential Equations in Population Dynamics. *Journal of Nonlinear Science*, *26*(1), 27–41. doi:10.100700332-015-9267-4

Korobeinikov, A. (2006). Lyapunov functions and global stability for SIR and SIRS epidemiological models with non-linear transmission. *Bulletin of Mathematical Biology*, *68*(3), 615–626. doi:10.100711538-005-9037-9 PMID:16794947

Krasovskii, N. N. (1963). *Stability of Motion: Application of Lyapunov's Second Method to Differential Systems and Equations with Time-Delay*. Stanford, CA: SU Press.

Lyapunov, A. M. (1949). Probl´eme G´en´eral de la Stabilit´e du Mouvement. *Ann. Math. Studies, 17*.

Michel, A. N., & Hu, B. (2000). Stability analysis of discontinuous dynamical systems using vector Lyapunov functions. *Circuits, Systems, and Signal Processing*, *19*(3), 221–243. doi:10.1007/BF01204576

Miller, R. K., & Michel, A. (1982). *Ordinary Differential Equations*. New York: Academic Press.

Razumikhin, B. S. (1956). Stability of delay systems. *Прикладня математика и механика*, *20*, 500–512.

ADDITIONAL READING

Aruğaslan, D. (2015). Dynamics of a harvested logistic type model with delay and piecewise constant argument. *Journal of Nonlinear Science and Applications*, *8*(5), 507–517. doi:10.22436/jnsa.008.05.06

Aruğaslan, D., & Yolcu, Ş. (2015). Bazı skaler impalsif diferansiyel denklem modellerinde kararlılık analizi. *Gaziosmanpaşa Bilimsel Araştırma Dergisi*, *11*, 70–82.

Cooke, K., Driessche, P., & Zou, X. (1999). Interaction of maturation delay and nonlinear birth in population and epidemic models. *Journal of Mathematical Biology*, *39*(4), 332–352. doi:10.1007002850050194 PMID:10550577

Greenspan, H. P. (1976). On the growth and stability of cell cultures and solid tumors. *Journal of Theoretical Biology*, *56*(1), 229–242. doi:10.1016/S0022-5193(76)80054-9 PMID:1263527

Volz, R. (1982). Global asymptotic stability of a periodic solution to an epidemic model. *Journal of Mathematical Biology*, *15*(3), 319–338. doi:10.1007/BF00275691 PMID:7153676

Wei, J., & Zou, X. (2006). Bifurcation analysis of a population model and the resulting SIS epidemic model with delay. *Journal of Computational and Applied Mathematics*, *197*(1), 169–187. doi:10.1016/j.cam.2005.10.037

Zhao, X.-Q., & Zou, X. (2001). Threshold dynamics in a delayed SIS epidemic model. *Journal of Mathematical Analysis and Applications*, *257*(2), 282–291. doi:10.1006/jmaa.2000.7319

KEY TERMS AND DEFINITIONS

Epidemic: The occurrence of more cases of disease, injury or other health condition than expected in a given area or among a specific group of persons during a particular period. The cases are usually presumed to have a common cause or to be related to one another in some way.

GPCA: Generalized piecewise constant argument is a deviation argument which is a piecewise function considered in differential equations with piecewise constant argument of generalized type.

Outbreak: Sometimes distinguished from an epidemic as more localized, or the term less likely to evoke public panic.

Population Dynamics: Population dynamics is the branch of life sciences that studies the size and age composition of populations as dynamical systems, and the biological and environmental processes driving them (such as birth and death rates, and by immigration and emigration). Example scenarios are ageing populations, population growth, or population decline.

SIS (Susceptible-Infected-Susceptible): It is a simple mathematical model of epidemics.

Chapter 10
The Parseval Equality and Expansion Formula for Singular Hahn–Dirac System

Bilender P. Allahverdiev
Süleyman Demirel University, Turkey

Hüseyin Tuna
Mehmet Akif Ersoy University, Turkey

ABSTRACT

This work studies the singular Hahn-Dirac system given by

$$\begin{pmatrix} 0 & -\frac{1}{q}D_{-\omega q^{-1},q^{-1}} \\ D_{\omega,q} & 0 \end{pmatrix}\begin{pmatrix} y_1 \\ y_2 \end{pmatrix}+\begin{pmatrix} p(x) & 0 \\ 0 & r(x) \end{pmatrix}\begin{pmatrix} y_1 \\ y_2 \end{pmatrix}=\mu\begin{pmatrix} y_1 \\ y_2 \end{pmatrix}.$$

Here μ is a complex spectral parameter, p(.) and r(.) are real-valued continuous functions at ω0, defined on [ω0,∞) and q∈(0,1), $\omega_0 := \omega/1-q$, ω>0, x∈[ω0,∞). The existence of a spectral function for this system is proved. Further, a Parseval equality and an expansion formula in eigenfunctions are proved in terms of the spectral function.

DOI: 10.4018/978-1-7998-0134-4.ch010

INTRODUCTION

The theory of Hahn difference operator was introduced by Hahn in 1949; see the papers (Hahn, 1949,1983). This operator provides a unifying structure for the study of the forward difference operator defined by

$$\Delta_{\omega} f(x) := \frac{f(\omega+x) - f(x)}{(\omega+x) - x},\ x \in \mathbb{R}$$

and the study of the quantum q-difference operator (Jackson, 1910) defined by

$$D_q f(x) := \frac{f(qx) - f(x)}{qx - x},\ x \neq 0.$$

Recently, Hahn difference operators are receiving an increase of interest due to their applications in the construction of families of orthogonal polynomials and approximation problems (see Alvarez-Nodarse, 2006; Dobrogowsa & Odzijewicz, 2006; Kwon, Lee, Park & Yoo, 1998; Lesky, 2005; Petronilho, 2007 and the references therein).

In the literature there exist some papers studying the Hahn difference operator. The theory of linear Hahn difference equations was developed in the paper (Hamza & Ahmed, 2013). In (Hamza & Ahmed, 2013), the authors also study the existence and uniqueness of the solution for initial value problems related to Hahn difference equations. Hamza and Makharesh (Hamza & Makheresh, 2016) investigated Leibniz's rule and Fubini's theorem associated with the Hahn difference operator. The nonlocal boundary value problem for nonlinear Hahn difference equation was developed in the paper (Sitthiwirattham, 2016). In 2018, (Annaby, Hamza & Makherseh, 2018), the regular Hahn-Sturm-Liouville problem

$$-\frac{1}{q} D_{-\omega q^{-1}, q^{-1}} D_{\omega,q} y(x) + v(x) y(x) = \lambda y(x),$$

$$a_1 y(\omega_0) + a_2 D_{-\omega q^{-1}, q^{-1}} y(\omega_0) = 0,$$

$$b_1 y(b) + b_2 D_{-\omega q^{-1}, q^{-1}} y(b) = 0,$$

has been studied, where $\omega_0 \leq x \leq b$, $\lambda \in \mathbb{C}$, $a_i, b_i \in \mathbb{R} := (-\infty, \infty)$, $i = 1,2$ and v(.) is a real-valued continuous function at ω_0, defined on $[\omega_0, b]$. Annaby et al. (Annaby, Hamza & Makherseh, 2018) defined a Hilbert space of ω,q- square summable functions. The authors discussed the formulation of the self-adjoint operator and the properties of the eigenvalues and the eigenfunctions. Furthermore, the authors constructed the Green's function and gave an eigenfunction expansion theorem. This equation is reduced to classic Sturm-Liouville problem when $\omega \to 0$ and $q \to 1$. The French mathematicians Sturm and Liouville were introduced first this problem in 1837 (Sturm & Liouville, 1837). Since then this field is an active field of research. For a deeper discussion of this theory we refer the reader to (Zettl, 2005; Amrein, Hinz & Pearson, 2005; Mukhtarov & Aydemir, 2018; Aydemir, Olgar, Mukhtarov & Muhtarov, 2018; Olgar, Mukhtarov & Aydemir, 2018; Aydemir & Mukhtarov, 2017; Allahverdiev, Eryılmaz & Tuna, 2017; Tuna, 2014, 2016; Tuna & Özek, 2017; Tuna & Eryılmaz, 2013).

In (Hira,2018), the author introduced the ω,q- analogy of the regular Dirac system

$$\begin{pmatrix} 0 & -\frac{1}{q} D_{-\omega q^{-1}, q^{-1}} \\ D_{\omega,q} & 0 \end{pmatrix} \begin{pmatrix} y_1 \\ y_2 \end{pmatrix} + \begin{pmatrix} p(x) & 0 \\ 0 & r(x) \end{pmatrix} \begin{pmatrix} y_1 \\ y_2 \end{pmatrix} = \mu \begin{pmatrix} y_1 \\ y_2 \end{pmatrix}, \; x \in (\omega_0, b),$$

where μ is a complex spectral parameter, p(.) and r(.) are real-valued continuous functions at ω_0, defined on $([\omega_0, b)$, $\omega_0 < b < \infty$ where $q \in (0,1)$ and $\omega > 0$. Hira investigated the existence and uniqueness of the solutions for this problem and gave its spectral properties. This system is reduced to classic Dirac system when $\omega \to 0$ and $q \to 1$. Now, we give some information related to the physical meaning of classic dirac system. The Dirac system is the most important system in quantum mechanics. This system formulates the fundamental physics of realistic quantum mechanics. It predicts the existence of antimatter and gives a description of the electron spin. The spectral properties of the Dirac systems have been considered in (Levitan & Sargsjan, 1991; Weidmann, 1987; Stone, 1926, 1930; Thaller, 1992).

In mathematical physics, when a partial differential equation is solved by the method of separation of variables, the problem of expanding an arbitrary function as a series of eigenfunctions is matched. Thus the spectral expansion theorems are essential for solving various problems in mathematics. The first paper for the spectral expansion problem were done by Weyl in 1910 (Weyl,1910). Since then, by several methods, a lot of authors have studied such problems. For instance, by the methods of integral equations, contour integration and finite difference (see Allahverdiev & Tuna, 2018a, 2018b, 2019a, 2019b, 2019c, 2019d, 2019e ; Berezanskii, 1968;

Levitan & Sargsjan, 1991; Guseinov, 2007, 2008; Levinson, 1951 ; Titchmarsh, 1962; Yosida 1950, 1960;).

It is the purpose of this paper to develop the spectral theory of singular Hahn-Dirac system. For this system, the existence of a spectral function is proved. A Parseval equality and an expansion formula in eigenfunctions are established.

PRELIMINARIES

In this section, some preliminary materials related to the Hahn calculus are provided. For more details, the reader may refer to (Annaby, Hamza & Aldwoah, 2012; Hahn, 1949,1983) and (Annaby, Hamza & Makharesh, 2018). For the purposes here, it will be assumed that $q\in(0,1)$ and $\omega>0$.

Let $\omega_0 := \omega/_{1-q}$ and I be a real interval containing ω_0.

Definition 1 (Hahn, 1949,1983). *Let $f : I \to \mathbb{R}$ be a function. The Hahn difference operator is defined by*

$$D_{\omega,q} f(x) = \begin{cases} \dfrac{f(\omega+qx)-f(x)}{\omega+(q-1)x}, & x \neq \omega_0, \\ f^{1}(\omega_0), & x = \omega_{0,} \end{cases}$$

provided that f is differentiable at ω_0. In this case, $D_{\omega,q} f$ is called the ω,q-derivative of f.

Remark 2 *The Hahn difference operator unifies two well known operators. When $q\to1$, the forward difference operator is obtained, which is defined by*

$$\Delta_\omega f(x) := \frac{f(\omega+x)-f(x)}{(\omega+x)-x}, \; x \in \mathbb{R}.$$

When $\omega\to0$, the Jackson q- difference operator is obtained, which is defined by

$$D_q f(x) := \frac{f(qx)-f(x)}{qx-x}, \; x \neq 0.$$

Furthermore, under appropriate conditions, it is obtained that

$$\lim_{q\to 1} D_{\omega,q} f(x) = f'(x).$$

Now some properties of the ω,q- derivative will be presented.

Theorem 3 (Annaby, Hamza & Makharesh, 2018) *Let* $f,g: I \to \mathbb{R}$ *be* ω,q- *differentiable at x∈I and h(x):=ω+qx. Then, we have*

$$D_{\omega,q}(af+bg)(x) = aD_{\omega,q} f(x) + bD_{\omega,q} g(x),\ a,b \in I,$$

$$D_{\omega,q}(fg)(x) = D_{\omega,q}(f(x))g(x) + f(\omega + xq)D_{\omega,q} g(x),$$

$$D_{\omega,q}\left(\frac{f}{g}\right)(x) = \frac{D_{\omega,q}(f(x))g(x) - f(x)D_{\omega,q} g(x)}{g(x)g(\omega+xq)},$$

$$D_{\omega,q} f(h^{-1}(x)) = D_{-\omega q^{-1},q^{-1}} f(x), h^{-1}(x) = q^{-1}(x-\omega),$$

for all x∈I. The ω,*q- integral of the function f can be defined as follows.*

Definition 4 (Jackson-Nörlund Integral (Annaby, Hamza & Makharesh, 2018)**.** *Let* $f: I \to \mathbb{R}$ *be a function and a,b,*ω_0*∈I. The* ω,*q- integral of the function f from a to b is defined by*

$$\int_a^b f(x)d_{\omega,q}x := \int_{\omega_0}^b f(x)d_{\omega,q}x - \int_{\omega_0}^a f(x)d_{\omega,q}x$$

where

$$\int_{\omega_0}^x f(t)d_{\omega,q}t := ((1-q)x-\omega)\sum_{n=0}^{\infty} q^n f(\omega\frac{1-q^n}{1-q} + xq^n),\ x \in I$$

provided that the series converges at x=a and x=b. In this case, f is called ω,*q- integrable on [a,b].*

Similarly, one can define the ω,*q*- integral of the function f over $[\omega_0,\infty)$ by

$$\int_{\omega_0}^{\infty} f(x)\, d_{\omega,q}x := \lim_{b\to\infty} \int_{\omega_0}^{b} f(x)\, d_{\omega,q}x .$$

The following properties of ω,q- integration can be found in (Annaby, Hamza & Makharesh, 2018).

Theorem 5 (Annaby, Hamza & Makharesh, 2018) *Let $f, g : I \to \mathbb{R}$ be ω,q-integrable on I and let a,b,c∈I, a<c<b and $\alpha, \beta \in \mathbb{R}$. Then, the following formulas hold:*

i) $D_{\omega,q}(af+bg)(x) = aD_{\omega,q}f(x) + bD_{\omega,q}g(x), a, b \in I$

ii) $D_{\omega,q}(fg)(x) = D_{\omega,q}\left(f(x)\right)g(x) + f(\omega + xq)D_{\omega,q}g(x)$

iii) $D_{\omega,q}\left(\dfrac{f}{q}\right)(x) = \dfrac{D_{\omega,q}\left(f(x)\right)q(x) - f(x)D_{\omega,q}g(x)}{g(x)g(\omega + xq)}$

iv) $D_{\omega,q}f\left(h^{-1}(x)\right) = D_{-\omega q^{-1},q^{-1}}f(x), h^{-1}(x) = q^{-1}(x-\omega)$

Now the ω,q- integration by parts is presented.

Lemma 6 (Annaby, Hamza & Makharesh, 2018) *Let $f, g : I \to \mathbb{R}$ be ω,q-integrable on I and let a,b∈I where a<b. Then, the following formula holds:*

$$\int_a^b f(x) D_{\omega,q} g(x)\, d_{\omega,q}x + \int_a^b g(\omega + qx) D_{\omega,q} f(x)\, d_{\omega,q}x = f(b)g(b) - f(a)g(a).$$

The next result is the fundamental theorem of Hahn calculus.

Theorem 7 (Annaby, Hamza & Makharesh, 2018) *Let $f : I \to \mathbb{R}$ be continuous at ω_0. Define*

$$F(x) := \int_{\omega_0}^{x} f(t)\, d_{\omega,q}t, \; x \in I.$$

Then F is continuous at ω_0. Moreover, $D_{\omega,q}F(x)$ exists for every x∈I and $D_{\omega,q}F(x)=f(x)$. Conversely,

$$\int_a^b D_{\omega,q} F(x) d_{\omega,q} x = f(b) - f(a).$$

Let $L^2_{\omega,q}(\omega_0,\infty)$ be the space of all complex-valued functions defined on $[\omega_0,\infty)$ such that

$$\|f\| := \left(\int_{\omega_0}^{\infty} |f(x)|^2 \, d_{\omega,q} x \right)^{1/2} < \infty.$$

The space $L^2_{\omega,q}(\omega_0,\infty)$ is a separable Hilbert space with the inner product

$$\langle f, g \rangle := \int_{\omega_0}^{\infty} f(x) \overline{g(x)} d_{\omega,q} x, \; f, g \in L^2_{\omega,q}(\omega_0,\infty),$$

(see Annaby, Hamza & Makharesh, 2018).

We introduce a convenient Hilbert space $\mathcal{H}_{\omega,q} = L^2_{\omega,q}((\omega_0,\infty); E)\left(E := \mathbb{C}^2\right)$ of vector-valued functions by using the inner product

$$\begin{aligned}(f,g) &:= \int_{\omega_0}^{\infty} (f(x), g(x))_E \, d_{\omega,q} x \\ &= \int_{\omega_0}^{\infty} f_1(x) \overline{g_1(x)} d_{\omega,q} x + \int_{\omega_0}^{\infty} f_2(x) \overline{g_2(x)} d_{\omega,q} x\end{aligned},$$

where

$$f(x) = \begin{pmatrix} f_1(x) \\ f_2(x) \end{pmatrix}, \; g(x) = \begin{pmatrix} g_1(x) \\ g_2(x) \end{pmatrix} \in \mathcal{H}_{\omega,q}.$$

Let $y(x) = \begin{pmatrix} y_1(x) \\ y_2(x) \end{pmatrix}, \; z(x) = \begin{pmatrix} z_1(x) \\ z_2(x) \end{pmatrix}.$

Then, the Wronskian of y(x) and z(x) is defined by the formula

$$W(y,z)(x)=y_1(x)z_2\left(h^{-1}(x)\right)-z_1(x)y_2\left(h^{-1}(x)\right),\ x\in[\omega_0,\infty).$$

Definition 8 *A matrix-valued function $M(x,t)$ of two variables with $\omega_0 \le x,t \le b$ is called the ω,q- Hilbert-Schmidt kernel if*

$$\int_{\omega_0}^{b}\int_{\omega_0}^{b}\left\|M(x,t)\right\|_E^2 d_{\omega,q}x d_{\omega,q}t<+\infty.$$

Theorem 9 (Kolmogorov & Fomin, 1970) *Let $\left(w_n\right)_{n\in\mathbb{N}}$ be a uniformly bounded sequence of real non-decreasing functions on a finite interval $a\le\mu\le b$, where $\mathbb{N}:=\{1,2,3,\ldots\}$. Then, there exist a subsequence $\left(w_{n_k}\right)_{k\in\mathbb{N}}$ and a non-decreasing function w such that*

$$\lim_{k\to\infty}w_{n_k}(\mu)=w(\mu),\ a\le\mu\le b.$$

Theorem 10 (Kolmogorov & Fomin, 1970) *Assume that $\left(w_n\right)_{n\in\mathbb{N}}$ is a real, uniformly bounded sequence of non-decreasing functions on a finite interval $a\le\mu\le b$, and suppose that*

$$\lim_{n\to\infty}w_n(\mu)=w(\mu),\ a\le\mu\le b.$$

If f is any continuous function on $a\le\mu\le b$, then

$$\lim_{n\to\infty}\int_a^b f(\mu)dw_n(\mu)=\int_a^b f(\mu)dw(\mu).$$

Theorem 11 (Naimark, 1968) *If*

$$\sum_{i,k=1}^{\infty}\left|a_{ik}\right|^2<\infty \tag{1}$$

then the operator A defined by the formula $A\{x_i\}=\{y_i\}$, $i=1,2,\ldots$, where

$$y_i = \sum_{k=1}^{\infty} a_{ik} x_k, \ i = 1,2,\ldots \tag{2}$$

is compact in the sequence space l^2.

Regular System

In this section, regular Hahn-Dirac system is studied.

The Hahn-Dirac system has the form

$$\begin{pmatrix} 0 & -\frac{1}{q} D_{-\omega q^{-1}, q^{-1}} \\ D_{\omega,q} & 0 \end{pmatrix} \begin{pmatrix} y_1 \\ y_2 \end{pmatrix} + \begin{pmatrix} p(x) & 0 \\ 0 & r(x) \end{pmatrix} \begin{pmatrix} y_1 \\ y_2 \end{pmatrix} = \mu \begin{pmatrix} y_1 \\ y_2 \end{pmatrix}, \ x \in (\omega_0, \infty). \tag{3}$$

Here μ is a complex spectral parameter, p(.) and r(.) are real-valued continuous functions at ω_0, defined on $[\omega_0,\infty)$ where $q \in (0,1)$ and $\omega > 0$.

The system (3) with the boundary conditions

$$y_1(\omega_0, \mu) \sin\beta + y_2(\omega_0, \mu) \cos\beta = 0, \ \beta \in \mathbb{R}, \tag{4}$$

$$y_2(\omega_0 + q^{-s}, \mu) \cos\alpha + y_1(\omega_0 + q^{-s}, \mu) \sin\alpha = 0, \ \alpha \in \mathbb{R}, s \in \mathbb{N}, \tag{5}$$

is a regular boundary-value problem.

Theorem 12 *The boundary value problem (3)-(5) has a compact resolvent operator, thus it has a purely discrete spectrum.*

Proof Denote by

$$\varphi_1(x, \mu) = \begin{pmatrix} \varphi_{11}(x, \mu) \\ \varphi_{12}(x, \mu) \end{pmatrix}$$

And

$$\varphi_2(x,\mu)=\begin{pmatrix}\varphi_{21}(x,\mu)\\ \varphi_{22}(x,\mu)\end{pmatrix}$$

respectively, the solution of the system (3) which satisfies the initial conditions

$$\varphi_{11}(\omega_0,\mu)=\cos\beta,\varphi_{12}(\omega_0,\mu)=-\sin\beta, \tag{6}$$

$$\varphi_{21}\left(\omega_0+q^{-s},\mu\right)=\cos\alpha,\varphi_{22}\left(\omega_0+q^{-s},\mu\right)=-\sin\alpha.$$

Let us define the *Green's matrix by the formula*

$$\mathrm{G}(\mathrm{x},\mathrm{t},\mu)=\frac{1}{W(\varphi_1,\varphi_2)}\begin{cases}\varphi_2(x,\mu)\varphi_1^T(t,\mu),\ t\le x\\ \varphi_1(x,\mu)\varphi_2^T(t,\mu),\ x<t\end{cases} \tag{7}$$

It will be proved that the function

$$y(x,\mu)=\int_{\omega_0}^{\omega_0+q^{-s}}G(x,t,\mu)f(t)d_{\omega,q}t \tag{8}$$

is the solution of the non-homogeneous system

$$-q^{-1}D_{-\omega q^{-1},q^{-1}}y_2+\{p(x)-\mu\}y_1=f_1(x), \tag{9}$$

$$D_{\omega,q}y_1+\{r(x)-\mu\}y_2=f_2(x), \tag{10}$$

where

$$f(.)=\begin{pmatrix}f_1(.)\\ f_2(.)\end{pmatrix}\in L^2_{\omega,q}((\omega_0,\omega_0+q^{-s});E),$$

which satisfies the boundary conditions (4)-(5).

It follows from (8) that

$$y_1(x,\mu)=\frac{1}{W(\varphi_1,\varphi_2)}\varphi_{21}(x,\mu)\int_{\omega_0}^{x}\left(\varphi_{11}(t,\mu)f_1(t)+\varphi_{12}(t,\mu)f_2(t)\right)d_{\omega,q}t$$
$$+\frac{1}{W(\varphi_1,\varphi_2)}\varphi_{11}(x,\mu)\int_{x}^{\omega_0+q^{-s}}\left(\varphi_{21}(t,\mu)f_1(t)+\varphi_{22}(t,\mu)f_2(t)\right)d_{\omega,q}t \tag{11}$$

$$y_2(x,\mu)=\frac{1}{W(\varphi_1,\varphi_2)}\varphi_{22}(x,\mu)\int_{\omega_0}^{x}\left(\varphi_{11}(t,\mu)f_1(t)+\varphi_{12}(t,\mu)f_2(t)\right)d_{\omega,q}t$$
$$+\frac{1}{W(\varphi_1,\varphi_2)}\varphi_{12}(x,\mu)\int_{x}^{\omega_0+q^{-s}}\left(\varphi_{21}(t,\mu)f_1(t)+\varphi_{22}(t,\mu)f_2(t)\right)d_{\omega,q}t \tag{12}$$

From (11), we have

$$\begin{aligned}
D_{\omega,q}y_1(x,\mu)=&\frac{1}{W(\varphi_1,\varphi_2)}D_{\omega,q}\varphi_{21}(x,\mu)\int_{\omega_0}^{x}\left(\varphi_{11}(t,\mu)f_1(t)+\varphi_{12}(t,\mu)f_2(t)\right)d_{\omega,q}t\\
&+\frac{1}{W(\varphi_1,\varphi_2)}D_{\omega,q}\varphi_{11}(x,\mu)\int_{x}^{\omega_0+q^{-s}}\left(\varphi_{21}(t,\mu)f_1(t)+\varphi_{22}(t,\mu)f_2(t)\right)d_{\omega,q}t\\
&+\frac{1}{W(\varphi_1,\varphi_2)}W(\varphi_1,\varphi_2)f_2(x)\\
=&\frac{1}{W(\varphi_1,\varphi_2)}\left\{-r(x)+\mu\right\}\varphi_{22}(x,\mu)\int_{\omega_0}^{x}\left(\varphi_{11}(t,\mu)f_1(t)+\varphi_{12}(t,\mu)f_2(t)\right)d_{\omega,q}t\\
&+\frac{1}{W(\varphi_1,\varphi_2)}\left\{-r(x)+\mu\right\}\varphi_{12}(x,\mu)\int_{x}^{\omega_0+q^{-s}}\left(\varphi_{21}(t,\mu)f_1(t)+\varphi_{22}(t,\mu)f_2(t)\right)d_{\omega,q}t+f_2(x)\\
=&\left\{-r(x)+\mu\right\}y_2(x,\mu)+f_2(x)
\end{aligned}$$

The validity of (9) is proved similarly. Hence the function $y(x,\mu)$ in (8) is the solution of the system (9)-(10). It is checked at once that (8) satisfies the boundary conditions (4)-(5). Without loss of generality, assume that $\mu=0$ is not an eigenvalue. Then, we have

$$G(x,t)=G(x,t,0)=\frac{1}{W(\varphi_1,\varphi_2)}\begin{cases}\varphi_2(x,0)\varphi_1^T(t,0), & t\le x\\ \varphi_1(x,0)\varphi_2^T(t,0), & x<t\end{cases} \tag{13}$$

Now, it will be shown that G(x,t) defined by (13) is a Hilbert-Schmidt kernel. By the upper half of the formula (13), we have

$$\int_{\omega_0}^{\omega_0+q^{-s}} d_{\omega,q}x \int_{\omega_0}^{x} \|G(x,t)\|_E^2 \, d_{\omega,q}t<\infty;$$

and by the lower half of (13), we have

$$\int_{\omega_0}^{\omega_0+q^{-s}} d_{\omega,q}x \int_{x}^{\omega_0+q^{-s}} \|G(x,t)\|_E^2 \, d_{\omega,q}t<\infty$$

since the inner integral exists and is a linear combination of the products $\varphi_{ij}(x,0)$ $\varphi_{kl}(t,0)(i,j,k,l=1,2)$, and these products belong to

$$L_{\omega,q}^2(\omega_0,\omega_0+q^{-s})\times L_{\omega,q}^2(\omega_0,\omega_0+q^{-s})$$

because each of the factors belongs to $L_{\omega,q}^2(\omega_0,\omega_0+q^{-s})$. Then, we obtain

$$\int_{\omega_0}^{\omega_0+q^{-s}} \int_{\omega_0}^{\omega_0+q^{-s}} \|G(x,t)\|_E^2 \, d_{\omega,q}x d_{\omega,q}t<\infty, \tag{14}$$

i.e., G(x,t) is a Hilbert-Schmidt kernel.

Now, it will be shown that the operator K defined by the formula

$$(Kf)(x)=\int_{\omega_0}^{\omega_0+q^{-s}} G(x,t)f(t)\,d_{\omega,q}t$$

is compact in $L_{\omega,q}^2((\omega_0,\omega_0+q^{-s});E)$.

Let $\psi_i=\psi_i(t)(i\in\mathbb{N}$ be a complete, orthonormal basis of $L_{\omega,q}^2((\omega_0,\omega_0+q^{-s});E)$. Since G(x,t) is a Hilbert-Schmidt kernel, we can define

$$x_i = (f, \psi_i) = \int_{\omega_0}^{\omega_0+q^{-s}} (f(t), \psi_i(t))_E \, d_{\omega,q}t,$$

$$y_i = (g, \psi_i) = \int_{\omega_0}^{\omega_0+q^{-s}} (g(t), \psi_i(t))_E \, d_{\omega,q}t,$$

$$a_{ik} = \int_{\omega_0}^{\omega_0+q^{-s}} \int_{\omega_0}^{\omega_0+q^{-s}} (G(x,t)\psi_i(x), \psi_k(t))_E \, d_{\omega,q}x d_{\omega,q}t.$$

Then, $L^2_{\omega,q}((\omega_0, \omega_0 + q^{-s}); E)$ is mapped isometrically into l^2. Consequently, the integral operator transforms to the operator defined by the formula (2) in the space l^2 by this mapping, and the condition (14) is translated into the condition (1). It follows from Theorem 11 that this operator is compact. Therefore, the original operator is compact and has a purely discrete spectrum.

Construction of the Spectral Function

In this section, the existence of a spectral function is proved.

Denote by $\Phi(x, \mu)$ the solution of the system (3) subjected to the initial conditions

$$\Phi_1(\omega_0, \mu) = \cos\beta, \; \Phi_2(\omega_0, \mu) = -\sin\beta.$$

Let us denote by $\mu_{m,q^{-s}}$ the eigenvalues of the problem (3), (4), (5), and by

$$\Phi_{m,q^{-s}}(x) = \begin{pmatrix} \Phi^{(1)}_{m,q^{-s}}(x) \\ \Phi^{(2)}_{m,q^{-s}}(x) \end{pmatrix} = \Phi(x, \mu_{m,q^{-s}}) = \begin{pmatrix} \Phi_1(x, \mu_{m,q^{-s}}) \\ \Phi_2(x, \mu_{m,q^{-s}}) \end{pmatrix}$$

the corresponding eigenfunctions which satisfy the conditions (4) and (5), where $(m \in \mathbb{Z} := \{0, \pm1, \pm2, \ldots\})$. If

$$f(x) = \begin{pmatrix} f_1(x) \\ f_2(x) \end{pmatrix}, \; \int_{\omega_0}^{\omega_0+q^{-s}} (f_1^2(x) + f_2^2(x)) d_{\omega,q}x < \infty,$$

and

$$\alpha^2_{m,q^{-s}} = \int_{\omega_0}^{\omega_0+q^{-s}} \left(\left(\Phi^{(1)}_{m,q^{-s}}(x) \right)^2 + \left(\Phi^{(2)}_{m,q^{-s}}(x) \right)^2 \right) d_{\omega,q}x,$$

i.e., if $f(.) \in L^2_{\omega,q}((\omega_0, \omega_0 + q^{-s}); \mathbb{R}^2)$, then we have

$$\begin{aligned} &\int_{\omega_0}^{\omega_0+q^{-s}} \left(f_1^2(x) + f_2^2(x) \right) d_{\omega,q}x \\ &= \sum_{m=-\infty}^{\infty} \frac{1}{\alpha^2_{m,q^{-s}}} \left\{ \int_{\omega_0}^{\omega_0+q^{-s}} \left(f_1(x)\Phi^{(1)}_{m,q^{-s}}(x) + f_2(x)\Phi^{(2)}_{m,q^{-s}}(x) \right) d_{\omega,q}x \right\}^2, \end{aligned} \tag{16}$$

which is called the *Parseval equality*.

Now let the non-decreasing step function $\varrho_{q^{-s}}$ on $\mathbb{R}$ be defined by

$$\varrho_{q^{-s}}(\mu) = \begin{cases} -\sum\limits_{\mu < \mu_{m,q^{-s}} < 0} \dfrac{1}{\alpha^2_{m,q^{-s}}}, & \text{for } \mu \le 0 \\ \sum\limits_{0 \le \mu_{m,q^{-s}} < \mu} \dfrac{1}{\alpha^2_{m,q^{-s}}}, & \text{for } \mu \ge 0. \end{cases}$$

Then, the equalities (16) can be written as

$$\int_{\omega_0}^{\omega_0+q^{-s}} \left(f_1^2(x) + f_2^2(x) \right) d_{\omega,q}x = \int_{-\infty}^{\infty} F_s^2(\mu)\, d\varrho_{q^{-s}}(\mu), \tag{17}$$

where

$$F_s(\mu) = \int_{\omega_0}^{\omega_0+q^{-s}} \left(f_1(x)\Phi_1(x,\mu) + f_2(x)\Phi_2(x,\mu) \right) d_{\omega,q}x.$$

It will be shown that the Parseval equality for the problem (3), (4) can be obtained from (17) by letting $s \to \infty$. For this aim, a lemma will be proved.

Lemma 13 *For any positive* κ, *there is a positive constant* $M=M(\kappa)$ *not depending on* q^{-s} *such that*

$$\bigvee_{\kappa}^{-\kappa}\left\{\varrho_{q^{-s}}(\mu)\right\}=\sum_{-\kappa<\mu_{m,q^{-s}}<\kappa}\frac{1}{\alpha_{m,q^{-s}}^{2}}=\varrho_{q^{-s}}(\kappa)-\varrho_{q^{-s}}(-\kappa)<M. \tag{18}$$

Proof Let $\sin\beta\neq 0$. Since $\Phi_2(x,\mu)$ is continuous on the region

$$\left\{(x,\mu):-\kappa\leq\mu\leq\kappa,\omega_0\leq x\leq\omega_0+a,a>0\right\},$$

by the condition $\Phi_2(\omega_0,\mu)=-\sin\beta$ there is a positive number h and near by 0 such that

$$\left(\frac{1}{h}\int_{\omega_0}^{\omega_0+h}\Phi_2(x,\mu)d_{\omega,q}x\right)^2>\frac{1}{2}(\sin\beta)^2. \tag{19}$$

Let us define

$$f_h(x)=\begin{pmatrix}f_{1h}(x)\\ f_{2h}(x)\end{pmatrix}$$

by

$$f_{1h}(x)=0,\ f_{2h}(x)=\begin{cases}\dfrac{1}{h}, & \omega_0\leq x<h\\ 0, x\geq h.\end{cases}$$

From (17), (18) and (19), we get

$$\int_{\omega_0}^{\omega_0+h}\left(f_{1h}^2(x)+f_{2h}^2(x)\right)d_{\omega,q}x=\frac{1}{h}\int_{-\infty}^{\infty}\left(\frac{1}{h}\int_{\omega_0}^{\omega_0+h}\Phi_2(x,\mu)d_{\omega,q}x\right)^2 d\varrho_{q^{-s}}(\mu)$$

$$\geq\int_{-\kappa}^{\kappa}\left(\frac{1}{h}\int_{\omega_0}^{\omega_0+h}\Phi_2(x,\mu)d_{\omega,q}x\right)^2 d\varrho_{q^{-s}}(\mu)>\frac{1}{2}(\sin\beta)^2\left\{\varrho_{q^{-s}}(\kappa)-\varrho_{q^{-s}}(-\kappa)\right\},$$

which proves the inequality (18).

If sinβ=0, then the function $f_h(x)=\begin{pmatrix} f_{1h}(x) \\ f_{2h}(x) \end{pmatrix}$ is defined by the formula

$$f_{1h}(x)=\begin{cases} \frac{1}{h}, & \omega_0 \le x < h, f_{2h}(x)=0. \\ 0, x \ge h. \end{cases}$$

Thus we obtain the inequality (18) by applying the Parseval equality.

Let ϱ be any non-decreasing function on -∞<μ<∞. Denote by $L_\varrho^2(\mathbb{R})$ the Hilbert space of all functions $f:\mathbb{R}\to\mathbb{R}$ which are measurable with respect to the Lebesque-Stieltjes measure defined by ϱ and such that

$$\int_{-\infty}^{\infty} f^2(\mu)\,d\varrho(\mu)<\infty$$

with the inner product

$$(f,g)_\varrho := \int_{-\infty}^{\infty} f(\mu)g(\mu)\,d\varrho(\mu).$$

The Parseval Equality and Spectral Expansion Theorem

The main result of this work is the following theorem.

Theorem 14 *For the singular Hahn-Dirac system (3)-(4), there exists a non-decreasing function $\varrho(\mu)$ on -∞<μ<∞ with the following properties:*

(i) *If* $f(.)=\begin{pmatrix} f_1(.) \\ f_2(.) \end{pmatrix}\in L_{\omega,q}^2((\omega_0,\omega_0+q^{-s});\mathbb{R}^2)$, *there exists a function* $F\in L_\varrho^2(\mathbb{R})$ *such that*

$$\lim_{s\to\infty}\int_{-\infty}^{\infty}\left\{F(\mu)-\int_{\omega_0}^{\omega_0+q^{-s}}\left(f_1(x)\Phi_1(x,\mu)+f_2(x)\Phi_2(x,\mu)\right)d_{\omega,q}x\right\}d\varrho(\mu)=0 \tag{20}$$

and the Parseval equality

$$\int_{\omega_0}^{\infty}\left(f_1^2(x)+f_2^2(x)\right)d_{\omega,q}x=\int_{-\infty}^{\infty}F^2(\mu)\,d\varrho(\mu) \tag{21}$$

holds.

(ii) *The integrals* $\int_{-\infty}^{\infty}F(\mu)\Phi_1(x,\mu)\,d\varrho(\mu)$ and $\int_{-\infty}^{\infty}F(\mu)\Phi_2(x,\mu)\,d\varrho(\mu)$ *converge to* f_1 *and* f_2 *in* $L^2_{\omega,q}(\omega_0,\infty)$, *respectively. That is,*

$$\lim_{n\to\infty}\int_{\omega_0}^{\infty}\left(f_1(x)-\int_{-n}^{n}F(\mu)\Phi_1(x,\mu)\,d\varrho(\mu)\right)^2 d_{\omega,q}x=0,$$

$$\lim_{n\to\infty}\int_{\omega_0}^{\infty}\left(f_2(x)-\int_{-n}^{n}F(\mu)\Phi_2(x,\mu)\,d\varrho(\mu)\right)^2 d_{\omega,q}x=0.$$

Note that the function ϱ is called a *spectral function* for the singular system (3)-(4).

Proof Assume that the function $f_\xi(x)=\begin{pmatrix} f_{1\xi}(x)\\ f_{2\xi}(x)\end{pmatrix}$ satisfies the following conditions.

1) $f_\xi(x)$ vanishes outside the interval $[\omega_0,\omega_0+q^{-\xi}]$, $q^{-\xi}<q^{-s}$.
2) The functions $f_\xi(x)$ and $D_{\omega,q}f_\xi(x)$ are continuous at ω_0.
3) $f_\xi(x)$ satisfies the boundary conditions (4) and (5).

If we apply to $f_\xi(x)$ the Parseval equality (17), then we obtain

$$\int_{\omega_0}^{\omega_0+q^{-\xi}}\left(f_{1\xi}^2(x)+f_{2\xi}^2(x)\right)d_{\omega,q}x=\int_{-\infty}^{\infty}F_\xi^2(\mu)\,d\varrho(\mu), \tag{22}$$

where

$$F_{\xi}(\mu)=\int_{\omega_0}^{\omega_0+q^{-\xi}}\left(f_{1\xi}(x)\Phi_1(x,\mu)+f_{2\xi}(x)\Phi_2(x,\mu)\right)d_{\omega,q}x. \tag{23}$$

Since $\Phi(x,\mu)$ satisfies the system (3), it is seen that

$$\Phi_1(x,\mu)=\frac{1}{\mu}\left(-q^{-1}D_{-\omega q^{-1},q^{-1}}\Phi_2(x,\mu)+p(x)\Phi_1(x,\mu)\right),$$

$$\Phi_2(x,\mu)=\frac{1}{\mu}\left(D_{\omega,q}\Phi_1(x,\mu)+r(x)\Phi_2(x,\mu)\right).$$

By (23), we get

$$F_{\xi}(\mu)=\frac{1}{\mu}\int_{\omega_0}^{\omega_0+q^{-s}}\begin{pmatrix}f_{1\xi}(x)\left(-q^{-1}D_{-\omega q^{-1},q^{-1}}\Phi_2(x,\mu)+p(x)\Phi_1(x,\mu)\right)\\+f_{2\xi}(x)\left(D_{\omega,q}\Phi_1(x,\mu)+r(x)\Phi_2(x,\mu)\right)\end{pmatrix}d_{\omega,q}x$$

Since $f_{\xi}(x)$ vanishes in a neighborhood of the point ω_0+q^{-s} and, $f_{\xi}(x)$ and $\Phi(x,\mu)$ satisfy the boundary condition (15), we obtain

$$F_{\xi}(\mu)=\frac{1}{\mu}\int_{\omega_0}^{\omega_0+q^{-s}}\begin{pmatrix}\Phi_1(x,\mu)\left(-q^{-1}D_{-\omega q^{-1},q^{-1}}f_{2\xi}(x)+p(x)f_{1\xi}(x)\right)\\+\Phi_2(x,\mu)\left(D_{\omega,q}f_{1\xi}(x)+r(x)f_{2\xi}(x)\right)\end{pmatrix}d_{\omega,q}x,$$

via ω,q- integration by parts. For any finite $\kappa>0$, by using (17), we have

$$\int_{|\mu|>\kappa}F_{\xi}^2(\mu)d\varrho_{q^{-s}}(\mu)$$

$$\leq\frac{1}{\kappa^2}\int_{|\mu|>\kappa}\left\{\int_{\omega_0}^{\omega_0+q^{-s}}\left(\Phi_1(x,\mu)\left(-q^{-1}D_{-\omega q^{-1},q^{-1}}f_{2\xi}(x)+p(x)f_{1\xi}(x)\right)\right.\right.$$
$$\left.\left.+\Phi_2(x,\mu)\left(D_{\omega,q}f_{1\xi}(x)+r(x)f_{2\xi}(x)\right)d_{\omega,q}x\right\}^2d\varrho_{q^{-s}}(\mu)$$

$$\leq \frac{1}{\kappa^2}\int_{-\infty}^{\infty}\left\{\int_{\omega_0}^{\omega_0+q^{-s}}\left(\Phi_1(x,\mu)\left(-q^{-1}D_{-\omega q^{-1},q^{-1}}f_{2\xi}(x)+p(x)f_{1\xi}(x)\right)\right.\right.$$
$$\left.\left.+\Phi_2(x,\mu)\left(D_{\omega,q}f_{1\xi}(x)+r(x)f_{2\xi}(x)\right)\right)d_{\omega,q}x\right\}^2 d\varrho_{q^{-s}}(\mu)$$
$$=\frac{1}{\kappa^2}\int_{\omega_0}^{\omega_0+q^{-\xi}}\left(\left(-q^{-1}D_{-\omega q^{-1},q^{-1}}f_{2\xi}(x)+p(x)f_{1\xi}(x)\right)^2+\left(D_{\omega,q}f_{1\xi}(x)+r(x)f_{2\xi}(x)\right)^2\right)d_{\omega,q}x.$$

From (22), we see that

$$\left|\int_{\omega_0}^{\omega_0+q^{-\xi}}\left(f_{1\xi}^2(x)+f_{2\xi}^2(x)\right)d_{\omega,q}x-\int_{-\kappa}^{\kappa}F_\xi^2(\mu)\,d\varrho_{q^{-s}}(\mu)\right|$$
$$\leq\frac{1}{\kappa^2}\int_{\omega_0}^{\omega_0+q^{-\xi}}\left(\left(-q^{-1}D_{-\omega q^{-1},q^{-1}}f_{2\xi}(x)+p(x)f_{1\xi}(x)\right)^2+\left(D_{\omega,q}f_{1\xi}(x)+r(x)f_{2\xi}(x)\right)^2\right)d_{\omega,q}x. \tag{24}$$

By Lemma 13, the set $\left\{\varrho_{q^{-s}}(\mu)\right\}$ is bounded. By using Theorems 9 and 10, we can find a sequence $\{q^{-s}\}$ such that the sequence $\varrho_{q^{-s}}(\mu)$ converges to a monotone function $\varrho(\mu)$ as $s\to\infty$. Passing to the limit with respect to $\{q^{-s}\}$ in (24), we get

$$\left|\int_{\omega_0}^{\omega_0+q^{-\xi}}\left(f_{1\xi}^2(x)+f_{2\xi}^2(x)\right)d_{\omega,q}x-\int_{-\kappa}^{\kappa}F_\xi^2(\mu)\,d\varrho(\mu)\right|$$
$$\leq\frac{1}{\kappa^2}\int_{\omega_0}^{\omega_0+q^{-\xi}}\left(\left(-q^{-1}D_{-\omega q^{-1},q^{-1}}f_{2\xi}(x)+p(x)f_{1\xi}(x)\right)^2+\left(D_{\omega,q}f_{1\xi}(x)+r(x)f_{2\xi}(x)\right)^2\right)d_{\omega,q}x.$$

Hence, letting $\kappa\to\infty$, we obtain

$$\int_{\omega_0}^{\omega_0+q^{-\xi}}\left(f_{1\xi}^2(x)+f_{2\xi}^2(x)\right)d_{\omega,q}x=\int_{-\infty}^{\infty}F_\xi^2(\mu)\,d\varrho(\mu).$$

Now, let f be an arbitrary function on $L^2_{\omega,q}((\omega_0,\infty);\mathbb{R}^2)$. It is known that there exists a sequence $\left\{f_\xi(x)\right\}$ of functions satisfying the conditions 1-3 and such that

$$\lim_{\xi\to\infty}\int_{\omega_0}^{\infty}\left\|f(x)-f_{\xi}(x)\right\|_E^2 d_{\omega,q}x=0.$$

Let

$$F_{\xi}(\mu)=\int_{\omega_0}^{\infty}\left\|f_{\xi}^{T}(x)\Phi(x,\mu)\right\|_E d_{\omega,q}x.$$

Then, we have

$$\int_{\omega_0}^{\infty}\left(f_{1\xi}^2(x)+f_{2\xi}^2(x)\right)d_{\omega,q}x=\int_{-\infty}^{\infty}F_{\xi}^2(\mu)d\varrho(\mu).$$

Since

$$\int_{\omega_0}^{\infty}\left\|f_{\xi_1}(x)-f_{\xi_2}(x)\right\|_E^2 d_{\omega,q}x\to 0 \text{ as } \xi_1,\xi_2\to\infty,$$

we have

$$\int_{-\infty}^{\infty}\left(F_{\xi_1}(\mu)-F_{\xi_2}(\mu)\right)^2 d\varrho(\mu)=\int_{\omega_0}^{\infty}\left\|f_{\xi_1}(x)-f_{\xi_2}(x)\right\|_E^2 d_{\omega,q}x\to 0$$

as $\xi_1,\xi_2\to\infty$. Consequently, there is a limit function F which satisfies

$$\int_{\omega_0}^{\infty}\left\|f(x)\right\|_E^2 d_{\omega,q}x=\int_{-\infty}^{\infty}F^2(\mu)d\varrho(\mu),$$

by the completeness of the space $L_{\varrho}^2(\mathbb{R})$.

The next goal is to show that the sequence

$$K_{\xi}(\mu)=\int_{\omega_0}^{\omega_0+q^{-\xi}}\left[f_1(x)\Phi_1(x,\mu)+f_2(x)\Phi_2(x,\mu)\right]d_{\omega,q}x$$

converges as $\xi\to\infty$ to F in the metric of the space $L^2_\varrho(\mathbb{R})$. Let g be another function in $L^2_{\omega,q}((\omega_0,\infty);\mathbb{R}^2)$. By similar arguments, let $G(\mu)$ be defined by g. It is clear that

$$\int_{\omega_0}^{\infty}\|f(x)-g(x)\|_E^2\, d_{\omega,q}x=\int_{-\infty}^{\infty}\{F(\mu)-G(\mu)\}^2\, d\varrho(\mu).$$

Set

$$g(x)=\begin{cases} f(x), & x\in\left[\omega_0,q^{-\xi}\right] \\ 0, & x\in\left(q^{-\xi},\infty\right). \end{cases}$$

Then, we have

$$\int_{-\infty}^{\infty}\{F(\mu)-K_\xi(\mu)\}^2\, d\varrho(\mu)=\int_{q^{-\xi}}^{\infty}\left(f_1^2(x)+f_2^2(x)\right)d_{\omega,q}x\to 0(\xi\to\infty),$$

which proves that K_ξ converges to F in $L^2_\varrho(\mathbb{R})$ as $\xi\to\infty$. This proves (i).

Now, (ii) will be proved. Suppose that

$$f(.)=\begin{pmatrix} f_1(.) \\ f_2(.) \end{pmatrix}, g(.)=\begin{pmatrix} g_1(.) \\ g_2(.) \end{pmatrix}\in L^2_{\omega,q}((\omega_0,\infty);\mathbb{R}^2)$$

and, $F(\mu)$ and $G(\mu)$ are their Fourier transforms, respectively. Then $F\pm G$ are the transforms of $f\pm g$. Consequently, by (21), we have

$$\int_{\omega_0}^{\infty}\left(\left[f_1(x)+g_1(x)\right]^2+\left[f_2(x)+g_2(x)\right]^2\right)d_{\omega,q}x=\int_{-\infty}^{\infty}\{F(\mu)+G(\mu)\}^2\, d\varrho(\mu),$$

$$\int_{\omega_0}^{\infty}\left(\left[f_1(x)-g_1(x)\right]^2+\left[f_2(x)-g_2(x)\right]^2\right)d_{\omega,q}x=\int_{-\infty}^{\infty}\{F(\mu)-G(\mu)\}^2\, d\varrho(\mu).$$

Subtracting the second relation from the first, we get

$$\int_{\omega_0}^{\infty}\left[f_1(x)g_1(x)+f_2(x)g_2(x)\right]^2 d_{\omega,q}x=\int_{-\infty}^{\infty}F(\mu)G(\mu)d\varrho(\mu) \tag{25}$$

which is called the *generalized Parseval equality*.

Set

$$f_\tau(x)=\begin{pmatrix}\int_{-\tau}^{\tau}F(\mu)\Phi_1(x,\mu)d\varrho(\mu)\\ \int_{-\tau}^{\tau}F(\mu)\Phi_2(x,\mu)d\varrho(\mu)\end{pmatrix},$$

where F is the function defined in (20). Let $g(.)=\begin{pmatrix}g_1(.)\\ g_2(.)\end{pmatrix}$ be a vector-function which is equal to zero outside the finite interval $\left[\omega_0,\omega_0+q^{-\vartheta}\right]$, where $\vartheta\in\mathbb{N}$. Thus we obtain

$$\begin{aligned}(f_\tau,g)&=\int_{\omega_0}^{\omega_0+q^{-\vartheta}}\left\{\int_{-\tau}^{\tau}F(\mu)\Phi_1(x,\mu)d\varrho(\mu)\right\}g_1(x)d_{\omega,q}x\\&+\int_{\omega_0}^{\omega_0+q^{-\vartheta}}\left\{\int_{-\tau}^{\tau}F(\mu)\Phi_2(x,\mu)d\varrho(\mu)\right\}g_2(x)d_{\omega,q}x\\&=\int_{-\tau}^{\tau}F(\mu)\left\{\int_{\omega_0}^{\omega_0+q^{-\vartheta}}\Phi_1(x,\mu)g_1(x)d_{\omega,q}x\right\}d\varrho(\mu)\\&+\int_{-\tau}^{\tau}F(\mu)\left\{\int_{\omega_0}^{\omega_0+q^{-\vartheta}}\Phi_2(x,\mu)g_2(x)d_{\omega,q}x\right\}d\varrho(\mu)\\&=\int_{-\tau}^{\tau}F(\mu)G(\mu)d\varrho(\mu)\end{aligned} \tag{26}$$

From (25), we get

$$(f,g)=\int_{-\infty}^{\infty}F(\mu)G(\mu)d\varrho(\mu). \quad (27$$

By (26) and (27), we have

$$(f_\tau - f,g)=\int_{|\mu|>\tau}F(\mu)G(\mu)d\varrho(\mu).$$

By using the Cauchy-Schwarz inequality, we obtain

$$|(f_\tau - f,g)|^2 \le \int_{|\mu|>\tau}F^2(\mu)d\varrho(\mu)\int_{|\mu|>\tau}G^2(\mu)d\varrho(\mu)\le \int_{|\mu|>\tau}F^2(\mu)d\varrho(\mu)\int_{-\infty}^{\infty}G^2(\mu)d\varrho(\mu).$$

If this inequality is applied to the function

$$g(x)=\begin{cases} f_\tau(x)-f(x), & x\in\left[\omega_0,\omega_0+q^{-\vartheta}\right], \\ 0, & x\in(\omega_0+q^{-\vartheta},\infty), \end{cases}$$

then we get

$$\|f_\tau - f\|^2 \le \int_{-\infty}^{\infty}F^2(\mu)d\varrho(\mu).$$

Letting $\tau\to\infty$ yields the desired results since the right-hand side does not depend on ϑ.

REFERENCES

Aldwoah, K. A. (2009). *Generalized time scales and associated difference equations* (Ph.D. Thesis). Cairo University.

Allahverdiev, B. P., Eryılmaz, A., & Tuna, H. (2017). Dissipative Sturm-Liouville operators with a spectral parameter in the boundary condition on bounded time scales. *Electronic Journal of Differential Equations*, *95*, 1–13.

Allahverdiev, B. P., & Tuna, H. (2018a). An expansion theorem for q-Sturm-Liouville operators on the whole line. *Turkish Journal of Mathematics, 42*, 1060–1071.

Allahverdiev B. P. & Tuna H. (2018b). Spectral expansion for the singular Dirac system with impulsive conditions. Turk J Math. 42.2527 – 2545.

Allahverdiev B. P. & Tuna H. (2019a). Eigenfunction expansion for singular Sturm-Liouville problems with transmission conditions, Electron. J. Differential Equations, Vol. 2019. No. 03. pp. 1-10.

Allahverdiev, B. P., & Tuna, H. (2019b). The spectral expansion for the Hahn–Dirac system on the whole line. *Turkish Journal of Mathematics, 43*, 1668–1687.

Allahverdiev B. P. & Tuna H. (2019c). On Expansion in Eigenfunction for q-Dirac Systems on the Whole Line, Mathematical Reports. Vol. 21(71), No. 3/2019.

Allahverdiev B. P. & Tuna H. (2019d). Eigenfunction Expansion in the Singular Case for Dirac Systems on Time Scales . Konuralp Journal of Mathematics, 7 (1).128-135.

Allahverdiev, B. P., & Tuna, H. (2019e). Eigenfunction expansion in the singular case for q-SturmLiouville operators. *Caspian Journal of Mathematical Sciences., 8*(2), 91–102.

Álvarez-Nodarse, R. (2006). On characterizations of classical polynomials. *Journal of Computational and Applied Mathematics, 196*(1), 320–337. doi:10.1016/j.cam.2005.06.046

Amrein, W. O., Hinz, A. M., & Pearson, D. B. (2005). *Sturm-Liouville theory, past and present. Basel.* Boston: Birkhauser Verlag. doi:10.1007/3-7643-7359-8

Annaby, M. H., Hamza, A. E., & Aldwoah, K. A. (2012). Hahn difference operator and associated Jackson--Nörlund integrals. *Journal of Optimization Theory and Applications, 154*(1), 133–153. doi:10.100710957-012-9987-7

Annaby, M. H., Hamza, A. E., & Makharesh, S. D. (2018). A Sturm-Liouville theory for Hahn difference operator. In X. Li & Z. Nashed (Eds.), Frontiers of Orthogonal Polynomials and Series (pp. 35–84). Singapore: World Scientific. doi:10.1142/9789813228887_0004

Aydemir, K., & Mukhtarov, O. S. (2017). Class of Sturm.Liouville Problems with Eigenparameter Dependent Transmission Conditions. *Numerical Functional Analysis and Optimization, 38*(10), 1260-1275.

Aydemir, K., Olgar, H., Mukhtarov, O. S., & Muhtarov, F. (2018). Differential operator equations with interface conditions in modifed direct sum spaces. *Filomat, 32*(3), 921–931. doi:10.2298/FIL1803921A

Berezanskii, J. M. (1968). *Expansions in Eigenfunctions of Selfadjoint Operators.* Providence, RI: Amer. Math. Soc.

Dobrogowska, A., & Odzijewicz, A. (2006). Second order $q-$ difference equations solvable by factorization method. *J. Comput. Appl. Math., 193*(1), 319-346.

Guseinov, G. S. (2007). Eigenfunction expansions for a Sturm-Liouville problem on time scales. *International Journal of Differential Equations, 2*(1), 93–104.

Guseinov, G. S. (2008). An expansion theorem for a Sturm-Liouville operator on semi-unbounded time scales. *Adv. Dyn. Syst. Appl., 3*(1), 147–160.

Hahn, W. (1949). Über orthogonalpolynome, die $q-$. *Mathematische Nachrichten, 2*(1-2), 4–34. doi:10.1002/mana.19490020103

Hahn, W. (1983). Ein beitrag zur theorie der orthogonalpolynome. *Monatshefte für Mathematik, 95*(1), 19–24. doi:10.1007/BF01301144

Hamza, A. E., & Ahmed, S. A. (2013). Theory of linear Hahn difference equations. *J. Adv. Math., 4*(2), 440–460.

Hamza, A. E., & Ahmed, S. A. (2013a). Existence and uniqueness of solutions of Hahn difference equations. *Adv. Difference Equations, 316*, 1–15.

Hamza, A. E., & Makharesh, S. D. (2016). Leibniz' rule and Fubinis theorem associated with Hahn difference operator. *Journal of Advanced Mathematical, 12*(6), 6335–6345.

Hira, F. (2018). *Dirac systems associated with Hahn difference operator.* arXiv:1806.00710v1

Jackson, F. H. (1910). Difference equations. *American Journal of Mathematics, 32*(4), 305–314. doi:10.2307/2370183

Kolmogorov, A. N., & Fomin, S. V. (1970). *Introductory Real Analysis* (R. A. Silverman, Trans.). New York: Dover Publications.

Kwon, K. H., Lee, D. W., Park, S. B., & Yoo, B. H. (1998). Hahn class orthogonal polynomials. *Kyungpook Math. J., 38*, 259–281.

Lesky, P. A. (2005). Eine Charakterisierung der klassischen kontinuierlichen, diskretenund Orthgonalpolynome. Aachen: Shaker.

Levinson, N. (1951). A simplified proof of the expansion theorem for singular second order linear differential equations. *Duke Math. J., 18*, 57-71.

Levitan, B. M., & Sargsjan, I. S. (1991). *Sturm-Liouville and Dirac Operators*. Dordrecht, The Netherlands: Kluwer Academic Publishers Group.

Mukhtarov, O. S., & Aydemir, K. (2018). Basis properties of the eigenfunctions of two-interval Sturm.Liouville problems. *Analysis and Mathematical Physics*, 1-20.

Naimark, M. A. (1969). Linear Differential Operators (2nd ed.). Frederick Ungar Publishing.

Olgar, H., Mukhtarov, O. S., & Aydemir, K. (2018). Some properties of eigenvalues and generalized eigenvectors of one boundary value problem. *Filomat, 32*(3), 911-920.

Petronilho, J. (2007). Generic formulas for the values at the singular points of some special monic classical $H_{q,\omega}$ – . *Journal of Computational and Applied Mathematics, 205*(1), 314–324. doi:10.1016/j.cam.2006.05.005

Sitthiwirattham, T. (2016). On a nonlocal boundary value problem for nonlinear second-order Hahn difference equation with two different derivatives. Advances in Difference Equations, 2016(1).

Stone, M. H. (1926). A comparison of the series of Fourier and Birkhoff. *Transactions of the American Mathematical Society, 28*(4), 695–761. doi:10.1090/S0002-9947-1926-1501372-6

Stone, M. H. (1932). *Linear Transformations in Hilbert Space and Their Application to Analysis*. Providence, RI: Amer. Math. Soc.

Sturm, C., & Liouville, J. (1837). Extrait d.un M'emoire sur le d'eveloppement des fonctions en s'eries dont les di¤erents termes sont assujettis. a satisfaire. a une meme equation differentielle lineaire, contenant un parametre variable. *Journal de Mathématiques Pures et Appliquées, 2*, 220–223.

Thaller, B. (1992). *The Dirac Equation, Berlin*. Heidelberg, Germany: Springer. doi:10.1007/978-3-662-02753-0

Titchmarsh, E. C. (1962). *Eigenfunction Expansions Associated with Second-Order Differential Equations. Part I* (2nd ed.). Oxford, UK: Clarendon Press. doi:10.1063/1.3058324

Tuna, H. (2014). Completeness of the rootvectors of a dissipative Sturm–Liouville operators on time scales. *Appl Math Comput., 228*, 108–115.

Tuna, H. (2016). Completeness Theorem for the Dissipative Sturm-Liouville Operator on Bounded Time Scales. *Indian Journal of Pure and Applied Mathematics, 47*(3), 535–544. doi:10.100713226-016-0196-1

Tuna, H., & Eryılmaz, A. (2013). Dissipative Sturm-Liouville Operators with Transmission Conditions. *Abstract and Applied Analysis, 2013*, 1–7. doi:10.1155/2013/248740

Tuna, H., & Özek, M. A. (2017). The one-dimensional Schrödinger operator on bounded time scales. *Mathematical Methods in the Applied Sciences, 40*(1), 78–83. doi:10.1002/mma.3966

Weidmann, J. (1987). *Spectral Theory of Ordinary Di¤erential Operators. Lecture Notes in Mathematics 1258*. Berlin: Springer.

Weyl, H. (1910). Über gewöhnliche Differentialgleichungen mit Singularitäten und die zugehörigen Entwicklungen willkürlicher Funktionen. *Mathematische Annalen, 68*(2), 220–269. doi:10.1007/BF01474161

Yosida, K. (1950). On Titchmarsh-Kodaira formula concerning Weyl-Stone eingenfunction expansion. *Nagoya Math. J., 1*, 49-58.

Yosida, K. (1960). *Lectures on Differential and Integral Equations*. New York: Springer.

Zettl, A. (2005). Sturm.Liouville Theory. In *Mathematical Surveys and Mono-graphs* (Vol. 121). American Mathematical Society.

Chapter 11
The Stability of an Epidemic Model With Piecewise Constant Argument by Lyapunov-Razumikhin Method

Nur Cengiz
Süleyman Demirel University, Turkey

Duygu Aruğaslan-Çinçin
Süleyman Demirel University, Turkey

ABSTRACT

The authors propose a nonlinear epidemic model by developing it with generalized piecewise constant argument (GPCA) introduced by Akhmet. The authors investigate invariance region for the considered model. For the taken model into consideration, they obtain a useful inequality concerning relation between the values of the solutions at the deviation argument and at any time for the epidemic model. The authors reach sufficient conditions for the existence and uniqueness of the solutions. Then, based on Lyapunov-Razumikhin method developed by Akhmet and Aruğaslan for the differential equations with generalized piecewise constant argument (EPCAG), sufficient conditions for the stability of the trivial equilibrium and the positive equilibrium are investigated. Thus, the theoretical results concerning the uniform stability of the equilibriums are given.

DOI: 10.4018/978-1-7998-0134-4.ch011

INTRODUCTION

In the present chapter, the main objective is to provide information on the stability analysis of the susceptible-infected-susceptible (SIS) model established by Cooke (1979). This model is developed by GPCA which is defined by Akhmet (2007a, 2007b). Firstly, invariance region for the considered equations with GPCA is investigated by the authors, respectively. The inequalities describing the relationship between the values received at this piecewise function and at any given time for the solutions of the models are given. This inequality is useful and important in the proofs of stability analysis in the sense of Lyapunov-Razumikhin method. It is aimed to obtain sufficient conditions guaranteeing the existence and uniqueness of the solutions of the proposed model. Afterwards, it is aimed to perform stability analysis with the help of Lyapunov-Razumikhin method developed by Akhmet and Aruğaslan (2009) (Aruğaslan, 2009) for EPCAG. Based on the relevant method, the conditions that guarantee the uniform stability and the uniform asymptotic stability of the trivial equilibrium and the positive equilibrium are presented. The obtained theoretical results depend on the parameters of the equation.

BACKGROUND

The mathematical evaluation of the problems encountered in real life and the interpretation of their past and future dynamics has been and continues to be a subject that attracted the attention of the scientific world. In this direction, it is possible to express these problems as mathematical models by differential equations. With the help of the qualitative theory of differential equations, informations about the behaviors of these models can be provided. However, the fact that these information presents distant results from the reality phenomenon strengthens the likelihood of misleading the behavior of real processes. In this situation, it is possible to reach the findings that will affect the life negatively by the models which do not fully reflect the dynamics of real life problems. In order to overcome this issue, it would be a step in the right direction to consider differential equations with deviation argument instead of ordinary differential equations. Because, differential equations with deviation argument improve models representing real life problems, and moreover they contribute to understanding how a situation in the past of these problems affect the current and subsequent state of them. In other words, the effects of a past value of the models that has been constructed and developed with the help of such equations on the current behavior can be observed. Therefore, the efforts to achieve the development of such equations have received considerable attention by many scientists. As a result of these efforts, the qualitative theory of such equations has shown a serious development process. It is possible to see that this process introduces differential equations which have arguments such as delay, piecewise constant argument, generalized piecewise constant argument (Cooke & Wiener, 1984; Akhmet 2007a, 2007b).

The use of such differential equations has led to the construction of remarkable models in many areas such as physics, engineering, medicine, biology, population dynamics, chemistry and economics. Thus, the scientific, technological, industrial and economic development of the society could be supported, and evaluations about the health of the society could be made possible. Of course, as time goes on, contributions to these supports and evaluations continue. Because, it is necessary to make different contributions to the changing problems of the developing world. In this respect, epidemic models can mathematically represent the problems encountered in the fields of medicine, biology and population dynamics. The negligence of the handling of such problems and the negligence of their scientific examination can affect the individuals and the development of the world with serious consequences. Because, the events affecting the health of the society directly concern people. It is not difficult to observe this fact. In order to understand the effects of the outbreaks affecting societies and cultures on populations, it is sufficient to examine the statistical numerals held in history. For this purpose, the following delayed SIS model

$$y'(t) = by(t-T)\left[1-y(t)\right]-cy(t). \tag{1}$$

was established by Cooke (1979). Here, b,c are positive and constant parameters corresponding to infection and recovery rates, respectively. The proportion of humans in the community who are infectious at time t and the proportion who are susceptible are denoted by $y(t)$ and $S(t)$ respectively. Then, Cooke (1979) takes $y(t)+S(t)=1$ by assuming that the infection in humans confers negligible immunity and does not result in death or isolation. At the end of this time expressed in this assumption, it is assumed that the vector can infect a susceptible human (Cooke, 1979; Huang et al., 2016). Additionally, denote by $z(t)$ the number of infectious vectors in the community at time t according to Cooke (1979). The population is evaluated in two classes: susceptible and infectious. This evaluation supports homogeneous mixing of the vector and human populations (Busenberg & Cooke, 1980). The model (1) with delay is quite valuable epidemic model in population dynamics. So, this model has attracted the attention of many scientists.

According to Cooke (1979), the equation (1) with delay has representation of the proportion of infective individuals by considering the following assumptions arising from the dynamic of the spread of a communicable disease:

(A) The infection is transmitted by a vector like mosquito to individuals.
(B) The infection in the individual gives an immunity, but does not cause death or isolation.
(C) The population is fixed without considering the change in births, deaths and immigration to the population.
(D) When a susceptible vector is infected by a person, there is a fixed time T during which the infectious agent develops in the vector. At the end of this time, the vector can infect a susceptible human.
(E) Human and vector populations have a homogeneous mixing.
(F) The rate of recovery of infected people is positive constant c.
(G) Th vector population is very large and $z(t)$ is simply proportional to $y(t\text{-}T)$. The infection is transmitted by a vector like mosquito to individuals.

In ligth of above descriptions and assumptions, the multiple of $S(t)z(t)$ means the number of new infections per unit time, and so the time-dependent change of the proportion of infectious humans in the community is expressed by the differential equation (1) (Cooke, 1979).

In addition, differential equations have a theoretical background to examine the development of a process. The examination of structures only developed by ordinary differential equations may not yield the expected results. As a remedy to this situation, differential equations with deviation arguments can be used. Because, such equations allow the models to be established more closely to their nature. Moreover, the dynamics of the models can be analyzed in more detail by them. In this respect, EPCAG introduced by Akhmet is of great importance in order to understand the problems encountered in many areas better (Akhmet, 2007a, 2007b). The deviation argument defined in these differential equations is a piecewise function that provides certain properties. This piecewise function makes it possible to choose arbitrary argument and the difference between the arguments is arbitrary. Furthermore, this argument is a discontinuous argument. This includes a remarkable situation in the modeling of actual processes. The greatest integer function, which is considered as the deviation argument in the differential equations with piecewise constant argument, permits to take always one unit of change between the arguments. However, EPCAG have a more general form of deviation argument than the argument contained in the differential equations with piecewise constant argument. Such equations have enabled many theoretical and practical studies to be carried out (Akhmet, 2008a, 2008b, 2010, 2011, 2014; Akhmet & Aruğaslan, 2009; Akhmet, Aruğaslan, & Yılmaz, 2010a, 2010b; Aruğaslan, 2009; Aruğaslan & Cengiz, 2017, 2018; Aruğaslan & Güzel, 2015). Besides, the construction of models with such differential equations brings about the process of examining the qualitative behaviors of these models. In this respect, there is an extensive literature on stability analysis of models. In this sense, the stability of the systems without reaching their solutions can be examined in the light of Lyapunov-Razumikhin (1956) and Lyapunov-Krasovskii (1963) methods. While it is desired to approach the phenomenon of reality, the structures of the models become more complex. In this case, the usefulness of Lyapunov methods is obvious and noteworthy. Furthermore, the introduction of Lyapunov-Razumikhin method by Akhmet and Aruğaslan (2009) for EPCAG is noteworthy (Aruğaslan, 2009). Because stability analysis with other methods for such equations brings about more intensive operations and calculations unlike Lyapunov-Razumikhin method for EPCAG.

As the subject of the present chapter, the model (1) is developed by EPCAG which is introduced by Akhmet (2007a, 2007b). Thus, under favour of the GPCA, the change between arguments can be arbitrarily chosen and so the model can be examined by a more natural approach. Basically, the authors aim to perform the stability examination for the developed model by GPCA. Then, the stability examinations with the help of Lyapunov-Razumikhin method for EPCAG (Akhmet & Aruğaslan, 2009) are performed for the epidemic model with GPCA. This method enables stability analysis to be performed more easily by requiring less operation.

MAIN FOCUS OF THE CHAPTER

Let R,N,N_0 and R^+ be sets of all real numbers, natural numbers, non-negative integers and non-negative real numbers, respectively, i.e., $R=(-\infty,\infty)$, $N_0=\{1,2,\ldots\}$ and $R^+=[0,\infty)$. Fix a real-valued sequence $\{\theta_i\}$, $i\in N_0$ such that $0=\theta_0<\theta_1<\ldots<\theta_i<\ldots$ with $\theta_i\rightarrow\infty$ as $i\rightarrow\infty$. In the present chapter, the authors address the SIS model with GPCA:

$$x'(t)=bx(\beta(t))\left[1-x(t)\right]-cx(t). \tag{2}$$

Here, in the equation (2), $x\in R$, $t\in R^+$, $\beta(t)=\theta_i$ if $t\in[\theta_i,\theta_{i+1})$, $i\in N_0$. For the sequence $\{\theta_i\}$, $i\in N_0$, let $0\leq\theta_i<\theta_{i+1}$ for all $i\in N_0$ and $\theta_i\rightarrow\infty$ as $i\rightarrow\infty$. Let us assume without loss of generality that $\theta_i<t_0\leq\theta_{i+1}$ .for some $i\in N_0$. In the nonlinear epidemic model (2), b,c are positive and constant parameters corresponding to infection and recovery rates, respectively. Equation (2) has a representation of the proportion of infective individuals by considering the expressions and the assumptions (A)-(H) in Background above.

As the main goal, the authors investigate the stability of the trivial equilibrium and the positive equilibrium of (2). While investigating the stability, they take Lyapunov-Razumikhin method developed by Akhmet and Aruğaslan (2009) for EPCAG into consideration. For this aim, as the first step, the authors examine positive invariance region for the considered model. Then, the authors indicate an important auxiliary result given a relation between the values of solutions at the deviation argument $\beta(t)$ and at any time t of the proposed model. This auxiliary result is useful in the proofs of the stability theory within the scope of Lyapunov-Razumikhin method. Next, the authors obtain sufficient conditions guaranteeing the existence and uniqueness of the solutions of the epidemic model with GPCA. By taking notice of the Lyapunov-Razumikhin method for EPCAG (Akhmet & Aruğaslan, 2009), the authors investigate the the uniform stability of the trivial equilibrium and the positive equilibrium. During all these investigations, the nature of the solutions is evaluated within the biologically meaningful range [0,1] as required by the examination performed for the positive invariance region.

Note that the equation (2) has the trivial equilibrium $x^*=0$ and the positive equilibrium $x^* = 1 - \frac{c}{b}, c < b$. If the positive equilibrium $x^* = 1 - \frac{c}{b}, c < b$ of (2) is transformed into the trivial equilibrium, by $y=x-x^*$ then the following equation is reached:

$$y'(t) = -by(t)\left[1 + y(\beta(t))\right] + cy(\beta(t)). \tag{3}$$

Now, let us describe crucial sets of functions:

$$\mathcal{K} = \left\{a \in \boldsymbol{C}\left(R^+, R^+\right) : \text{strictly increasing and } a(0) = 0\right\},.$$

$$\mathcal{M} = \left\{d \in \boldsymbol{C}\left(R^+, R^+\right) : d(0) = 0,\ d(s) > 0 \text{ for } s > 0\right\},.$$

which will be used in the stability examinations.

The definitions related to solutions of the nonlinear epidemic models (2) and (3) are given by Definition 1 and Definiton 2, respectively.

Definiton 1 A function $x(t)$ is a solution of (3) ((4)) on R^+ if:

(i) $x(t)$ is continuous on R^+.

(ii) the derivative $x'(t)$.exists for $t \in R^+$ with the possible exception of the points θ_i, $i \in N_0$, where one-sided derivatives exist;

(iii) equation (3) ((4)) is satisfied by $x(t)$.on each interval $(\theta_i, \theta_{i+1}), i \in N_0$, .and it holds for the right derivative of $x(t)$.at the points θ_i, $i \in N_0$.

Definiton 2 A function $y(t)$.is a solution of (5) ((6)) on R^+ if:

(i) $y(t)$ is continuous on R^+.

(ii) the derivative $y'(t)$.exists for $t \in R^+$ with the possible exception of the points θ_i, $i \in N_0$, where one-sided derivatives exist;

(iii) equation (5) ((6)) is satisfied by $y(t)$ on each interval $(\theta_i, \theta_{i+1}), i \in N_0$, .and it holds for the right derivative of $y(t)$ at the points θ_i, $i \in N_0$.

THE THEORETICAL RESULTS FOR (2)

The following assumptions will be needed throughout the chapter:

(C1) There exists a positive numbers $\underline{\theta}, \overline{\theta}$ .such that $\underline{\theta} \leq \theta_{i+1} - \theta_i \leq \overline{\theta}, i \in N_0$;

(C2) $\overline{\theta}\left(b+c+(b+c)\left(1+\overline{\theta}(b+c)\right)e^{\overline{\theta}(b+c)}\right) < 1$;

(C3) $3\overline{\theta}(c+b)e^{(c+b)\overline{\theta}} < 1$;

(C4) $b \leq c$;

(C5) $c < b$.

The Results Concerning the Positive Invariance

Now, the results concerning positive invariance for (2) are given by the lemma and theorems below. So, this results give an information about the invariance region of the solutions for (3). Take just solutions $x(t)$ with $0 \leq x(\theta_0) = x_0 \leq 1$.

Lemma 1 The equation (2) with $x(\theta_0) = x_0$ is equivalent to the following integral equation

$$x(t) = e^{-c(t-\theta_0)}x_0 + \int_{\theta_0}^{t} e^{-c(t-s)}bx(\beta(s))\left[1-x(s)\right]ds, t \in [\theta_0, \alpha). \tag{4}$$

Proof: *Necessity.* Let $x(t)$.be the solution of (2) with $x(\theta_0) = x_0$. Based on Definiton 1, (4) satisfies the equation (2) on each interval $[\theta_i, \theta_{i+1}), i \in N_0$. For $t \in [\theta_0, \theta_1)$,. the solution is

$$x(t) = e^{-c(t-\theta_0)}x_0 + \int_{\theta_0}^{t} e^{-c(t-s)}bx(\beta(s))\left[1-x(s)\right]ds..$$

Letting $t \to \theta_1$, by the continuity of the solutions, it is obtained

$$x_1 = x(\theta_1) = e^{-c(\theta_1-\theta_0)}x_0 + \int_{\theta_0}^{\theta_1} e^{-c(\theta_1-s)}bx(\beta(s))[1-x(s)]ds.$$

Therefore, (4) holds on $[\theta_0,\theta_1]$. Assume that (4) is valid on the interval $[\theta_0,\theta_k]$ for some $k \geq 1$. Then, for $t \in [\theta_k, \theta_{k+1})$, .it is true that

$$x(t) = e^{-c(t-\theta_k)}x_k + \int_{\theta_k}^{t} e^{-c(t-s)}bx(\beta(s))[1-x(s)]ds$$
$$= e^{-c(t-\theta_0)}x_0 + \int_{\theta_0}^{t} e^{-c(t-s)}bx(\beta(s))[1-x(s)]ds.$$

Letting $t \to \theta_{k+1}$. it can be seen that

$$x_{k+1} = x(\theta_{k+1}) = e^{-c(\theta_{k+1}-\theta_0)}x_0 + \int_{\theta_0}^{\theta_{k+1}} e^{-c(\theta_{k+1}-s)}bx(\beta(s))[1-x(s)]ds.$$

Thus, (4) holds on $[\theta_0, \theta_{k+1}]$. Based on induction method, it can be observed for all $t \geq \theta_0$.

Sufficiency. Let $x(t)$ be the solution of (2). Fix $i \in N_0$ and consider the interval $[\theta_i, \theta_{i+1})$. Differentiating (4), it can be seen that $x(t)$ holds (2). Letting $t \to \theta_i +$ and considering that $x(\beta(t))$.is right continuous function, it is seen that $x(t)$ satisfies (2) on $[\theta_i, \theta_{i+1})$.

Theorem 1 If $x:[\theta_0,\alpha) \to R$ is a solution of (2) for $\theta_0 < t < \theta_{i+1} \leq \alpha$ .atisfying the initial condition $0 \leq x(\theta_0) = x_0 \leq 1$ and $b \leq c$ holds, then the set $\Omega = \{x \in R: 0 \leq x \leq 1\}$ is positively invariant for (2). Moreover, if $0 < x_0 \leq 1$ then $0 < x(t) \leq 1$. If $0 < x_0 < 1$ then $0 < x(t) < 1$.

Proof: Let us assume without loss of generality that $\theta_i < t_0 \leq \theta_{i+1}$ .for some $i \in N_0$. Let $x(t):[\theta_0,\alpha) \to R$.be a solution of (2) throught the initial condition $x(\theta_0) = x_0$ satisfying $0 \leq x_0 \leq 1$. Then, the solution of (3) is equivalent to the following integral equation

$$x(t)=e^{-c(t-\theta_0)}x_0+\int_{\theta_0}^{t}e^{-c(t-s)}bx(\beta(s))\left[1-x(s)\right]ds, t\in[\theta_0,\alpha).$$

Now, suppose that x(t) does not satisfy the inequality $0\leq x(t)\leq 1$ for $0\leq\theta_0\leq t<\theta_{i+1}\leq\alpha, i\in N$. By continuity of $x(t)$, there exists a largest number μ. $0\leq\theta_0\leq\mu<\theta_{i+1}\leq\alpha$, $i\in N$ such that $0\leq x(t)\leq 1$ for $0\leq t\leq\mu$ and either

(i) $x(\mu)=0$ and $x(t)<0$ on $(\mu,\mu+\varepsilon)$ for some $\varepsilon>0$; or
(ii) $x(\mu)=1$ and $x(t)>1$ on $(\mu,\mu+\varepsilon)$ for some $\varepsilon>0$.

First, consider the case (i). This situation shows that $bx_0[1-x(t)]\geq 0$ and $-cx(t)>0$. Thus, while $0\leq\theta_0\leq t\leq\mu<\theta_1$ it is seen that

$$x'(t)=bx_0\left[1-x(t)\right]-cx(t)\geq -cx(t)>0,.$$

and then $x(t)$ is non-decreasing on $(\mu,\mu+\varepsilon)$ which is a contradiction. In detail, it can be seen that

$$x(t)=e^{-c(t-\theta_0)}x_0+\int_{\theta_0}^{t}e^{-c(t-s)}bx_0\left[1-x(s)\right]ds\geq e^{-c\bar{\theta}}x_0\geq 0.$$

on $[\theta_0,\theta_1)$. For $t=\theta_1$,

$$x(\theta_1)=x_1=e^{-c(\theta_1-\theta_0)}x_0+\int_{\theta_0}^{\theta_1}e^{-c(\theta_1-s)}bx_0\left[1-x(s)\right]ds\geq e^{-c\bar{\theta}}x_0\geq 0..$$

Therefore, the inequality $0\leq x(t)$ holds on $[\theta_0,\theta_1]$ when $0\leq x(\theta_0)=x_0\leq 1$. Moreover, by performing the operation on each interval $\left[\theta_i,\theta_{i+1}\right)$, $i=1,2,3,\ldots$, in a similiar manner, it can be proved that $x(t)$ satisfies $0\leq x(t)$ for all t in $[\theta_0,\alpha)$ while $0\leq x(\theta_0)=x_0\leq 1$.

Second, consider the case (ii). This case implies that $-cx(t)<-c$, $x_0[1-x(t)]\leq 0$. Thus, while $0\leq\theta_0\leq t\leq\mu<\theta_1$ it is seen that

$$x'(t)=bx_0\left[1-x(t)\right]-cx(t)<0-c<0.$$

and then $x(t)$ is non-increasing on $(\mu,\mu+\varepsilon)$ which is a contradiction. In detail, it can be seen on $[\theta_0,\theta_1)$ that

$$x(t)=e^{-c(t-\theta_0)}x_0+\int_{\theta_0}^{t}e^{-c(t-s)}bx_0\left[1-x(s)\right]ds\le e^{-c(t-\theta_0)}x_0\le e^{-c\underline{\theta}}x_0\le 1,.$$

which is a contradiction. For $t=\theta_1$,

$$x(\theta_1)=x_1=e^{-c(\theta_1-\theta_0)}x_0+\int_{\theta_0}^{\theta_1}e^{-c(\theta_1-s)}bx_0\left[1-x(s)\right]ds\le e^{-c(\theta_1-\theta_0)}x_0\le e^{-c\underline{\theta}}x_0\le 1..$$

Therefore, the inequality $x(t)\leq 1$ holds on $[\theta_0,\theta_1]$ when $0\leq x(\theta_0)=x_0\leq 1$. Moreover, by performing the operations on each interval $\left[\theta_i,\theta_{i+1}\right)$, $i=1,2,3,\ldots$, in a similiar manner, it can be proved that $x(t)$ satisfies $x(t)\leq 1$ for all t in $[\theta_0,\alpha)$ while $0\leq x(\theta_0)=x_0\leq 1$.

Thus, performing the operation on each interval $\left[\theta_i,\theta_{i+1}\right)$, $i=1,2,3,\ldots$, in a similiar manner, proves that $x(t)$ satisfies $0\leq x(t)\leq 1$ for all t in $[\theta_0,\alpha)$ while $0\leq x(\theta_0)=x_0\leq 1$.

Moreover, suppose that $0<x_0\leq 1$. By the above conclusion, it is known that x(t) satisfy the inequality $0\leq x(t)\leq 1$.for $\theta_0\leq t<\alpha$. From now on, it is shown that x(t) remains strictly positive. In the contrary, let $\bar{t}$ be the first point where $x(\bar{t})=0$ on $[\theta_0,\theta_1)$. Then, it is obvious that

$$x'(\bar{t})=bx_0\left[1-x(\bar{t})\right]-cx(\bar{t})=bx_0>0,.$$

which is a contradiction. Besides, it can be seen that x(t) remains strictly positive:

$$x(t)=e^{-c(t-\theta_0)}x_0+\int_{\theta_0}^{t}e^{-c(t-s)}bx_0\left[1-x(s)\right]ds\ge e^{-c\bar{\theta}}x_0>0\,.$$

on $[\theta_0,\theta_1)$. For $t=\theta_1$, it can be easily seen that $x_1>0$. Therefore, the inequality $0<x(t)\leq 1$ holds on $[\theta_0,\theta_1]$ when $0<x_0\leq 1$. Moreover, by performing the operations on each interval $\left[\theta_i,\theta_{i+1}\right)$, $i=1,2,3,\ldots$, in a similiar manner, it can be proved that $x(t)$ satisfies $0<x(t)\leq 1$ for all t in $[\theta_0,\alpha)$ while $0<x(\theta_0)=x_0\leq 1$.

Next, suppose that $0<x_0<1$. In this situation, it is shown that x(t) remains strictly less than one. On the contrary, let $\bar{t}$ .be the first point where $x(\bar{t})=1$ .on $[\theta_0,\theta_1)$. Hence, for $t\in[\theta_0,\theta_1)$.

$$x'(\bar{t})=bx_0\left[1-x(\bar{t})\right]-cx(\bar{t})=-c<0,.$$

which means that $x(t)$ does not exceed the value $x=1$, and it can be seen that x(t) remains strictly less than one:

$$x(t)=e^{-c(t-\theta_0)}x_0+\int_{\theta_0}^{t}e^{-c(t-s)}bx_0\left[1-x(s)\right]ds<e^{-c(t-\theta_0)}x_0+\int_{\theta_0}^{t}e^{-c(t-s)}bds$$

$$\leq e^{-c(t-\theta_0)}x_0+\int_{\theta_0}^{t}e^{-c(t-s)}cds=e^{-c(t-\theta_0)}x_0+x_0-e^{-c(t-\theta_0)}x_0<1,$$

and so $x(t)<1$ on $[\theta_0,\theta_1)$. For $t=\theta_1$, it can be easily seen that $x_1<1$. Therefore, the inequality $0<x(t)<1$ holds on $[\theta_0,\theta_1]$. Moreover, by performing the operation on each interval $[\theta_i,\theta_{i+1})$, $i=1,2,3,\ldots$, in a similiar manner, it can be proved that $x(t)$ satisfies $0<x(t)<1$ for all t in $[\theta_0,\alpha)$ while $0<x(\theta_0)=x_0<1$.

Besides, for (3), it is obvious that the invariance region Ω changes as $\Omega^*=\left\{y\in R:-1+\frac{c}{b}\leq y\leq\frac{c}{b}\right\}$ while transforming the positive equilibrium into trivial equilibrium.

The Auxiliary Results for the Solutions of (2)

Now, for (2), the lemmas containing an auxiliary result which is noteworthy for the proof of stability analysis in the sense of Lyapunov-Razumikhin method will be given.

Lemma 3 Let the assumptions (C1) and (C2) be satisfied. Then, the following inequality

$$x(\beta(t))\leq \hbar_1 x(t). \tag{5}$$

holds for all $t\geq 0$ $x\in\Omega$ and (2) where $\hbar_1=\left\{1-\bar{\theta}\left(b+(b+c)(1+\bar{\theta}b)e^{\bar{\theta}(b+c)}\right)\right\}^{-1}$.

Proof: Let us fix $t \in R^+$. Then there exists $k \in N_0$ such that $t \in [\theta_k, \theta_{k+1})$. By integral equation, the solution $x(t)$ of (2) can be written as

$$x(t) = x(\theta_k) + \int_{\theta_k}^{t} \left(bx(\theta_k) - bx(\theta_k)x(s) - cx(s) \right) ds, t \in [\theta_k, \theta_{k+1}),$$

and thus it is true that

$$|x(t)| \leq |x(\theta_k)| + \left| \int_{\theta_k}^{t} \left(b|x(\theta_k)| + b|x(\theta_k)||x(s)| + c|x(s)| \right) ds \right|.$$

By $x \in \Omega$ and then based on the Gronwall-Bellman Lemma,

$$x(t) \leq x(\theta_k) + \bar{\theta} bx(\theta_k) + \left| \int_{\theta_k}^{t} (b+c)x(s) ds \right| \leq (1 + \bar{\theta} b) x(\theta_k) e^{\bar{\theta}(b+c)} \tag{6}$$

is obtained. Besides, for $t \in [\theta_k, \theta_{k+1})$. it can be written

$$x(\theta_k) = x(t) - \int_{\theta_k}^{t} \left(bx(\theta_k) - bx(\theta_k)x(s) - cx(s) \right) ds.$$

The last equality gives

$$|x(\theta_k)| = x(\theta_k) = x(t) + \left| \int_{\theta_k}^{t} \left(bx(\theta_k) + bx(\theta_k)x(s) + cx(s) \right) ds \right|,$$

and then since $x \in \Omega$.

$$x(\theta_k) \leq x(t) + \bar{\theta} bx(\theta_k) + \left| \int_{\theta_k}^{t} (b+c)x(s) ds \right|$$

is obtained. The inequality (6) gives that

$$x(\theta_k) \le x(t) + \bar{\theta} b x(\theta_k) + \left| \int_{\theta_k}^{t} (b+c)(1+\bar{\theta} b) x(\theta_k) e^{\bar{\theta}(b+c)} ds \right|$$

$$= x(t) + \bar{\theta}\left(b + (b+c)(1+\bar{\theta} b) e^{\bar{\theta}(b+c)}\right) x(\theta_k).$$

Then, the following inequality

$$x(\theta_k) \le \left\{1 - \bar{\theta}\left(b + (b+c)(1+\bar{\theta} b) e^{\bar{\theta}(b+c)}\right)\right\}^{-1} x(t)$$

is reached. It follows from the condition (C2) that the inequality (5) for $t \in [\theta_k, \theta_{k+1})$. Hence, (5) holds for all $t \ge 0$.

Similiar result for (3) can be shown by the following lemma.

Lemma 4 Let the assumptions (C1) and (C2) be satisfied. Then, the following inequality

$$|y(\beta(t))| \le \hbar_2 |y(t)| \tag{7}$$

holds for all $t \ge 0$ $y \in \Omega^*$ and (3), where $\hbar_2 = \left\{1 - \bar{\theta}\left(c + (b+c)(1+\bar{\theta} c) e^{\bar{\theta}(b+c)}\right)\right\}^{-1}$.

The Existence and Uniqueness of the Solutions of (2)

Now, sufficient conditions for the existence and uniqueness of the solutions of (2) with GPCA shall be given.

Firstly, note that $g(u,v) := bv[1-u] - cu = bv - u(bv+c)$ is a continuous function and has continuous partial derivatives for $u, v \in \Omega$. First partial derivatives of the function $g(u,v)$.give that

$$\left|\frac{\partial g}{\partial u}\right| = |-(bv+c)| \le b|v| + c \le b + c, \quad \left|\frac{\partial g}{\partial v}\right| = |b - ub| = b|1-u| \le b,$$

for $u, v \in \Omega$ he Lipschitz constant can be chosen as $\ell = b + c$ .by assuming that ℓ .is sufficiently small.

Lemma 5 Let (C1)-(C3) be satisfied and $i \in N_0$ be fixed. Then for every $(\xi, x_0) \in [\theta_i, \theta_{i+1}] \times \Omega$, .there exists a unique solution $x(t) = x(t, \xi, x_0)$ .of (2) on $[\theta_i, \theta_{i+1}]$..

Proof: *Existence.* Fix $i \in N_0$ and assume without loss of generality that $\theta_i \leq \xi \leq \theta_{i+1}$..Define a norm $\left|x(t)\right|_0 = \max_{[\theta_i, \xi]} \left|x(t)\right|$. Take $x^0(t)=x^0$ and define a sequence

$$x^{p+1}(t) = x^0 + \int_{\xi}^{t} \left(bx^p(\theta_i) - bx^p(\theta_i)x^p(s) - cx^p(s)\right) ds, \; p \geq 0, t \in [\theta_i, \theta_{i+1}).$$

Or consider $g(u,v) = bv[1-u]-cu = bv-u(bv+c)$ and, by taking $g(x^p(s), x^p(\theta_i)) = bx^p(\theta_i) - bx^p(\theta_i)x^p(s) - cx^p(s)$ define a sequence

$$x^{p+1}(t) = x^0 + \int_{\xi}^{t} g\left(x^p(s), x^p(\theta_i)\right) ds, \; p \geq 0, \; t \in [\theta_i, \theta_{i+1}).$$

Then, for $p=0$ it is true that

$$\begin{aligned} \left|x^1(t) - x^0(t)\right| &\leq \max_{[\theta_i, \xi]} \left|x^1(t) - x^0(t)\right| = \left|x^1(t) - x^0(t)\right|_0 \\ &\leq \max_{[\theta_i, \xi]} \left| \int_{\xi}^{t} \left(b\left|x^0(\theta_i)\right|\left|1 - x^0(s)\right| + c\left|x^0(s)\right|\right) ds \right|. \\ &\leq \bar{\theta}(b+c)\left|x^0\right| \leq 2\bar{\theta}(b+c)x^0. \end{aligned} \tag{8}$$

Second, for $p=0$ and $p=1$ it can be written

$$\begin{aligned} \left|x^2(t) - x^1(t)\right|_0 &= \max_{[\theta_i, \xi]} \left| \int_{\xi}^{t} \left| g\left(x^1(s), x^1(\theta_i)\right) - g\left(x^0(s), x^0(\theta_i)\right) \right| ds \right| \\ &\leq \max_{[\theta_i, \xi]} \left| (b+c) \int_{\xi}^{t} \left(\left|x^1(\theta_i) - x^0(\theta_i)\right| + \left|x^1(s) - x^0(s)\right| \right) ds \right| \\ &\leq (b+c)\bar{\theta}\left|x^1(\theta_i) - x^0(\theta_i)\right| + \max_{[\theta_i, \xi]} \left| (b+c) \int_{\xi}^{t} \left|x^1(s) - x^0(s)\right| ds \right|. \end{aligned}$$

Using (8), for $t=\theta_1$, the inequality $\left|x^1(\theta_i)-x^0(\theta_i)\right|_0 \le 2\bar{\theta}(b+c)x^0$ .is reached and then

$$\left|x^2(s)-x^1(s)\right|_0 \le 2(b+c)^2\bar{\theta}^2x^0 + \max_{[\theta_i,\xi]}\left|\int_\xi^t 2\bar{\theta}(b+c)^2x^0ds\right| \le 2^2(b+c)^2\bar{\theta}^2x^0$$

is obtained. Next, $\left|x^{p+1}(t)-x^p(t)\right|_0$ .can be evaluated for the value p=2,3,... similiarly. Then, the induction method result in

$$\left|x^{p+1}(t)-x^p(t)\right|_0 \le \left(2(b+c)\bar{\theta}\right)^{p+1}x^0.$$

Then, (C3) implies that the sequence $x^p(t)$ is convergent and its limit $x(t)$ satisfies

$$x(t)=x^0+\int_\xi^t\left(bx(\theta_i)-bx(\theta_i)x(s)-cx(s)\right)ds$$

on $t\in[\theta_i,\xi]$. The existence is proved.

Uniqueness: Let $x_j(t)=x(t,\xi,x_j^0), x_j(\xi)=x_j^0, j=1,2,$ .denote the solutions of (3) where $\theta_i \le \xi \le \theta_{i+1}$. Now, it will be shown that $x_1(t)\neq x_2(t)$ while $x_1^0 \neq x_2^0$ .for every $t\in[\theta_i,\theta_{i+1}]$. .For all $t\in[\theta_i,\theta_{i+1}]$. let define $g(u,v)=bv[1-u]-cu=bv-u(bv+c)$, and so the solutions $x_1(t)$ and $x_2(t)$ satisfy the following integral equations

$$x_1(t)=x_1^0+\int_\xi^t g\left(x_1(s),x_1(\theta_i)\right)ds \text{ and } x_2(t)=x_2^0+\int_\xi^t g\left(x_2(s),x_2(\theta_i)\right)ds,$$

respectively. It is true that

$$\begin{aligned}\left|x_1(t)-x_2(t)\right| &\le \left|x_1^0-x_2^0\right|+\int_{\xi}^{t}\left|g\left(x_1(s),x_1(\theta_i)\right)-g\left(x_2(s),x_2(\theta_i)\right)\right|ds\\ &\le \left|x_1^0-x_2^0\right|+\int_{\xi}^{t}(b+c)\left(\left|x_1(s)-x_2(s)\right|+\left|x_1(\theta_i)-x_2(\theta_i)\right|\right)ds\\ &\le \left|x_1^0-x_2^0\right|+(b+c)\bar{\theta}\left|x_1(\theta_i)-x_2(\theta_i)\right|+\int_{\xi}^{t}(b+c)\left|x_1(s)-x_2(s)\right|ds.\end{aligned}$$

Then, the Gronwall-Bellman inequality gives

$$\left|x_1(t)-x_2(t)\right| \le \left\{\left|x_1^0-x_2^0\right|+(b+c)\bar{\theta}\left|x_1(\theta_i)-x_2(\theta_i)\right|\right\}e^{(b+c)\bar{\theta}}. \tag{9}$$

For $t=\theta_i$, it is obvious that

$$\left|x_1(\theta_i)-x_2(\theta_i)\right| \le \left\{\left|x_1^0-x_2^0\right|+(b+c)\bar{\theta}\left|x_1(\theta_i)-x_2(\theta_i)\right|\right\}e^{(b+c)\bar{\theta}}$$

and so

$$\left|x_1(\theta_i)-x_2(\theta_i)\right| \le \left\{1-(b+c)\bar{\theta}e^{(b+c)\bar{\theta}}\right\}^{-1}\left|x_1^0-x_2^0\right|e^{(b+c)\bar{\theta}}. \tag{10}$$

Substituting (10) into (9), the following inequality is reached:

$$\begin{aligned}&\left|x_1(t)-x_2(t)\right| \le \left\{1+(b+c)\bar{\theta}\left\{1-(b+c)\bar{\theta}e^{(b+c)\bar{\theta}}\right\}^{-1}e^{(b+c)\bar{\theta}}\right\}\left|x_1^0-x_2^0\right|e^{(b+c)\bar{\theta}}\\ &=\left\{1-(b+c)\bar{\theta}e^{(b+c)\bar{\theta}}\right\}^{-1}e^{(b+c)\bar{\theta}}\left|x_1^0-x_2^0\right|.\end{aligned} \tag{11}$$

Now, on the contrary, assume that there exists $t^* \in \left[\theta_i,\theta_{i+1}\right]$.such that $x_1(t^*)=x_2(t^*)$. Then,

$$x_1^0-x_2^0=\int_{\xi}^{t^*}-\left(g\left(x_1(s),x_1(\theta_i)\right)-g\left(x_2(s),x_2(\theta_i)\right)\right)ds,$$

so

$$\left|x_1^0 - x_2^0\right| \le \int_{\xi}^{t^*} (b+c)\left(\left|x_1(s) - x_2(s)\right| + \left|x_1(\theta_i) - x_2(\theta_i)\right|\right) ds.$$

By the last inequality and then (11)

$$\left|x_1^0 - x_2^0\right| \le \int_{\xi}^{t^*} 2(b+c)\left\{1-(b+c)\bar{\theta}e^{(b+c)\bar{\theta}}\right\}^{-1} e^{(b+c)\bar{\theta}} \left|x_1^0 - x_2^0\right| ds$$

$$\le 2\bar{\theta}(b+c)\left\{1-(b+c)\bar{\theta}e^{(b+c)\bar{\theta}}\right\}^{-1} e^{(b+c)\bar{\theta}} \left|x_1^0 - x_2^0\right| < \left|x_1^0 - x_2^0\right|.$$

Then, based on the assumption (C3), a contradiction is obtained. The uniqueness is proved..

Thus, the following theorem gives sufficient conditions that guarantee the existence and uniqueness of the solutions for (2) based on the paper (Akhmet & Aruğaslan, 2009).

Theorem 4 Assume that the conditions (C1)-(C3) hold true. Then for every $(t_0,x_0)\in R^+\times\Omega$, there exists a unique solution $x(t)=x(t,t_0,x_0)$ of (2) on R^+ in the sense of Definition 1 such that $x(t_0)=x_0$. Second, note that $g^*(u,v):= -bu[1+v]+cv$ is a continuous function and has continuous partial derivatives for $u,v\in\Omega^*$. First partial derivatives of the function $g^*(u,v)$ give that

$$\left|\frac{\partial g^*}{\partial u}\right| = \left|-b(v+1)\right| \le c+b, \quad \left|\frac{\partial g^*}{\partial v}\right| = \left|c-ub\right| \le b,$$

for $u,v\in\Omega^*$. The Lipschitz constant can be chosen as $\ell^* = c+b$ by assuming that ℓ^* is sufficiently small. Thus, the following lemma and theorem give sufficient conditions that guarantee the existence and uniqueness of the solutions for (3) based on the paper (Akhmet & Aruğaslan, 2009).

Lemma 6 Let (C1)-(C3) be satisfied and $i\in N_0$ be fixed. Then for every $(\xi, y_0)\in[\theta_i,\theta_{i+1}]\times\Omega^*$, there exists a unique solution $y(t=y(t,\xi,y_0)$ of (3) on $[\theta_i,\theta_{i+1}]$.

Theorem 5 Assume that the conditions (C1)-(C3) hold true. Then for every $(t_0,y_0)\in R^+\times\Omega^*$, there exists a unique solution $y(t)=y(t,t_0,y_0)$ of (3) on R^+ in the sense of Definition 2 such that $y(t_0)=y_0$.

The Stability Analysis of the Solutions for (2)

From now on, sufficient conditions guaranteeing the uniform stability for the trivial equilibirum of (2) and of (3) (i.e., for the positive equilibrium of (2)) will be presented. While investigating these conditions, Lyapunov-Razumikhin method developed by Akhmet and Aruğaslan (2009) will be taken into consideration.

Theorem 6 Assume that the conditions (C1)-(C4) are satisfied. Then, the trivial equilibrium of (2) is uniformly stable in Ω.

Proof: Based on Definition 1.4 in paper which contains Lyapunov-Razumikhin method developed by Akhmet and Aruğaslan (2009), construct the following positive definite Lyapunov function

$$V(x(t))=x^2(t) \tag{12}$$

which is positive definite. Functions $u, v \in \mathcal{K}$ can be found such that $u(|x|)\leq V(x)\leq v(|x|)$. The evaluation for the derivative of (12) for $t\neq\theta_i$, $i\in N_0$, results in

$$V'\left(x(t),x\left(\beta(t)\right)\right)=2x(t)bx\left(\beta(t)\right)\left[1-x(t)\right]-2cx^2(t).$$

Then, for $x\in\Omega$, it can be obtained the following inequality

$$V'\left(x(t),x\left(\beta(t)\right)\right)\leq 2x(t)bx\left(\beta(t)\right)-2cx^2(t)\leq 2bx^2(t)-2cx^2(t)$$

whenever $x(\beta(t))\leq x(t)$. So, the conclusion

$$V'\left(x(t),x\left(\beta(t)\right)\right)\leq -2(c-b)x^2(t)$$

is reached. Thus, based on the results given in Theorem 2.4 established by Akhmet and Aruğaslan (2009), it follows from the condition (C4) that the trivial equilibrium of (2) is uniformly stable.

Theorem 7 Assume that the conditions (C1)-(C3), (C5) are satisfied. Then, the trivial equilibrium of (3) (i.e., the positive equilibrium of (2)) is uniformly stable in Ω^*.

Proof: Based on Definition 1.4 in paper which contains Lyapunov-Razumikhin method (Aruğaslan, 2009), let us construct the following positive definite Lyapunov function

$$V(y(t))=y^2(t) \tag{13}$$

It can be found functions $u, v \in \mathcal{K}$ such that $u(|y|) \leq V(y) \leq v(|y|)$. Now, let us evaluate the derivative of (13) for $t \neq \theta_i$, $i \in N_0$.

$$V'(y(t), y(\beta(t))) = -2by(t)y(t)\left[1 + y(\beta(t))\right] + 2cy(t)y(\beta(t)).$$

Then, for $y \in \Omega^*$,

$$V'(y(t), y(\beta(t))) \leq -2cy^2(t) + 2c|y(t)||y(\beta(t))| \leq -2cy^2(t) + 2cy^2(t)$$

whenever $|y(\beta(t))| \leq |y(t)|$. Thus, it is seen that $V'(y(t), y(\beta(t))) \leq 0$.

So, by Theorem 2.4 in (Akhmet & Aruğaslan, 2009), it is seen that the trivial equilibrium of (3) is uniformly stable which it means the positive equilibrium of (2) is uniformly stable.

FUTURE RESEARCH DIRECTIONS

In this chapter, the model is developed by the piecewise function $\beta(t)$ as deviating argument. As a subject of future research, this model can be developed by DEPCAG taking the piecewise function $\gamma(t)$ as deviating argument, or by functional differential equations. Alternatively, different models from the model studied in the present chapter may be subjected to similar analysis by developing the models with the help of DEPCAG or of functional differential equations.

CONCLUSION

The present chapter addresses a nonlinear epidemic equation modeled by differential equations with GPCA. It is obvious that the model studied in the present chapter are a remarkable model in the scope of population dynamics. Modeling this epidemic equation by a generalized piecewise constant argument that makes it possible to have knowledge of effects of its past behaviors on the present behaviour is a sign of the value of the results achieved in the chapter. Because, the effect of a past value of real life problems on current behavior can be very serious. Even, the past situation of systems can change their current situation seriously. Therefore, the argument chosen when building models are very important. In this respect, the deviation argument considered in model in the chapter is remarkable and it is obvious that it contributes to the development of model. In the chapter, the fact that model contains such an argument makes it difficult to attain an explicit equation for its behavior. Therefore, analyzing it without reaching a explicit form of solution has a facilitating effect. In this direction, the analysis of the relevant equation is performed with the help of Lyapunov-Razumikhin method without the need to reach its exact solution. It is seen that this method developed by Akhmet and Aruğaslan (2009) for EPCAG is very useful. Because, for the analysis of the nonlinear epidemic model with GPCA, computations and operations performed in the sense of this method can be preferred convenience according to other methods in the literature.

ACKNOWLEDGMENT

This research received no specific grant from any funding agency in the public, commercial, or not-for-profit sectors.

REFERENCES

Akhmet, M. U. (2007a). Integral manifolds of differential equations with piecewise constant argument of generalized type. *Nonlinear Analysis: Theory, Methods and Applications*, *66*, 367–383.

Akhmet, M. U. (2007b). On the reduction principle for differential equations with piecewise constant argument of generalized type. *Journal of Mathematical Analysis and Applications*, *336*(1), 646–663. doi:10.1016/j.jmaa.2007.03.010

Akhmet, M. U. (2008a). Asymptotic behavior of solutions of differential equations with piecewise constant arguments. *Applied Mathematics Letters, 21*(9), 951–956. doi:10.1016/j.aml.2007.10.008

Akhmet, M. U. (2008b). Stability of differential equations with piecewise constant arguments of generalized type. *Nonlinear Analysis, 68*(4), 794–803. doi:10.1016/j.na.2006.11.037

Akhmet, M. U. (2010). *Principles of Discontinuous Dynamical Systems*. New York: Springer. doi:10.1007/978-1-4419-6581-3

Akhmet, M. U. (2011). *Nonlinear Hybrid Continuous Discrete-Time Models*. Amsterdam: Atlantis Press. doi:10.2991/978-94-91216-03-9

Akhmet, M. U. (2014). Quasilinear retarded differential equations with functional dependence on piecewise constant argument. *Communications on Pure and Applied Analysis, 13*(2), 929–947. doi:10.3934/cpaa.2014.13.929

Akhmet, M. U., & Aruğaslan, D. (2009). Lyapunov-Razumikhin method for differential equations with piecewise constant argument. *Discrete and Continuous Dynamical Systems. Series A, 25*(2), 457–466. doi:10.3934/dcds.2009.25.457

Akhmet, M. U., Aruğaslan, D., & Yılmaz, E. (2010a). Stability analysis of recurrent neural networks with piecewise constant argument of generalized type. *Neural Networks, 23*(7), 805–811. doi:10.1016/j.neunet.2010.05.006 PMID:20605400

Akhmet, M. U., Aruğaslan, D., & Yılmaz, E. (2010b). Stability in cellular neural networks with piecewise constant argument. *Journal of Computational and Applied Mathematics, 233*(9), 2365–2373. doi:10.1016/j.cam.2009.10.021

Arnold, L., & Schmalfuss, B. (2001). Lyapunov's second method for random dynamical systems. *Journal of Differential Equations, 177*(1), 235–265. doi:10.1006/jdeq.2000.3991

Aruğaslan, D. (2009). *Süreksizlikleri Olan Diferansiyel Denklemler ve Populasyon Dinamiği*. Ankara, Turkey: Orta Doğu Teknik Üniversitesi.

Aruğaslan, D., & Cengiz, N. (2017). Green's Function and Periodic Solutions of a Spring-Mass System in which the Forces are Functionally Dependent on Piecewise Constant Argument. *Süleyman Demirel University Journal of Natural and Applied Sciences, 21*(1), 266–278.

Aruğaslan, D., & Cengiz, N. (2018). Existence of periodic solutions for a mechanical system with piecewise constant forces. *Hacettepe Journal of Mathematics and Statistics, 47*(3), 521–538.

Arugaslan, D., & Güzel, L. (2015). Stability of the logistic population model with generalized piecewise constant delays. Adv. Difference Equ., 2015(173).

Busenberg, S., & Cooke, K. L. (1980). The effect of integral conditions in certain equations modelling epidemics and population growth. *Journal of Mathematical Biology*, *10*(1), 13–32. doi:10.1007/BF00276393 PMID:7205075

Challamel, N., & Gilles, G. (2007). Stability and dynamics of a harmonically excited elastic-perfectly plastic oscillator. *Journal of Sound and Vibration*, *301*(3-5), 608–634. doi:10.1016/j.jsv.2006.10.014

Cooke, K. L. (1979). Stability analysis for a vector disease model. *The Rocky Mountain Journal of Mathematics*, *7*, 253–263.

Cooke, K. L., & Wiener, J. (1984). Retarded differential equations with piecewise constant delays. *Journal of Mathematical Analysis and Applications*, *99*(1), 265–297. doi:10.1016/0022-247X(84)90248-8

Gopalsamy, K., & Liu, P. (1998). Persistence and global stability in a population model. *Journal of Mathematical Analysis and Applications*, *224*(1), 59–80. doi:10.1006/jmaa.1998.5984

Greenspan, H. P. (1976). On the growth and stability of cell cultures and solid tumors. *Journal of Theoretical Biology*, *56*(1), 229–242. doi:10.1016/S0022-5193(76)80054-9 PMID:1263527

Hahn, W. (1967). *Stability of Motion*. Berlin: Springer-Verlag. doi:10.1007/978-3-642-50085-5

Huang, G., Liu, A., & Fory's, U. (2016). Global Stability Analysis of Some Nonlinear Delay Differential Equations in Population Dynamics. *Journal of Nonlinear Science*, *26*(1), 27–41. doi:10.100700332-015-9267-4

Huebel, L. (2005). *Analysis of solid tumor growth models: Mechanisms of volume loss and slowed growth rate. B.S.* Texas State University.

Korobeinikov, A. (2006). Lyapunov functions and global stability for SIR and SIRS epidemiological models with non-linear transmission. *Bulletin of Mathematical Biology*, *68*(3), 615–626. doi:10.100711538-005-9037-9 PMID:16794947

Krasovskii, N. N. (1963). *Stability of Motion: Application of Lyapunov's Second Method to Differential Systems and Equations with Time-Delay*. Stanford, CA: SU Press.

Lyapunov, A. M. (1949). Probl´eme G´en´eral de la Stabilit´e du Mouvement. *Ann. Math. Studies, 17.*

Michel, A. N., & Hu, B. (2000). Stability analysis of discontinuous dynamical systems using vector Lyapunov functions. *Circuits, Systems, and Signal Processing, 19*(3), 221–243. doi:10.1007/BF01204576

Miller, R. K., & Michel, A. (1982). *Ordinary Differential Equations*. New York: Academic Press.

Razumikhin, B. S. (1956). Stability of delay systems. *Прикладня математика и механика, 20*, 500–512.

ADDITIONAL READING

Aruğaslan, D. (2015). Dynamics of a harvested logistic type model with delay and piecewise constant argument. *Journal of Nonlinear Science and Applications, 8*(5), 507–517. doi:10.22436/jnsa.008.05.06

Aruğaslan, D., & Yolcu, Ş. (2015). Bazı skaler impalsif diferansiyel denklem modellerinde kararlılık analizi. *Gaziosmanpaşa Bilimsel Araştırma Dergisi, 11*, 70–82.

Cooke, K., Driessche, P., & Zou, X. (1999). Interaction of maturation delay and nonlinear birth in population and epidemic models. *Journal of Mathematical Biology, 39*(4), 332–352. doi:10.1007002850050194 PMID:10550577

Volz, R. (1982). Global asymptotic stability of a periodic solution to an epidemic model. *Journal of Mathematical Biology, 15*(3), 319–338. doi:10.1007/BF00275691 PMID:7153676

Wei, J., & Zou, X. (2006). Bifurcation analysis of a population model and the resulting SIS epidemic model with delay. *Journal of Computational and Applied Mathematics, 197*(1), 169–187. doi:10.1016/j.cam.2005.10.037

Zhao, X.-Q., & Zou, X. (2001). Threshold dynamics in a delayed SIS epidemic model. *Journal of Mathematical Analysis and Applications, 257*(2), 282–291. doi:10.1006/jmaa.2000.7319

KEY TERMS AND DEFINITIONS

Epidemic: The occurrence of more cases of disease, injury or other health condition than expected in a given area or among a specific group of persons during a particular period. Usually, the cases are presumed to have a common cause or to be related to one another in some way.

GPCA: Generalized piecewise constant argument is a deviation argument which is a piecewise function considered in differential equations with piecewise constant argument of generalized type.

Outbreak: Sometimes distinguished from an epidemic as more localized, or the term less likely to evoke public panic.

Population Dynamics: Population dynamics is the branch of life sciences that studies the size and age composition of populations as dynamical systems, and the biological and environmental processes driving them (such as birth and death rates, and by immigration and emigration). Example scenarios are ageing populations, population growth, or population decline.

SIS: A model is a simple mathematical model of epidemics.

Compilation of References

Abazari, R., & Abazari, M. (2012). Numerical simulation of generalized Hirota–Satsuma coupled KdV equation by RDTM and comparison with DTM. *Communications in Nonlinear Science and Numerical Simulation, 17*(2), 619–629. doi:10.1016/j.cnsns.2011.05.022

Ablowitz, M. J., & Clarkson, P. A. (1990). *Nonlinear evolution equations and inverse scattering transform*. Cambridge University Press.

Abu-Omar, A., & Kittaneh, F. (2015). Notes on some spectral radius and numerical radius inequalities. *Studia Mathematica, 272*(2), 97–109. doi:10.4064m227-2-1

Acemoglu, D., & Verdier, T. (2000). The choice between market failures and corruption. The American Economic Review, 90(1), 194–211. doi:10.1257/aer.90.1.194 doi:10.1257/aer.90.1.194

Agamaliyev, A., & Nabiyev, A. (2005). On eigenvalues of some boundary value problems for a polynomial pencil of Sturm-Liouville equation. *J. Applied Math. and Computations, 165*(3), 503–515. doi:10.1016/j.amc.2004.04.116

Agarwal, R., Echambadi, R., Franco, A. M., & Sarkar, M. B. (2004). Knowledge transfer through inheritance: Spin-out generation, development, and survival. Academy of Management Journal, 47(4), 501–522.

Akhmet, M. U. (2007a). Integral manifolds of differential equations with piecewise constant argument of generalized type. *Nonlinear Analysis: Theory, Methods and Applications, 66*, 367–383.

Akhmet, M. U. (2007b). On the reduction principle for differential equations with piecewise constant argument of generalized type. *Journal of Mathematical Analysis and Applications, 336*(1), 646–663. doi:10.1016/j.jmaa.2007.03.010

Akhmet, M. U. (2008a). Asymptotic behavior of solutions of differential equations with piecewise constant arguments. *Applied Mathematics Letters, 21*(9), 951–956. doi:10.1016/j.aml.2007.10.008

Akhmet, M. U. (2008b). Stability of differential equations with piecewise constant arguments of generalized type. *Nonlinear Analysis, 68*(4), 794–803. doi:10.1016/j.na.2006.11.037

Akhmet, M. U. (2010). *Principles of Discontinuous Dynamical Systems*. New York: Springer. doi:10.1007/978-1-4419-6581-3

Akhmet, M. U. (2011). *Nonlinear Hybrid Continuous Discrete-Time Models*. Amsterdam: Atlantis Press. doi:10.2991/978-94-91216-03-9

Akhmet, M. U. (2014). Quasilinear retarded differential equations with functional dependence on piecewise constant argument. *Communications on Pure and Applied Analysis*, *13*(2), 929–947. doi:10.3934/cpaa.2014.13.929

Akhmet, M. U., & Aruğaslan, D. (2009). Lyapunov-Razumikhin method for differential equations with piecewise constant argument. *Discrete and Continuous Dynamical Systems. Series A*, *25*(2), 457–466. doi:10.3934/dcds.2009.25.457

Akhmet, M. U., Aruğaslan, D., & Yılmaz, E. (2010a). Stability analysis of recurrent neural networks with piecewise constant argument of generalized type. *Neural Networks*, *23*(7), 805–811. doi:10.1016/j.neunet.2010.05.006 PMID:20605400

Akhmet, M. U., Aruğaslan, D., & Yılmaz, E. (2010b). Stability in cellular neural networks with piecewise constant argument. *Journal of Computational and Applied Mathematics*, *233*(9), 2365–2373. doi:10.1016/j.cam.2009.10.021

Aktosun, T., Klaus, M., & van der Mee, C. (1998). Wave scattering in one dimension with absorption. *Journal of Mathematical Physics*, *39*(4), 1957–1992. doi:10.1063/1.532271

Aldwoah, K. A. (2009). *Generalized time scales and associated difference equations* (Ph.D. Thesis). Cairo University.

Ali, A. R. (2008). Does auditor industry specialization matter? Evidence from the bond market. Journal of Audit Practice, 5(3), 44–72.

Allahverdiev B. P. & Tuna H. (2018b). Spectral expansion for the singular Dirac system with impulsive conditions. Turk J Math. 42.2527 – 2545.

Allahverdiev B. P. & Tuna H. (2019a). Eigenfunction expansion for singular Sturm-Liouville problems with transmission conditions, Electron. J. Differential Equations, Vol. 2019. No. 03. pp. 1-10.

Allahverdiev B. P. & Tuna H. (2019c). On Expansion in Eigenfunction for q-Dirac Systems on the Whole Line, Mathematical Reports. Vol. 21(71), No. 3/2019.

Allahverdiev B. P. & Tuna H. (2019d). Eigenfunction Expansion in the Singular Case for Dirac Systems on Time Scales . Konuralp Journal of Mathematics, 7 (1).128-135.

Allahverdiev, B. P., Eryılmaz, A., & Tuna, H. (2017). Dissipative Sturm-Liouville operators with a spectral parameter in the boundary condition on bounded time scales. *Electronic Journal of Differential Equations*, *95*, 1–13.

Allahverdiev, B. P., & Tuna, H. (2018a). An expansion theorem for q-Sturm-Liouville operators on the whole line. *Turkish Journal of Mathematics*, *42*, 1060–1071.

Allahverdiev, B. P., & Tuna, H. (2019b). The spectral expansion for the Hahn–Dirac system on the whole line. *Turkish Journal of Mathematics*, *43*, 1668–1687.

Allahverdiev, B. P., & Tuna, H. (2019e). Eigenfunction expansion in the singular case for q-SturmLiouville operators. *Caspian Journal of Mathematical Sciences.*, *8*(2), 91–102.

Alonso, M. (1980). Schrödinger spectral problems with energy-dependent potentials as sources of nonlinear Hamiltonian evolution equations. *Journal of Mathematical Physics*, *21*(9), 2342–2349. doi:10.1063/1.524690

Alparslan Gök, S. (2011). Set-valued solution concepts using interval-type payoffs for interval games. *Journal of Mathematical Economics, 47*(4-5), 621-626.

Alparslan Gök, S. Z., Branzei, R., & Tijs, S. (2008). *Cores and Stable Sets for Interval-Valued Games*. CentER Discussion Paper; Vol. 2008-17. Tilburg: Operations Research.

Alparslan Gök, S. Z., Branzei, R., & Tijs, S. (2009a). Convex interval games. *Journal of Applied Mathematics and Decision Sciences*, *2009*, 342089. doi:10.1155/2009/342089

Alparslan Gök, S. Z., Miquel, S., & Tijs, S. (2009b). Cooperative under interval uncertainty. *Mathematical Methods of Operations Research*, *69*(1), 99–109. doi:10.100700186-008-0211-3

Alvarez-Nodarse, R. (2003). *Polinomios hipergemétricos y q-polinomios. Monografías del Seminario Matemático "García de Galdeano" Número 26*. Spain: Prensas Universitarias de Zaragoza.

Alvarez-Nodarse, R. (2006). On characterizations of classical polynomials. *Journal of Computational and Applied Mathematics*, *196*(1), 320–337. doi:10.1016/j.cam.2005.06.046

Amrein, W. O., Hinz, A. M., & Pearson, D. B. (2005). *Sturm-Liouville theory, past and present. Basel*. Boston: Birkhauser Verlag. doi:10.1007/3-7643-7359-8

Andrews, G. E., Askey, R., & Roy, R. (1999). *Special functions, Encyclopedia of Mathematics and Its Applications*. Cambridge, UK: The University Press.

Annaby, M. H., Hamza, A. E., & Makharesh, S. D. (2018). A Sturm-Liouville theory for Hahn difference operator. In X. Li & Z. Nashed (Eds.), Frontiers of Orthogonal Polynomials and Series (pp. 35–84). Singapore: World Scientific. doi:10.1142/9789813228887_0004

Annaby, M. H., Hamza, A. E., & Aldwoah, K. A. (2012). Hahn difference operator and associated Jackson--Nörlund integrals. *Journal of Optimization Theory and Applications*, *154*(1), 133–153. doi:10.100710957-012-9987-7

Aparicio, J. (2010). *Cooperative logistics games. In SCIYO* (p. 129). COM.

Applebaum, D. (2009). *Lévy Processes and Stochastic Calculus* (2nd ed.). Cambridge, UK: Cambridge University Press.

Arafa, A. A. M., Rida, S. Z., & Mohamed, H. (2011). Homotopy analysis method for solving biological population model. *Communications in Theoretical Physics, 56*(5), 797–800. doi:10.1088/0253-6102/56/5/01

Arnold, L., & Schmalfuss, B. (2001). Lyapunov's second method for random dynamical systems. *Journal of Differential Equations, 177*(1), 235–265. doi:10.1006/jdeq.2000.3991

Aronzajn, N. (1950). Theory of reproducing kernels. *Transactions of the American Mathematical Society, 68*(3), 337–404. doi:10.1090/S0002-9947-1950-0051437-7

Aruğaslan, D., & Güzel, L. (2015). Stability of the logistic population model with generalized piecewise constant delays. Adv. Difference Equ., 2015(173).

Aruğaslan, D. (2009). *Süreksizlikleri Olan Diferansiyel Denklemler ve Populasyon Dinamiği.* Ankara, Turkey: Orta Doğu Teknik Üniversitesi.

Aruğaslan, D., & Cengiz, N. (2017). Green's Function and Periodic Solutions of a Spring-Mass System in which the Forces are Functionally Dependent on Piecewise Constant Argument. *Süleyman Demirel University Journal of Natural and Applied Sciences, 21*(1), 266–278.

Aruğaslan, D., & Cengiz, N. (2018). Existence of periodic solutions for a mechanical system with piecewise constant forces. *Hacettepe Journal of Mathematics and Statistics, 47*(3), 521–538.

Askey, R. A., Atakishiyev, N. M., & Suslov, S. K. (1993). An analog of the Fourier transformation for a q-harmonic oscillator. *Symmetries in Science, VI*, 57–63. doi:10.1007/978-1-4899-1219-0_5

Askey, R. A., & Suslov, S. K. (1993). The q-harmonic oscillator and an analogue of the Charlier polynomials. *Journal of Physics. A, Mathematical and General, 26*(15), L693–L698. doi:10.1088/0305-4470/26/15/014

Askey, R. A., & Suslov, S. K. (1993). The q-harmonic oscillator and the Al-Salam and Carlitz polynomials. *Letters in Mathematical Physics, 29*(2), 123–132. doi:10.1007/BF00749728

Atakishiyev, N. M., & Suslov, S. K. (1991). Difference analogs of the harmonic oscillator. *Theoretical and Mathematical Physics, 85*, 442–444.

Atakishiyev, N. M., & Suslov, S. K. (1991). Realization of the q-harmonic oscillator. *Theoretical and Mathematical Physics, 87*, 1055–1062.

Aydemir, K., & Mukhtarov, O. S. (2017). Class of Sturm.Liouville Problems with Eigenparameter Dependent Transmission Conditions. *Numerical Functional Analysis and Optimization, 38*(10), 1260-1275.

Aydemir, K., Olgar, H., Mukhtarov, O. S., & Muhtarov, F. (2018). Differential operator equations with interface conditions in modifed direct sum spaces. *Filomat, 32*(3), 921–931. doi:10.2298/FIL1803921A

Aziz, H., Cahan, C., Gretton, C., Kilby, P., Mattei, N., & Walsh, T. (2016). A study of proxies for shapley allocations of transport costs. *Journal of Artificial Intelligence Research*, *56*, 573–611. doi:10.1613/jair.5021

Bai, G., Mao, J., & Lu, G. (2004). Grey transportation problem. *Kybernetes*, *33*(2), 219–224. doi:10.1108/03684920410514148

Bakherad, M. (2018). Some Berezin Number Inequalities for Operator Matrices. *Czechoslovak Mathematical Journal*, *68*(143), 997–1009. doi:10.21136/CMJ.2018.0048-17

Bakherad, M., & Garayev, M. T. (2019). Berezin number inequalities for operators. *Concr. Oper.*, *6*(1), 33–43. doi:10.1515/conop-2019-0003

Bannai, E. (1990). Orthogonal polynomials in coding theory and algebraic combinatorics. In Theory and Practice (pp. 25–54). NATO, ASI Series.

Bayracı, S., & Ünal, G. (2014). Stochastic interest rate volatility modeling with a continuous-time GARCH(1, 1) model. *Journal of Computational and Applied Mathematics*, *259*, 464–473. doi:10.1016/j.cam.2013.10.017

Behme, A., Klüppelberg, C., & Mayr, K. (2014). Asymmetric COGARCH processes. *Journal of Applied Probability*, *51A*(ASpec.), 161–173. doi:10.1239/jap/1417528473

Beltrametti, E. G., & Bugajski, S. (1995). A classical extensions of quantum mechanics. *Journal of Physics. A, Mathematical and General*, *28*(:12), 3329–3343. doi:10.1088/0305-4470/28/12/007

Berezanskii, J. M. (1968). *Expansions in Eigenfunctions of Selfadjoint Operators*. Providence, RI: Amer. Math. Soc.

Berezin, F. A. (1972). Covariant and contravariant symbols for operators, Math. *USSR-Izv.*, *6*(5), 1117–1151. doi:10.1070/IM1972v006n05ABEH001913

Berezin, F. A. (1974). Quantization. *Math. USSR Izvestija*, *8*(5), 1109–1165. doi:10.1070/IM1974v008n05ABEH002140

Berg, C., & Ismail, M. E. H. (1996). Q-Hermite polynomials and classical orthogonal Polynomials. *Can. J. Math.*, *48*(1), 43–63. doi:10.4153/CJM-1996-002-4

Bergelson, V., Blass, A., & Hindman, N. (1994). Partition theorems for spaces of variable words. *Proceedings of the London Mathematical Society*, *68*(3), 449–476. doi:10.1112/plms3-68.3.449

Berger, C. A., & Coburn, L. A. (1987). Toeplitz operators on the Segal-Bargmann space. *Transactions of the American Mathematical Society*, *301*(2), 813–829. doi:10.1090/S0002-9947-1987-0882716-4

Bergman, S. (1922). Über die Entwicklung der harmischen Funktionen der Ebene und des Rümes nach Orthogonal Functionen. *Mathematische Annalen*, *86*(3-4), 238–271. doi:10.1007/BF01457987

Black, F. (1976). Studies of Stock Price Volatility Changes. *Proceedings from the American Statistical Association, Business and Economic Statistics Section*, 177-181.

Bochner, S. (1929). Über sturm-liouvilleesche polynomsysteme. *Mathematische Zeitschrift, 29*(1), 730–736. doi:10.1007/BF01180560

Bonner, S. E., Palmrose, Z.-V., & Young, S. M. (1998). Fraud type and auditor litigation: An analysis of SEC accounting and auditing enforcement releases. The Accounting Review, 503–532.

Borm, Hamers, & Hendrickx. (2001). Operations research games: A survey. *Top, 9*(2), 139.

Borzov, V. V., & Damaskinsky, E. V. (2006). Generalized coherent states for q-oscillator connected with q-Hermite polynomials. *Journal of Mathematical Sciences, 132*(1), 26–36. doi:10.100710958-005-0472-9

Brandt, W. (1926). Über eine Verallgemeinerung des Gruppengriffes. *Mathematische Annalen, 96*(1), 360–366. doi:10.1007/BF01209171

Branzei, R., Branzei, O., Gök, S. Z. A., & Tijs, S. (2010a). Cooperative interval games: A survey. *Central European Journal of Operations Research, 18*(3), 397–411. doi:10.100710100-009-0116-0

Branzei, R., Gök, S. Z. A., & Branzei, O. (2011). SZ Alparslan Gök, and O. Branzei. *Cooperative games under interval uncertainty: On the convexity of the interval undominated cores. Central European Journal of Operations Research, 19*(4), 523–532. doi:10.100710100-010-0141-z

Branzei, R., Tijs, S., & Alparslan Gök, S. Z. (2010b). How to handle interval solutions for cooperative interval games. *International Journal of Uncertainty, Fuzziness and Knowledge-based Systems, 18*(2), 123–132. doi:10.1142/S0218488510006441

Brickey, K. F. (2003). From Enron to WorldCom and beyond: Life and crime after Sarbanes-Oxley. Wash. ULQ, 81, 357.

Brown, R. (1987). From groups to groupoids: A brief survey. *Bulletin of the London Mathematical Society, 19*(2), 113–134. doi:10.1112/blms/19.2.113

Bugajski, S. (1996). Fundamentals of fuzzy probability theory. *International Journal of Theoretical Physics, 35*(11), 2229–2244. doi:10.1007/BF02302443

Busenberg, S., & Cooke, K. L. (1980). The effect of integral conditions in certain equations modelling epidemics and population growth. *Journal of Mathematical Biology, 10*(1), 13–32. doi:10.1007/BF00276393 PMID:7205075

Caputo, M. (1967). Linear models of dissipation whose Q is almost frequency independent—II. *Geophysical Journal International, 13*(5), 529–539. doi:10.1111/j.1365-246X.1967.tb02303.x

Caputo, M., & Fabrizio, M. (2015). A new definition of fractional derivative without singular kernel. *Progress in Fractional Differentiation and Applications, 1*(2), 73–85.

Carruth, P. J. (2011). *Earnings Management: The Role Of Accounting Professionals.* Academic Press.

Challamel, N., & Gilles, G. (2007). Stability and dynamics of a harmonically excited elastic-perfectly plastic oscillator. *Journal of Sound and Vibration*, *301*(3-5), 608–634. doi:10.1016/j.jsv.2006.10.014

Chevalley, C. (1956). *Fundemental Concepts of Algebra*. New York: Academic Press, Inc.

Chihara, T. S. (1978). *An Introduction to orthogonal polynomials*. New York: Gordon and Breach.

Coburn, L. A. (2004). A Lipschitz estimate for Berezin's operator calculus. *Proceedings of the American Mathematical Society*, *133*(01), 127–131. doi:10.1090/S0002-9939-04-07476-3

Cooke, K. L. (1979). Stability analysis for a vector disease model. *The Rocky Mountain Journal of Mathematics*, *7*, 253–263.

Cooke, K. L., & Wiener, J. (1984). Retarded differential equations with piecewise constant delays. *Journal of Mathematical Analysis and Applications*, *99*(1), 265–297. doi:10.1016/0022-247X(84)90248-8

Cryer, C. W. (1970). Rodrigues formula and the classical orthogonal polynomials. *Bol. Un. Mat. Ital.*, *25*(3), 1–11.

Çule, M., & Fulton, M. (2009). Business culture and tax evasion: Why corruption and the unofficial economy can persist. Journal of Economic Behavior & Organization, 72(3), 811–822. doi:10.1016/j.jebo.2009.08.005 doi:10.1016/j.jebo.2009.08.005

Demiray, H. (2004). A travelling wave solution to the KdV–Burgers equation. *Applied Mathematics and Computation*, *154*(3), 665–670. doi:10.1016/S0096-3003(03)00741-0

Deng, J. (1988). *Grey Forecasting and Decision-Making*. Wuhan, China: Huazhong University of Science and Technology Press.

Deng, J.-L. (1982). Control problems of grey systems. *Systems & Control Letters*, *1*(5), 288–294. doi:10.1016/S0167-6911(82)80025-X

Deng, J.-L. (1990). *Grey system theory tutorial*. Wuhan: Huazhong University of Science and Technology Press.

Dobrogowska, A., & Odzijewicz, A. (2006). Second order $q-$ difference equations solvable by factorization method. *J. Comput. Appl. Math.*, *193*(1), 319-346.

Dokuchaev, M., Exel, R., & Piccione, P. (2000). Partial Representations and Partial Group Algebras. *Journal of Algebra*, *226*, 505–532. doi: 10.1006/jabr.1999.8204

Dollery, B. E., & Wallis, J. L. (1997). Market failure, government failure, leadership and public policy. Journal of Interdisciplinary Economics, 8(2), 113–126. doi:10.1177/02601079X9700800202 doi:10.1177/02601079X9700800202

Doty, J. (2013). *The Role of the Audit in the Global Economy*. at the ICAS Aileen Beattie Memorial Lecture, London, UK.

Dragomir, S.S. (2008a). Grüss' type inequalities for functions of selfadjoint operators in Hilbert spaces. Preprint. *RGMIA Res. Rep. Coll., 11*(e), Art. 11.

Dragomir, S.S. (2008b). Some new Grüss' type Inequalities for functions of selfadjoint operators in Hilbert spaces. Preprint. *RGMIA Res. Rep. Coll., 11*(e), Art. 12.

Dragomir, S. S. (2004). *Discrete Inequalities of the Cauchy-Bunyakovsky-Schwarz Type*. NY: Nova Science Publishers.

Dragomir, S. S. (2006). Reverse inequalities for the numerical radius of linear operators in Hilbert spaces. *Bulletin of the Australian Mathematical Society, 73*(2), 255–262. doi:10.1017/S0004972700038831

Dragomir, S. S. (2007). A survey of some recent inequalities for the norm and numerical radius of operators in Hilbert spaces. *Banach Journal of Mathematical Analysis, 1*(2), 154–175. doi:10.15352/bjma/1240336213

Dragomir, S. S. (2007). Inequalities for the norm and the numerical radius of linear operators in Hilbert spaces. *Demonstratio Math., 40*(2), 411–417.

Dragomir, S. S. (2011). Some Slater type inequalities for convex functions of selfadjoint operators in Hilbert spaces. *Revista de la Unión Matemática Argentina, 52*(1), 109–120.

El-Sayed, A. M. A., Behiry, S. H., & Raslan, W. E. (2010). Adomian's decomposition method for solving an intermediate fractional advection–dispersion equation. *Computers & Mathematics with Applications (Oxford, England), 59*(5), 1759–1765. doi:10.1016/j.camwa.2009.08.065

El-Sayed, A. M. A., & Gaber, M. (2006). The Adomian decomposition method for solving partial differential equations of fractal order in finite domains. *Physics Letters. [Part A], 359*(3), 175–182. doi:10.1016/j.physleta.2006.06.024

Engle, R. F. (1982). Autoregressive conditional heteroscedasticity with estimates of the variance of United Kingdom inflation. *Econometrica, 50*(4), 987–1007. doi:10.2307/1912773

Engliš, M. (1994). Functions invariant under the Berezin transform. *Journal of Functional Analysis, 121*(1), 233–254. doi:10.1006/jfan.1994.1048

Engliš, M. (1995). Toeplitz operators and the Berezin transform on H^2. *Linear Algebra and Its Applications, 223/224*, 171–204. doi:10.1016/0024-3795(94)00056-J

Engliš, M. (1999). Compact Toeplitz operators via the Berezin transform on bounded symmetric domains. *Integral Equations and Operator Theory, 33*(4), 426–455. doi:10.1007/BF01291836

Erbe, L., Goodrich, C. S., Jia, B., & Peterson, A. (2016). Survey of the qualitative properties of fractional difference operators: Monotonicity, convexity, and asymptotic behavior of solutions. *Advances in Difference Equations, 2016*(1), 43. doi:10.118613662-016-0760-3

Fairchild, R. (2008). *Does audit tenure lead to more fraud? A game theoretic approach.* Available at: https://papers.ssrn.com/sol3/papers.cfm?abstract_id=993400

Fan, E. (2000). Extended tanh-function method and its applications to nonlinear equations. *Physics Letters. [Part A], 277*(4), 212–218. doi:10.1016/S0375-9601(00)00725-8

Fine, N. J. (1988). Basic hypergeometric series and applications. American Mathematical Society.

Foulis, D., & Bennett, M. K. (1994). Effect algebras and unsharp quantum logics. *Foundations of Physics, 24*(10), 1331–1352. doi:10.1007/BF02283036

Frisk, M., Göthe-Lundgren, M., Jörnsten, K., & Rönnqvist, M. (2010). Cost allocation in collaborative forest transportation. *European Journal of Operational Research, 205*(2), 448–458. doi:10.1016/j.ejor.2010.01.015

Fuchs, L. (1970). *Infinite Abelian Groupd.* New York: Academic Press, Inc.

Furuta, T., Mićić Hot, J., Pečarić, J., & Seo, Y. (2005). *Mond-Pečarić Method in Operator Inequalities. Inequalities for Bounded Selfadjoint Operators on a Hilbert Space.* Zagreb: Element.

Garayev, M. T., Gürdal, M., & Okudan, A. (2016). Hardy-Hilbert's inequality and a power inequality for Berezin numbers for operators. *Mathematical Inequalities & Applications, 3*(3), 883–891. doi:10.7153/mia-19-64

Garayev, M. T., Gürdal, M., & Saltan, S. (2017). Hardy Type Inequality For Reproducing Kernel Hilbert Space Operators and Related Problems. *Positivity, 21*(4), 1615–1623. doi:10.100711117-017-0489-6

Garayev, M., Saltan, S., & Gundogdu, D. (2018). On the Inverse Power Inequality for the Berezin Number of Operators. *J. Math. Inequal., 12*(4), 997–1003. doi:10.7153/jmi-2018-12-76

Gasymov, M. G., & Guseinov, G. Sh. (1981). Determining the diffusion operator from spectral data [Russian]. *Akad. Nauk Azerbaijan SSR Dokl., 37*(2), 19–23.

Gasymow, M.G., Gasymow. (1982). On the spectral theory of differential operators polnomially depending on a parameter [Russian]. *Uspekhi Matematicheskhi Nauk, 37*(4), 99.

Goktas, U., & Hereman, W. (1997). Symbolic computation of conserved densities for systems of nonlinear evolution equations. *Journal of Symbolic Computation, 24*(5), 591–622. doi:10.1006/jsco.1997.0154

Goldhaber, J. (1970). *And Ehrlich, G., Algebra.* New York: The Macmillan Company.

Gopalsamy, K., & Liu, P. (1998). Persistence and global stability in a population model. *Journal of Mathematical Analysis and Applications, 224*(1), 59–80. doi:10.1006/jmaa.1998.5984

Gorenflo, R., & Mainardi, F. (2000). *Essentials of fractional calculus.* Academic Press.

Gorenstein, D. (1968). *Finite Groups.* New York: Harper and Row, Publishers.

Greenberg, M. D. (1978). *Foundation of Applied Mathematics.* Prentice–Hall Inc.

Greenspan, H. P. (1976). On the growth and stability of cell cultures and solid tumors. *Journal of Theoretical Biology, 56*(1), 229–242. doi:10.1016/S0022-5193(76)80054-9 PMID:1263527

Gul, F. A., Zhou, G., & Zhu, X. (2013). Investor protection, firm informational problems, big N auditors, and cost of debt around the world. Auditing, 32(3), 1–30. doi:10.2308/ajpt-50462 doi:10.2308/ajpt-50462

Guo, S., Mei, L., Li, Y., & Sun, Y. (2012). The improved fractional sub-equation method and its applications to the space–time fractional differential equations in fluid mechanics. *Physics Letters. [Part A]*, *376*(4), 407–411. doi:10.1016/j.physleta.2011.10.056

Guseinov, G. S. (1985). On spectral analysis of a quadratic pencil of Stırm-Liouville operators. *Soviet Math. Dokl.*, *32*(3), 859–862.

Guseinov, G. S. (2007). Eigenfunction expansions for a Sturm-Liouville problem on time scales. *International Journal of Differential Equations*, *2*(1), 93–104.

Guseinov, G. S. (2008). An expansion theorem for a Sturm-Liouville operator on semi-unbounded time scales. *Adv. Dyn. Syst. Appl.*, *3*(1), 147–160.

Guseinov, I. M. (1997). On a transformation operatör. *J. Mat. Notes*, *62*(2), 172–180. doi:10.1007/BF02355905

Guseinov, I. M., Nabiev, A. A., & Pashayev, R. T. (2000). Transformation operators and asymptotic formulas for the eigenvalues of a polynomial pencil of Sturm-Liouville operators. *Sib. Math. J.*, *41*(3), 453–464. doi:10.1007/BF02674102

Gustafson, K. E., & Rao, D. K. M. (1997). *Numerical Range*. Springer-Verlag.

Hahn, W. (1935). Über die jacobischen polynome und zwei verwandte polynomklassen. *Mathematische Zeitschrift*, *39*(1), 634–638. doi:10.1007/BF01201380

Hahn, W. (1949). Über orthogonalpolynome, die $q-$. *Mathematische Nachrichten*, *2*(1-2), 4–34. doi:10.1002/mana.19490020103

Hahn, W. (1967). *Stability of Motion*. Berlin: Springer-Verlag. doi:10.1007/978-3-642-50085-5

Hahn, W. (1983). Ein beitrag zur theorie der orthogonalpolynome. *Monatshefte für Mathematik*, *95*(1), 19–24. doi:10.1007/BF01301144

Hajmohamadi, M., Lashkaripour, R., & Bakherad, M. (n.d.). Improvements of Berezin number inequalities. *Linear and Multilinear Algebra*. doi: 10.1080/03081087.2018.1538310

Hall, M. (1970). *The Theory of Groups*. New York: Academic Press, Inc.

Hamza, A. E., & Ahmed, S. A. (2013). Theory of linear Hahn difference equations. *J. Adv. Math.*, *4*(2), 440–460.

Hamza, A. E., & Ahmed, S. A. (2013a). Existence and uniqueness of solutions of Hahn difference equations. *Adv. Difference Equations*, *316*, 1–15.

Hamza, A. E., & Makharesh, S. D. (2016). Leibniz' rule and Fubinis theorem associated with Hahn difference operator. *Journal of Advanced Mathematical*, *12*(6), 6335–6345.

Haug, S. (2006). *Exponential COGARCH and other continuous time models with applications to high frequency data* (Dissertation). Technische Universitat München.

Haug, S., Klüppelberg, C., Lindner, A., & Zapp, M. (2007). Moment Estimating the COGARCH (1,1) model - a first go. *The Econometrics Journal, 10*(2), 320–341. doi:10.1111/j.1368-423X.2007.00210.x

Hazewinkel, M. (1994). Groupoid. In *Encyclopedia of Mathematics*. Kluwer Academic Publishers.

He, J. H., & Wu, X. H. (2006). Exp-function method for nonlinear wave equations. *Chaos, Solitons, and Fractals, 30*(3), 700–708. doi:10.1016/j.chaos.2006.03.020

Herstein, I. (1965). *Topics in Algebra*. Waltham, MA: Addison-Wesley, Publishing Company, Inc.

Hettich, R., & Kortanek, K. O. (1993). Semi-infinite programming: Theory, methods, and applications. *SIAM Review, 35*(3), 380–429. doi:10.1137/1035089

Hilfer, R. (2000). *Applications of Fractional Calculus in Physics* (Vol. 128). Singapore: World Scientific Publishing. doi:10.1142/3779

Hira, F. (2018). *Dirac systems associated with Hahn difference operator*. arXiv:1806.00710v1

Hosking, J. R. (1981). Fractional Differencing. *Biometrika, 68*(1), 165–176. doi:10.1093/biomet/68.1.165

Huang, G., Liu, A., & Fory's, U. (2016). Global Stability Analysis of Some Nonlinear Delay Differential Equations in Population Dynamics. *Journal of Nonlinear Science, 26*(1), 27–41. doi:10.100700332-015-9267-4

Huebel, L. (2005). *Analysis of solid tumor growth models: Mechanisms of volume loss and slowed growth rate. B.S.* Texas State University.

Hungerford, T. W. (1974). Algebra. Springer-Verlag.

Iacus, S. M., Mercuri, L., & Rroji, E. (2015a). *Discrete time approximation of a COGARCH(p,q) model and its estimation.* arXiv preprint arXiv:1511.00253

Iacus, S. M., Mercuri, L., & Rroji, E. (2015b). Estimation and Simulation of a COGARCH(p,q) model in the YUIMA project. arXiv preprint arXiv:1505.03914

Infante, D., & Smirnova, J. (2009). Rent-seeking under a weak institutional environment. Economics Letters, 104(3), 118–121. doi:10.1016/j.econlet.2009.04.022 doi:10.1016/j.econlet.2009.04.022

Jackson, F. H. (1910). Difference equations. *American Journal of Mathematics, 32*(4), 305–314. doi:10.2307/2370183

James, K. S. (1989). Linear programming and its applications. Springer-Verlag World Publishing Corp.

Jarad, F., Abdeljawad, T., & Baleanu, D. (2012). Caputo-type modification of the Hadamard fractional derivatives. *Advances in Difference Equations*, *2012*(1), 142. doi:10.1186/1687-1847-2012-142

Jaulent, M., Jean, C. (1976). The inverse scattering problem for the one dimensional Schrödinger equation with an energy-dependent potential I, II. *Ann. Inst. Henri Poincare, 25*, 105-137.

Jaulent, M. (1972). On an inverse scattering problem with an energy dependent potential. *Ann. Inst. Henri Poincare*, *27*(4), 363–378.

Jaulent, M., & Jean, C. (1982). Solution of a Schrödinger inverse scattering problem with a polynomial spectral dependence in the potential. *Journal of Mathematical Physics*, *23*(2), 258–266. doi:10.1063/1.525347

Jekel, S. (2013). *Partial Groups*. Northeastern University. Doi: 10.13140/2.1.1409.3127

Jordanov, R. (1984). About some spectral properties of the Schrödinger equation with an energy dependent potential generating fully integrable Hamiltonian systems. *Ann. de L'universite de Sofia 'Kliment Ohridski'*, *78*, 1–29.

Jumarie, G. (2006). Modified Riemann-Liouville derivative and fractional Taylor series of nondifferentiable functions further results. *Computers & Mathematics with Applications (Oxford, England)*, *51*(9-10), 1367–1376. doi:10.1016/j.camwa.2006.02.001

Kamimura, Y. (2007). An inversion formula in energy dependent scattering. *Journal of Integral Equations and Applications*, *19*(4), 473–512. doi:10.1216/jiea/1192628620

Karaev, M. T. (2006). Berezin symbol and invertibility of operators on the functional Hilbert spaces. *Journal of Functional Analysis*, *238*(1), 181–192. doi:10.1016/j.jfa.2006.04.030

Karaev, M. T. (2013). Reproducing Kernels and Berezin symbols Techniques in Various Questions of Operator Theory. *Complex Analysis and Operator Theory*, *7*(4), 983–1018. doi:10.100711785-012-0232-z

Karaev, M. T., Gürdal, M., & Yamancı, U. (2013). Special operator classes and their properties. *Banach Journal of Mathematical Analysis*, *7*(2), 74–85. doi:10.15352/bjma/1363784224

Karaev, M. T., Gürdal, M., & Yamancı, U. (2014). Some results related with Berezin symbols and Toeplitz operators. *Mathematical Inequalities & Applications*, *17*(3), 1031–1045. doi:10.7153/mia-17-76

Kilbas, A. A., Srivastava, H. M., & Trujillo, J. J. (2006). *Theory of appliciations of fractional differential equations*. Academic Press.

Kittaneh, F. (2005). Numerical radius inequalities for Hilbert space operators. *Studia Mathematica*, *168*(1), 73–80. doi:10.4064m168-1-5

Kittaneh, F., Moslehian, M. S., & Yamazaki, T. (2015). Cartesian decomposition and numerical radius inequalities. *Linear Algebra and Its Applications*, *471*, 46–53. doi:10.1016/j.laa.2014.12.016

Klüppelberg, C., Maller, R., & Szimayer, A. (2011). The COGARCH: A Review, with News on Option Pricing and Statistical Inference. In *Surveys in Stochastic Processes. Proc. Of the 33rd SPA Conference in Berlin* (pp. 29–58). EMS Publishing House.

Klüppelberg, C., Lindner, A., & Maller, R. A. (2004). Continous time GARCH process driven by a Lévy process: Stationarity and second order behavior. *Journal of Applied Probability*, *41*(3), 601–622. doi:10.1239/jap/1091543413

Koekoek, R., Lesky, P. A., & Swarttouw, R. F. (2010). *Hypergeometric orthogonal polynomials and their q-analogues. Springer Monographs in Mathematics*. Berlin: Springer-Verlag. doi:10.1007/978-3-642-05014-5

Kolmogorov, A. N., & Fomin, S. V. (1970). *Introductory Real Analysis* (R. A. Silverman, Trans.). New York: Dover Publications.

Koornwinder, T. H. (1994). Compact quantum groups and q-special functions. Representations of lie groups and quantum groups, 46–128.

Koornwinder, T. H. (1995). Orthogonal polynomials in connection with quantum groups. Orthogonal polynomials, theory and practice, 257–292.

Korobeinikov, A. (2006). Lyapunov functions and global stability for SIR and SIRS epidemiological models with non-linear transmission. *Bulletin of Mathematical Biology*, *68*(3), 615–626. doi:10.100711538-005-9037-9 PMID:16794947

Krasovskii, N. N. (1963). *Stability of Motion: Application of Lyapunov's Second Method to Differential Systems and Equations with Time-Delay*. Stanford, CA: SU Press.

Kudryashov, N. A. (1990). Exact solutions of the generalized Kuramoto-Sivashinsky equation. *Physics Letters. [Part A]*, *147*(5-6), 287–291. doi:10.1016/0375-9601(90)90449-X

Kurosh, A. G. (1960). *The Theory of Groups* (Vols. 1–2). New York: Chelsea Publishing Company.

Kwon, K. H., Lee, D. W., Park, S. B., & Yoo, B. H. (1998). Hahn class orthogonal polynomials. *Kyungpook Math. J.*, *38*, 259–281.

Kyoto. (1997a). *Kyoto contract*. Retrieved from http://www.unfccc.org/resource/convkp.html

Kyoto. (1997c). *Kyoto protocol*. Retrieved from http://vitalgraphics.grida.no/kyoto

Lang, S. (1965). *Algebra*. Reading, MA: Addison-Wesley, Publishing Company, Inc.

Lesky, P. A. (2005). Eine Charakterisierung der klassischen kontinuierlichen, diskretenund Orthgonalpolynome. Aachen: Shaker.

Levinson, N. (1951). A simplified proof of the expansion theorem for singular second order linear differential equations. *Duke Math. J.*, *18*, 57-71.

Levitan, B. M., & Sargsjan, I. S. (1991). *Sturm-Liouville and Dirac Operators*. Dordrecht, The Netherlands: Kluwer Academic Publishers Group.

Liu, C., & Dai, Z. (2008). Exact soliton solutions for the fifth-order Sawada–Kotera equation. *Applied Mathematics and Computation*, *206*(1), 272–275. doi:10.1016/j.amc.2008.08.028

Liu, S., Fu, Z., Liu, S., & Zhao, Q. (2001). Jacobi elliptic function expansion method and periodic wave solutions of nonlinear wave equations. *Physics Letters. [Part A]*, *289*(1), 69–74. doi:10.1016/S0375-9601(01)00580-1

Li, Z. B., & He, J. H. (2010). Fractional complex transform for fractional differential equations. *Mathematical and Computational Applications*, *15*(5), 970–973. doi:10.3390/mca15050970

Llorca, N (2004). *On the Owen set of transportation situations. In Essays in Cooperative Games* (pp. 215–228). Boston, MA: Springer. doi:10.1007/978-1-4020-2936-3_17

Lu, B. (2012). Bäcklund transformation of fractional Riccati equation and its applications to nonlinear fractional partial differential equations. *Physics Letters. [Part A]*, *376*(28-29), 2045–2048. doi:10.1016/j.physleta.2012.05.013

Lyapunov, A. M. (1949). Probl´eme G´en´eral de la Stabilit´e du Mouvement. *Ann. Math. Studies, 17*.

Macfarlane, A. J. (1989). On q-analogues of the quantum harmonic oscillator and the quantum group SU_q. *Journal of Physics. A, Mathematical and General*, *22*(21), 4581–4588. doi:10.1088/0305-4470/22/21/020

Maksudov, F. G., & Guseinov, G. Sh. (1986). On the solution of the inverse scattering problem for the quadratic pencil of the Schrödinger equation on the full-line. *Dokl. Akad. Nauk USSR*, *289*(1), 42–46.

Maller, R. A., Mueller, G., & Szimayer, A. (2008). GARCH modelling in continuous time for irregularly spaced time series data. *Bernoulli*, *14*(2), 519–542. doi:10.3150/07-BEJ6189

Mandelbrot, B. (1963). The Variation of Certain Speculative Prices. *The Journal of Business*, *36*(4), 394–419. doi:10.1086/294632

Marcellan, F., Branquinho, A., & Petronilho, J. (1994). Classical orthogonal polynomials: A functional approach. *Acta Applicandae Mathematicae*, *34*, 283–303. doi:10.1007/BF00998681

Marchenko, V. A. (1997). *Sturm-Liouville's operators and their applications*. Kiev: Naukova Dumko.

Marín, J. M., Rodríguez-Bernal, M. T., & Romero, E. (2016). *ABC and Hamiltonian Monte-Carlo methods in COGARCH models*. UC3M Working Papers, Statistics and Econometrics, 16-01.

Marín, J. M., Rodríguez-Bernal, M. T., & Romero, E. (2015). Data cloning estimation of GARCH and COGARCH models. *Journal of Statistical Computation and Simulation*, *85*(9), 1818–1831. doi:10.1080/00949655.2014.903948

Marriage, M. (2019). Auditors promote 'mythology' on fraud detection, claims regulator. *Financial Times*. Retrieved from https://www.ft.com/content/43ab3668-2959-11e9-88a4-c32129756dd8

McCarthy, C. A. (1967). Cp. *Israel Journal of Mathematics*, *5*(4), 249–271. doi:10.1007/BF02771613

Medem, J. C., Álvarez-Nodarse, R., & Marcellán, F. (2001). On the q-polynomials: A distributional study. *Journal of Computational and Applied Mathematics*, *135*(2), 157–196. doi:10.1016/S0377-0427(00)00584-7

Medio, A., & Lines, M. (2001). Nonlinear dynamics: A primer. Cambridge University Press. doi:10.1017/CBO9780511754050 doi:10.1017/CBO9780511754050

Michel, A. N., & Hu, B. (2000). Stability analysis of discontinuous dynamical systems using vector Lyapunov functions. *Circuits, Systems, and Signal Processing*, *19*(3), 221–243. doi:10.1007/BF01204576

Miller, G. S. (2006). The press as a watchdog for accounting fraud. Journal of Accounting Research, 44(5), 1001–1033. doi:10.1111/j.1475-679X.2006.00224.x doi:10.1111/j.1475-679X.2006.00224.x

Miller, R. K., & Michel, A. (1982). *Ordinary Differential Equations*. New York: Academic Press.

Mond, B., & Pečarić, J. (1993). Convex inequalities in Hilbert space. *Houston Journal of Mathematics*, *19*, 405–420.

Moore, D. A., Tetlock, P. E., Tanlu, L., & Bazerman, M. H. (2006). Conflicts of interest and the case of auditor independence: Moral seduction and strategic issue cycling. Academy of Management Review, 31(1), 10–29. doi:10.5465/amr.2006.19379621 doi:10.5465/amr.2006.19379621

Mukhtarov, O. S., & Aydemir, K. (2018). Basis properties of the eigenfunctions of two-interval Sturm.Liouville problems. *Analysis and Mathematical Physics*, 1-20.

Müller, G. (2010). MCMC estimation of the COGARCH (1, 1) model. *Journal of Financial Econometrics*, *8*(4), 481–510. doi:10.1093/jjfinec/nbq029

Nabiev, A. A. (2006). Inverse scattering problem for the Schrödinger-type equation with a polynomial energy-dependent potential. *Inverse Problems*, *22*(6), 2055–2068. doi:10.1088/0266-5611/22/6/009

Nabiev, A. A., & Guseinov, I. M. (2005). On the Jost solutions of the Schrödinger-type equations with a polynomial energy-dependent potential. *Inverse Problems*, *22*(1), 55–67. doi:10.1088/0266-5611/22/1/004

Nabiev, I. M. (2004). The inverse spectral problem for the diffusion operator on an interval [Russian]. *Mat. Fiz.Anal. Geom.*, *11*(3), 302–313.

Naimark, M. A. (1969). Linear Differential Operators (2nd ed.). Frederick Ungar Publishing.

Nikiforov, A., Suslov, S. K., & Uvarov, V. (1991). *Classical orthogonal polynomials of a discrete variable*. Berlin: Springer-Verlag. doi:10.1007/978-3-642-74748-9

Nikiforov, A., & Uvarov, V. (1986). Classical orthogonal polynomials of a discrete variable on nonuniform lattices. *Letters in Mathematical Physics*, *11*(1), 27–34. doi:10.1007/BF00417461

Nikiforov, A., & Uvarov, V. (1988). *Special functions of mathematical physics*. Basel: Birkhäuser. doi:10.1007/978-1-4757-1595-8

Nordgren, E., & Rosenthal, P. (1994). Boundary values of Berezin symbols, Nonselfadjoint operators and related topics. Oper. Theory Adv. Appl., 73.

Odibat, Z., & Momani, S. (2008). A generalized differential transform method for linear partial differential equations of fractional order. *Applied Mathematics Letters*, *21*(2), 194–199. doi:10.1016/j.aml.2007.02.022

Olgar, H., Mukhtarov, O. S., & Aydemir, K. (2018). Some properties of eigenvalues and generalized eigenvectors of one boundary value problem. *Filomat, 32*(3), 911-920.

Owen, G. (1975). On the core of linear production games. *Mathematical Programming*, *9*(1), 358–370. doi:10.1007/BF01681356

Özener, O. Ö., & Ergun, Ö. (2008). Allocating costs in a collaborative transportation procurement network. *Transportation Science*, *42*(2), 146–165. doi:10.1287/trsc.1070.0219

Palancı, O., Alparslan Gök, S. Z., Ergün, S., & Weber, G. W. (2015). Cooperative grey games and the grey Shapley value. *Optimization*, *64*(8), 1657–1668. doi:10.1080/02331934.2014.956743

Palancı, O., Alparslan Gök, S. Z., Olgun, M. O., & Weber, G.-W. (2016). Transportation interval situations and related games. *OR-Spektrum*, *38*(1), 119–136. doi:10.100700291-015-0422-y

Petronilho, J. (2007). Generic formulas for the values at the singular points of some special monic classical $H_{q,\omega}$ – . *Journal of Computational and Applied Mathematics*, *205*(1), 314–324. doi:10.1016/j.cam.2006.05.005

Pickl, S. W., & Gerhard-W., W. (2002). *On Optimal Control of Heating Processes. In Operations Research Proceedings 2001*. Berlin: Springer.

Pickl, St., & Weber, G. W. (2000). An algorithmic approach by linear programming problems in generalized semi-infinite optimization, J. *Computing Techniques*, *5*(3), 62–82.

Platen, E., & Bruti-Liberati, N. (2010). *Numerical Solution of Stochastic Differential Equations with Jumps in Finance*. Springer-Verlag Berlin Heidelberg. doi:10.1007/978-3-642-13694-8

Podlubny, I. (1999).. . *Fractional Differential Equations, Academic Press*, *198*, 10–200.

Potter, M. J. A. (2000). *Berezin symbols and operator theory*. Thesis.

Puerto, J., Fernández, F. R., & Hinojosa, Y. (2008). Partially ordered cooperative games: Extended core and Shapley value. *Annals of Operations Research*, *158*(1), 143–159. doi:10.100710479-007-0242-9

Pulido, Sánchez-Soriano, & Llorca. (2002). Game theory techniques for university management: an extended bankruptcy model. *Annals of Operations Research, 109*(1-4), 129-142.

Pulido, M., Borm, P., Hendrickx, R., Llorca, N., & Sánchez-Soriano, J. (2008). Compromise solutions for bankruptcy situations with references. *Annals of Operations Research, 158*(1), 133–141. doi:10.100710479-007-0241-x

Razumikhin, B. S. (1956). Stability of delay systems. *Прикладня математика и механика, 20*, 500–512.

Rotman, J. (1973). *The Theory of Groups* (2nd ed.). Boston: Allyn and Bacon, Inc.

Routh, R. J. (1885). On some properties of certain solutions of a differential equation of the second order. *Proceedings of the London Mathematical Society, 16*, 245–261.

Saitoh, S. (1988). *Theory of reproducing kernels and its applications*. Pitman Research Notes in Mathematics Series 189.

Saitoh, S., & Sawano, Y. (2016). *Theory of reproducing kernels and applications*. Springer. doi:10.1007/978-981-10-0530-5

Samko, S. G., Kilbas, A. A., & Marichev, O. M. (1987). *Integral and derifatives of fractional order and its applications*. Minsk: Nauka and Tekhnika.

Sánchez-Soriano, J. (2003). The pairwise egalitarian solution. *European Journal of Operational Research, 150*(1), 220–231. doi:10.1016/S0377-2217(02)00503-9

Sánchez-Soriano, J. (2006). Pairwise solutions and the core of transportation situations. *European Journal of Operational Research, 175*(1), 101–110. doi:10.1016/j.ejor.2005.04.033

Sánchez-Soriano, J., Llorca, N., Meca, A., Molina, E., & Pulido, M. (2002). An integrated transport system for Alacant's students. *Annals of Operations Research, 109*(1-4), 41–60. doi:10.1023/A:1016387715917

Sánchez-Soriano, J., López, M. A., & Garc. (2001). On the core of transportation games. *Mathematical Social Sciences, 41*(2), 215–225. doi:10.1016/S0165-4896(00)00057-3

Sandholm, W. H. (2010). Population games and evolutionary dynamics. MIT Press.

Sattari, M., Moslehian, M. S., & Yamazaki, T. (2015). Some generalized numerical radius inequalities for Hilbert space operators. *Linear Algebra and Its Applications, 470*, 216–227. doi:10.1016/j.laa.2014.08.003

Sattinger, D. H., & Szmigielski, J. (1995). Energy dependent scattering theory. *Differential and Integral Equations, 8*, 945–959.

Schrijver, A. (1986). *Theory of Linear and Integer Programming*. Chichester, UK: John Wiley and Sons.

Schwartz, L. (1964). Sous-espaces hilbertiens d'espaces vectoriels topologiques et noyaux associés (noyaux reproduisants). *Journal d'Analyse Mathématique*, *13*(1), 115–256. doi:10.1007/BF02786620

Scott, W. R. (1964). *Group Theory*. Englewood Cliffs, NJ: Prentice-Hall, Inc.

Sitthiwirattham, T. (2016). On a nonlocal boundary value problem for nonlinear second-order Hahn difference equation with two different derivatives. Advances in Difference Equations, 2016(1).

Slater, M. S. (1981). A companion inequality to Jensen's inequality. *Journal of Approximation Theory*, *32*(2), 160–166. doi:10.1016/0021-9045(81)90112-X

Soons, D. E., Slikker, M., & Fransoo, J. C. (2011). *The determination and division of benefits among partners of a horizontal cooperation in transportation*. Technische Universiteit Eindhoven.

Stone, M. H. (1932). *Linear Transformations in Hilbert Space and Their Application to Analysis*. Providence, RI: Amer. Math. Soc.

Stone, M. H. (1926). A comparison of the series of Fourier and Birkhoff. *Transactions of the American Mathematical Society*, *28*(4), 695–761. doi:10.1090/S0002-9947-1926-1501372-6

Sturm, C., & Liouville, J. (1837). Extrait d.un M'emoire sur le d'eveloppement des fonctions en s'eries dont les di¤erents termes sont assujettis. a satisfaire. a une meme equation differentielle lineaire, contenant un parametre variable. *Journal de Mathématiques Pures et Appliquées*, *2*, 220–223.

Szegö, G. (1939). *Orthogonal polynomials*. American Mathematical Society Colloquium Publication Volume XXIII.

Szegö, S. (1921). Über orthogonale Polynome, die zu einer gegebenen Kurve der Komplexen Ebenegehören. *Mathematische Zeitschrift*, *9*(3-4), 218–270. doi:10.1007/BF01279030

Thaller, B. (1992). *The Dirac Equation, Berlin*. Heidelberg, Germany: Springer. doi:10.1007/978-3-662-02753-0

Theys, C., Dullaert, W., & Notteboom, T. (2008). *Analyzing cooperative networks in intermodal transportation: a game-theoretic approach*. In Nectar Logistics and Freight Cluster Meeting, Delft, The Netherlands.

Titchmarsh, E. C. (1962). *Eigenfunction Expansions Associated with Second-Order Differential Equations. Part I* (2nd ed.). Oxford, UK: Clarendon Press. doi:10.1063/1.3058324

Tricomi F. (1955). Vorlesungenüber orthogonalreihen. *Grundlehren der mathematischen wissenschafen,* 76.

Tsay, R. S. (2012). *Analysis of Financial Time Series*. Wiley and Sons.

Tsutsmi, M. (1981). On the inverse scattering problem for the one -dimensional Schrödinger equation with an energy dependent potential. *Journal of Mathematical Analysis and Applications*, *83*(1), 316–350. doi:10.1016/0022-247X(81)90266-3

Tuna, H. (2014). Completeness of the rootvectors of a dissipative Sturm–Liouville operators on time scales. *Appl Math Comput., 228*, 108–115.

Tuna, H. (2016). Completeness Theorem for the Dissipative Sturm-Liouville Operator on Bounded Time Scales. *Indian Journal of Pure and Applied Mathematics, 47*(3), 535–544. doi:10.100713226-016-0196-1

Tuna, H., & Eryılmaz, A. (2013). Dissipative Sturm-Liouville Operators with Transmission Conditions. *Abstract and Applied Analysis, 2013*, 1–7. doi:10.1155/2013/248740

Tuna, H., & Özek, M. A. (2017). The one-dimensional Schrödinger operator on bounded time scales. *Mathematical Methods in the Applied Sciences, 40*(1), 78–83. doi:10.1002/mma.3966

Turner, L. E. (2001). *Quality, transparency, accountability.* Paper presented at the Speech to the Financial Executives Institute-AICPA's "Benchmarking the Quality of Earnings" Conference, New York, NY.

Van der Mee, C., & Pivovarchik, V. (2001). Inverse scattering for a Schrödinger equation with energy dependent potential. *Journal of Mathematical Physics, 42*(1), 158–181. doi:10.1063/1.1326921

Van der Waerden, B. L. (n.d.). *Algebra* (7th ed.; 2 vols.). New York: Frederick Ungar Publishing Co.

Van Rijckeghem, C., & Weder di Mauro, B. (2001). Bureaucratic corruption and the rate of temptation: Do wages in the civil service affect corruption, and by how much? Journal of Development Economics, 65(2), 307–331. doi:10.1016/S0304-3878(01)00139-0 doi:10.1016/S0304-3878(01)00139-0

Vilenkin, N. J., & Klimyk, A. U. (1998). Representations of lie groups and special functions. *Bulletin of the American Mathematical Society, 35*(3), 265–270.

Wang, M. (1995). Solitary wave solutions for variant Boussinesq equations. *Physics Letters. [Part A], 199*(3-4), 169–172. doi:10.1016/0375-9601(95)00092-H

Wang, M. (1996). Exact solutions for a compound KdV-Burgers equation. *Physics Letters. [Part A], 213*(5-6), 279–287. doi:10.1016/0375-9601(96)00103-X

Wang, M., Li, X., & Zhang, J. (2007). Various exact solutions of nonlinear Schrödinger equation with two nonlinear terms. *Chaos, Solitons, and Fractals, 31*(3), 594–601. doi:10.1016/j.chaos.2005.10.009

Wang, M., Zhou, Y., & Li, Z. (1996). Application of a homogeneous balance method to exact solutions of nonlinear equations in mathematical physics. *Physics Letters. [Part A], 216*(1-5), 67–75. doi:10.1016/0375-9601(96)00283-6

Wazwaz, A. M. (2010). *Partial differential equations and solitary waves theory.* Springer Science & Business Media.

Weidmann, J. (1987). *Spectral Theory of Ordinary Di¤erential Operators. Lecture Notes in Mathematics 1258.* Berlin: Springer.

Weyl, H. (1910). Über gewöhnliche Differentialgleichungen mit Singularitäten und die zugehörigen Entwicklungen willkürlicher Funktionen. *Mathematische Annalen*, *68*(2), 220–269. doi:10.1007/BF01474161

Wu, Y., Geng, X., Hu, X., & Zhu, S. (1999). A generalized Hirota–Satsuma coupled Korteweg–de Vries equation and Miura transformations. *Physics Letters. [Part A]*, *255*(4-6), 259–264. doi:10.1016/S0375-9601(99)00163-2

Xu, G. (1999). *Handbook of Operations Research Fundamentals*. Beijing, China: Science Press.

Xu, T., Li, J., Zhang, H. Q., Zhang, Y. X., Yao, Z. Z., & Tian, B. (2007). New extension of the tanh-function method and application to the Whitham–Broer–Kaup shallow water model with symbolic computation. *Physics Letters. [Part A]*, *369*(5-6), 458–463. doi:10.1016/j.physleta.2007.05.047

Yamancı, U., Garayev, M. T., & Çelik, C. (2019). Hardy-Hilbert type inequality in reproducing kernel Hilbert space: Its applications and related results. *Linear and Multilinear Algebra*, *67*(4), 830–842. doi:10.1080/03081087.2018.1490688

Yamancı, U., & Gürdal, M. (2017). On numerical radius and Berezin number inequalities for reproducing kernel Hilbert space. *New York Journal of Mathematics*, *23*, 1531–1537.

Yamancı, U., Gürdal, M., & Garayev, M. T. (2017). Berezin Number Inequality for Convex Function in Reproducing Kernel Hilbert Space. *Filomat*, *31*(18), 5711–5717. doi:10.2298/FIL1718711Y

Ye, M., & Simunic, D. A. (2016). *The impact of PCAOB-type regulations on auditors under different legal systems*. Rotman School of Management Working Paper (2697268).

Yosida, K. (1950). On Titchmarsh-Kodaira formula concerning Weyl-Stone eingenfunction expansion. *Nagoya Math. J.*, *1*, 49-58.

Yosida, K. (1960). *Lectures on Differential and Integral Equations*. New York: Springer.

Yurko, V. A. (2000). An inverse problem for differential operator pencils, *Mat. Sb.*, *191*(10), 137-160; English transl. *Sbornik Mathematics*, *191*(10), 1561–1586. doi:10.1070/SM2000v191n10ABEH000520

Yusufoğlu, E., & Bekir, A. (2007). A travelling wave solution to the Ostrovsky equation. *Applied Mathematics and Computation*, *186*(1), 256–260. doi:10.1016/j.amc.2006.07.099

Zaremba, S. (1907). L'equation biharminique et une class remarquable de fonctions foundamentals harmoniques. Bull. *Int. de l'Academie des Sci. de Cracovie*, *39*, 147–196.

Zassenhaus, H. (1958). *The Theory of Groups*. New York: Chelsea Publishing Company.

Zettl, A. (2005). Sturm.Liouville Theory. In *Mathematical Surveys and Mono-graphs* (Vol. 121). American Mathematical Society.

Zhang, J. L., Wang, M. L., & Li, X. Z. (2006). The subsidiary ordinary differential equations and the exact solutions of the higher order dispersive nonlinear Schrödinger equation. *Physics Letters. [Part A]*, *357*(3), 188–195. doi:10.1016/j.physleta.2006.03.081

Zhang, S., Tong, J. L., & Wang, W. (2008). A generalized-expansion method for the mKdV equation with variable coefficients. *Physics Letters. [Part A], 372*(13), 2254–2257. doi:10.1016/j.physleta.2007.11.026

Zhang, S., & Zhang, H. Q. (2011). Fractional sub-equation method and its applications to nonlinear fractional PDEs. *Physics Letters. [Part A], 375*(7), 1069–1073. doi:10.1016/j.physleta.2011.01.029

Zimmerman, J. L. (2015). The role of accounting in the twenty-first century firm. Accounting and Business Research, 45(4), 485–509. doi:10.1080/00014788.2015.1035549 doi:10.1080/00014788.2015.1035549

About the Contributors

Bilender P. Allahverdiev is a Professor at Department of Mathematics in Süleyman Demirel University (Turkey), he obtained the PhD Degree in Baku (Azarbaijan).

Sakina Alwhishi is an instructor at El Mergib University, Libya. She has received her B.S. in Mathematics at El Mergib University, in 2006. She completed her Master of Science at Atilim University, Turkey, in 2017 under the supervision of Assoc. Prof. Dr. Rezan Sevinik Adıgüzel. Her research areas include the discrete q-orthogonal polynomials.

Yakup Arı is an Assistant Professor in the Department of Economics and Finance at Alanya Alaaddin Keykubat University. Having graduated with a bachelor's degree in Mathematics, he pursued an MBA degree in Finance and a PhD degree in Financial Economics – all of which at Yeditepe University with full scholarships. He worked as a statistical consultant at several private consultancy firms in Istanbul. He teaches courses in mathematical economics, technical analysis, probability and statistics, biostatistics and econometrics. His primary research interest lies in the area of time series analysis, stochastic models driven by Lévy processes, the Bayesian approach in statistics and econometrics, in addition to statistical methods in Engineering and Social Sciences.

Mustafa Ekici works as a lecturer at Department of Mathematics and Science Education of Usak University in Usak-Turkey. He holds his PhD at Applied mathematics in Gazi University (Ankara- Turkey) and BSc (Gazi University, Turkey) in Mathematics. His research interests are Fractional differential equations, Solitons and Nonlinear wave equations. He has co-authored research papers at various journals and conference proceedings.

Mert Erinç is a PhD Candidate at Bocconi University, Department of Accounting. Received his BA from Marmara University in Business Administration and Management, and his MAs from Bogazici University and Galatasaray University in Economics. Held a visiting position at George Washington University. His primary areas of research are audit market regulation and audit transparency.

Rezan Sevinik Adıgüzel completed her Ph.D. in mathematics in 2010 at Middle East Technical University (METU) in Turkey. She receieved her B. S. Degree in mathematics at Ankara University in 2002. Her primary research interest is classical ortghogonal polynomials of discrete variable.

Hüseyin Tuna is an Associated Professor at Department of Mathematics in Mehmet Akif Ersoy University (Turkey), he obtained the PhD Degree in Isparta (Turkey). His research interest lies in differential/difference equations.

Mehmet Turan is an Assoc. Prof. Dr. of the Mathematics Department at Atilim University. He has received his B.S. in Mathematics at Middle East Technical University in 2002 and his Ph.D. from the same university in 2009. His research areas include differential equations and approximation theory.

Index

I

L

M

O

P

R

S

T

U

V

www.ingramcontent.com/pod-product-compliance
Lightning Source LLC
LaVergne TN
LVHW080030210926
848481LV00011B/3148